Materials Development for Direct Write Technologies

MATERIALS RESEARCH SOCIETY
SYMPOSIUM PROCEEDINGS VOLUME 624

Materials Development for Direct Write Technologies

Symposium held April 24–26, 2000, San Francisco, California, U.S.A.

EDITORS:

Douglas B. Chrisey
Naval Research Laboratory
Washington, D.C., U.S.A.

Daniel R. Gamota
Motorola Inc.
Schaumburg, Illinois, U.S.A.

Henry Helvajian
Aerospace Corporation
Los Angeles, California, U.S.A.

David Paul Taylor
Aerospace Corporation
Los Angeles, California, U.S.A.

Materials Research Society
Warrendale, Pennsylvania

CAMBRIDGE UNIVERSITY PRESS
Cambridge, New York, Melbourne, Madrid, Cape Town,
Singapore, São Paulo, Delhi, Mexico City

Cambridge University Press
32 Avenue of the Americas, New York NY 10013-2473, USA

Published in the United States of America by Cambridge University Press, New York

www.cambridge.org
Information on this title: www.cambridge.org/9781107413030

Materials Research Society
506 Keystone Drive, Warrendale, PA 15086
http://www.mrs.org

First published 2001
First paperback edition 2013

Single article reprints from this publication are available through
University Microfilms Inc., 300 North Zeeb Road, Ann Arbor, MI 48106

CODEN: MRSPDH

ISBN 978-1-107-41303-0 Paperback

CONTENTS

Preface xi

Introduction xiii

Materials Research Society Symposium Proceedings xv

RELEVANCE OF DIRECT
WRITE PROCESSING

Commercial Applications and Review for Direct Write Technologies 3
    Kenneth H. Church, Charlotte Fore, and Terry Feeley

*Real-World Applications of Laser Direct Writing 9
    D.J. Ehrlich, Richard Aucoin, M.J. Burns, Kenneth Nill, and
    Scott Silverman

POWDER OR DROPLET BASED
DIRECT WRITE PROCESSING

Charged Molten Metal Droplet Deposition as a Direct Write
Technology 17
    M. Orme, J. Courter, Q. Liu, J. Zhu, and R. Smith

Manufacture of Microelectronic Circuitry by Drop-on-Demand
Dispensing of Nanoparticle Liquid Suspensions 23
    J.B. Szczech, C.M. Megaridis, D.R. Gamota, and J. Zhang

Deposition of Ceramic Materials Using Powder and Precursor
Vehicles Via Direct Write Processing 29
    P.D. Rack, J.M. Fitz-Gerald, A.C. Geiculescu, H.J. Rack,
    A. Piqué, R.C.Y. Auyeung, and D.B. Chrisey

Role of Powder Production Route in Direct Write Applications 35
    M.B. (Arun) Ranade, Z. Serpil Gonen, and Bryan W. Eichhorn

*Invited Paper

v

**Material Systems Used by Micro Dispensing and Ink Jetting
Technologies** 41
Jie Zhang, Irina Shmagin, James Skinner, John Szczech,
and Daniel Gamota

**Offset Printing of Liquid Microstructures for High Resolution
Lithography** 47
Scott M. Miller, Anton A. Darhuber, Sandra M. Troian, and
Sigurd Wagner

**Sol-Gel-Derived 0-3 Composite Materials for Direct Write
Electronics Applications** 53
Steven M. Coleman, Robert L. Parkhill, Robert M. Taylor,
and Edward T. Knobbe

Direct Write Metallizations for Ag and Al 59
C.J. Curtis, A. Miedaner, T. Rivkin, J. Alleman, D.L. Schulz,
and D.S. Ginley

**Ink Jet Deposition of Ceramic Suspensions: Modeling and
Experiments of Droplet Formation** 65
N. Reis and B. Derby

**Electrostatic Printing: A Versatile Manufacturing Process for
the Electronics Industries** 71
Robert H. Detig

LASER DIRECT WRITE TECHNIQUES

***A UV Direct Write Approach for Formation of Embedded Structures
in Photostructurable Glass-Ceramics** 79
P.D. Fuqua, D.P. Taylor, H. Helvajian, W.W. Hansen, and
M.H. Abraham

***Direct Patterning of Hydrogenated Amorphous Silicon by Near
Field Scanning Optical Microscopy** 87
Russell E. Hollingsworth, William C. Bradford,
Mary K. Herndon, Joseph D. Beach, and Reuben T. Collins

Laser Direct Write of Materials for Microelectronics Applications 99
K.M.A. Rahman, D.N. Wells, and M.T. Duignan

*Invited Paper

*Laser Guided Direct Writing 107
    Michael J. Renn

***Photo-Induced Large Area Growth of Dielectrics With
Excimer Lamps** 115
    Ian W. Boyd and Jun-Ying Zhang

**Laser Direct Write of Conducting and Insulating Tracks in
Silicon Carbide** 127
    D.K. Sengupta, N.R. Quick, and A. Kar

Laser Processing of Parmod™ Functional Electronic Materials 135
    Paul H. Kydd, David L. Richard, Douglas B. Chrisey, and
    Kenneth H. Church

***Matrix Assisted Pulsed Laser Evaporation Direct Write (MAPLE DW):
A New Method to Rapidly Prototype Active and Passive Electronic
Circuit Elements** 143
    J.M. Fitz-Gerald, D.B. Chrisey, A. Piqu, R.C.Y. Auyeung,
    R. Mohdi, H.D. Young, H.D. Wu, S. Lakeou, and R. Chung

ION AND ELECTRON BEAM
DIRECT WRITE TECHNIQUES

**Direct Focused Ion Beam Writing of Printheads for Pattern
Transfer Utilizing Microcontact Printing** 157
    David M. Longo and Robert Hull

Ion Beam Induced Chemical Vapor Deposition of Dielectric Materials 163
    H.D. Wanzenboeck, A. Lugstein, H. Langfischer, E. Bertagnolli,
    and M. Gritsch, and H. Hutter

Focused Electron Beam Induced Deposition of Gold and Rhodium 171
    P. Hoffmann, I. Utke, F. Cicoira, B. Dwir, K. Leifer,
    E. Kapon, and P. Doppelt

*Invited Paper

OTHER TECHNIQUES FOR
DIRECT WRITE PROCESSING

*Thermal Spray Techniques for Fabrication of Meso-Electronics
and Sensors 181
   S. Sampath, H. Herman, A. Patel, R. Gambino, R. Greenlaw,
   and E. Tormey

Electrical Characteristics of Thermal Spray Silicon Coatings 189
   S.Y. Tan, R.J. Gambino, R. Goswami, S. Sampath, and
   H. Herman

Direct Write Techniques for Fabricating Unique Antennas 195
   Robert M. Taylor, Kenneth H. Church, James Culver, and
   Steve Eason

Synthesis and Radical Polymerization of Pyrocarbonate Functionalized
Monomers: Application to Positive-Tone Photoresists 199
   S.O. Vansteenkiste, Y. Martelé, E.H. Schacht, M. Van Damme,
   H. van Aert, and J. Vermeersch

Pattern Writing by Implantation in a Large-Scale PSII System With
Planar Inductively Coupled Plasma Source 205
   Lingling Wu, Hongjun Gao, and Dennis M. Manos

Lateral Dye Distribution With Ink-Jet Dye Doping of Polymer
Organic Light Emitting Diodes 211
   Conor F. Madigan, Thomas R. Hebner, J.C. Sturm,
   Richard A. Register, and Sandra Troian

PRINTING METHODS FOR
DIRECT WRITE PROCESSING

High Resolution Copper Lines by Direct Imprinting 219
   C.M. Hong, X. Sun, S. Wagner, and S.Y. Chou

All-Printed Inorganic Logic Elements Fabricated by
Liquid Embossing 225
   Colin Bulthaup, Eric Wilhelm, Brian Hubert, Brent Ridley,
   and Joe Jacobson

*Invited Paper

*Rapid Prototyping of Patterned Multifunctional Nanostructures 231
Hongyou Fan, Gabriel P. López, and C. Jeffrey Brinker

CONSIDERATIONS FOR
DIRECT WRITE PROCESSING

Numerical Simulation of Laser Induced Substrate Heating for
Direct Write of Mesoscopic Integrated Conformal Electronics (MICE) 243
S. Lowry, J.C. Sheu, Robert Stewart, and Robert Parkhill

Thermal Stability and Analysis of Laser Deposited Platinum Films 249
G.J. Berry, J.A. Cairns, M.R. Davidson, Y.C. Fan,
A.G. Fitzgerald, A.H. Fzea, J. Lobban, P. McGivern,
J. Thomson, and W. Shaikh

CFD Modeling of Laser Guided Deposition for Direct Write
Fabrication 255
J.C. Sheu, M.G. Giridharan, and S. Lowry

Computational Modeling of Direct Print Microlithography 261
A.A. Darhuber, S.M. Miller, S.M. Troian, and S. Wagner

Using Convective Flow Splitting for the Direct Printing of
Fine Copper Lines 267
T. Cuk, S.M. Troian, C.M. Hong, and S. Wagner

Calculation of Hamaker Constants in Nonaqueous Fluid Media 275
Nelson Bell and Duane Dimos

Author Index 281

Subject Index 283

*Invited Paper

PREFACE

The goal of Symposium V, "Materials Development for Direct Write Technologies," held April 24–26 at the 2000 MRS Spring Meeting in San Francisco, California, was to identify and develop new materials approaches based on the direct write technique (transfer method) and to demonstrate the required electronic or other device performance (chem/bio sensors, phosphor display, FET, etc.). Many different CAD/CAM approaches exist to direct write or transfer material patterns and each technique has its own merits and shortcomings. Many approaches were presented including plasma spray, laser particle guidance, MAPLE DW, laser CVD, micropen, inkjet, e-beam, focused ion beam, and several novel liquid or droplet microdispensing approaches. One common theme to all techniques is their dependence on high-quality starting materials, typically with specially tailored chemistries and/or rheological properties (viscosity, density and surface tension).

Typical starting materials, sometimes termed "pastes" or "inks", can include combinations of powders, nanopowders, flakes, organic precursors, binders, vehicles, solvents, dispersants, surfactants, etc. This wide variety of materials with applications as conductors, resistors, and dielectrics are being developed especially for low-temperature deposition (< 300–400°C). This will allow fabrication of passive electronic components and RF devices with the performance of conventional thick film materials, but on low-temperature flexible substrates, e.g., plastics, paper, fabrics, etc. Examples include silver, gold, palladium, and copper conductors, polymer thick film and ruthenium oxide-based resistors and metal titanate-based dielectrics. Fabricating high-quality crystalline materials at these temperatures is nearly impossible. One strategy is to form a high density packed powder combined with chemical precursors that form low melting point nanoparticles *in situ* to chemically weld the powder together. The chemistries used are wide ranging, but include various thermal, photochemical and vapor, liquid, and/or gas coreactants. The chemistries are careful to avoid carbon and hydroxide incorporation that will cause high losses at microwave frequencies or chemistries that are incompatible with other fabrication line processing steps.

To further improve the electronic properties for low-temperature processing, especially of the oxide ceramics, laser surface sintering is used to enhance particle-particle bonding. In most cases, individual direct write techniques make trade-offs between particle bonding chemistries that are amenable with the transfer process and direct write properties such as resolution or speed. The resolution of direct write lines can be on the micron scale, speeds can be greater than 100 mm/sec, and the electronic material properties are comparable to conventional screen-printed materials. Optimized materials for direct write technologies result in: deposition of finer features, minimal process variation, lower prototyping and production costs, higher manufacturing yields, decreased prototyping and production times, greater manufacturing flexibility, and reduced capital investments.

Douglas B. Chrisey
Daniel R. Gamota
Henry Helvajian
David Paul Taylor

October 2000

INTRODUCTION AND NOTES ON THE PROCEEDINGS

So far, direct write processing techniques have been relegated to niche applications. One reason for this is that these techniques are serial in nature. Another reason is that the cost of these techniques is tied to their acceptance in the marketplace. Presently, the specialty applications that use direct write processing techniques are important enough to justify this symposium. In the future, I believe that direct write processing will become more important—and possibly much more important.

In assembling these proceedings, I have used the paper of Ken Church et. al. to provide an overview. The breadth of these proceedings indicates only part of the range of applications for direct write processing. "Editing" of silicon integrated circuits, surface texturing hard drive surfaces and the drilling of via holes in printed circuit boards are important applications of laser direct write processing techniques in mature markets. Dan Ehrlich's paper on "Real World Applications" was moved to near the front because it provides another perspective on the current level of acceptance of direct write processing. A significant market already exists for these applications of direct write technology. However, these more mature direct write processing technologies were not emphasized in this symposium. Instead, the emphasis of this symposium was novel direct write processing technologies.

A significant fraction of the papers involve use of direct write processing to achieve the integration of board level electronic circuit elements. It is necessary to write conductors and other materials for wiring, capacitors and resistors. Many applications require more specialized components, so it is necessary to write structures to be used as batteries, antennas and other devices. The dream driving this research involves fabricating future electronic circuit boards—fully populated with components—by some turn-key direct write processing machine. Such a machine might be realized through a new kind of inkjet printing, a laser printer technology, or a micro-pen approach. Alternatively, the new technology to build such a turn-key direct write machine might arise from a very different area, such as the one of the new laser or particle beam techniques. Some of this effort has been influenced by the Mesoscopic Integrated Conformal Electronics (MICE) research initiative sponsored by DARPA under the direction of William Warren. These new technologies should bring about a revolution in the rapid prototyping of new designs.

Other papers of the symposium explore direct write processing technologies more consistent with MEMS (MicroElectroMechanical Systems) applications. The ability to handle non-silicon materials and to assemble different components into an integrated device are important factors in successfully creating a variety of MEMS devices. These MEMS devices are already built into inkjet printer heads, optical scanners, accelerometers for airbag deployment, and they are becoming even more ubiquitous. It is interesting to consider the future of micro-instrumentation ranging from biomolecular sequencers on a chip to pico-satellites the size of a hockey puck. I have become convinced that one of the key benefits of using direct write techniques for these applications is the ability to fabricate true three-dimensional structures.

Another group of papers in this symposium explores the possibility of using direct write techniques to process material on the nanometer scale. A variety of ion beam or electron beam milling techniques can be used to fabricate devices on this scale. Lasers might also be used to create nanometer scale structures, either by using vacuum ultraviolet wavelengths or using other tricks. The papers of this symposium use a contact printing scheme in combination with ion milling of the stamps to produce nanometer scale structures.

There is considerable overlap in categorizing the papers of this symposium. The papers have been roughly grouped by processing technique and area of major interest. Is laser printing with a variant of electrostatic ink properly grouped with laser techniques or with particle transfer methods? Does contact printing of liquids with a stamp belong in the same category as a micropen method involving painting a slurry of suspended particles onto a surface? The choices I have made in organizing the papers are rather arbitrary. With low temperature processing, there is often a need for post-processing or sintering which varies with the direct write materials chosen and the survivability of the substrate. It rapidly becomes clear that the ultimate success of most of these approaches to direct write technology depends on the nature and quality of the materials as much as the processing technique.

The Materials Development for Direct Write Technologies symposium contains papers that emphasize both processing techniques and materials. I can only hope that the printed version of the proceedings papers conveys some of the interest and cross-disciplinary character of the symposium itself.

David Paul Taylor
The Aerospace Corporation
Los Angeles, CA

MATERIALS RESEARCH SOCIETY SYMPOSIUM PROCEEDINGS

Volume 578— Multiscale Phenomena in Materials—Experiments and Modeling, I.M. Robertson, D.H. Lassila, R. Phillips, B. Devincre, 2000, ISBN: 1-55899-486-6

Volume 579— The Optical Properties of Materials, J.R. Chelikowsky, S.G. Louie, G. Martinez, E.L. Shirley, 2000, ISBN: 1-55899-487-4

Volume 580— Nucleation and Growth Processes in Materials, A. Gonis, P.E.A. Turchi, A.J. Ardell, 2000, ISBN: 1-55899-488-2

Volume 581— Nanophase and Nanocomposite Materials III, S. Komarneni, J.C. Parker, H. Hahn, 2000, ISBN: 1-55899-489-0

Volume 582— Molecular Electronics, S.T. Pantelides, M.A. Reed, J. Murday, A. Aviram, 2000, ISBN: 1-55899-490-4

Volume 583— Self-Organized Processes in Semiconductor Alloys, A. Mascarenhas, D. Follstaedt, T. Suzuki, B. Joyce, 2000, ISBN: 1-55899-491-2

Volume 584— Materials Issues and Modeling for Device Nanofabrication, L. Merhari, L.T. Wille, K.E. Gonsalves, M.F. Gyure, S. Matsui, L.J. Whitman, 2000, ISBN: 1-55899-492-0

Volume 585— Fundamental Mechanisms of Low-Energy-Beam-Modified Surface Growth and Processing, S. Moss, E.H. Chason, B.H. Cooper, T. Diaz de la Rubia, J.M.E. Harper, R. Murti, 2000, ISBN: 1-55899-493-9

Volume 586— Interfacial Engineering for Optimized Properties II, C.B. Carter, E.L. Hall, S.R. Nutt, C.L. Briant, 2000, ISBN: 1-55899-494-7

Volume 587— Substrate Engineering—Paving the Way to Epitaxy, D. Norton, D. Schlom, N. Newman, D. Matthiesen, 2000, ISBN: 1-55899-495-5

Volume 588— Optical Microstructural Characterization of Semiconductors, M.S. Unlu, J. Piqueras, N.M. Kalkhoran, T. Sekiguchi, 2000, ISBN: 1-55899-496-3

Volume 589— Advances in Materials Problem Solving with the Electron Microscope, J. Bentley, U. Dahmen, C. Allen, I. Petrov, 2000, ISBN: 1-55899-497-1

Volume 590— Applications of Synchrotron Radiation Techniques to Materials Science V, S.R. Stock, S.M. Mini, D.L. Perry, 2000, ISBN: 1-55899-498-X

Volume 591— Nondestructive Methods for Materials Characterization, G.Y. Baaklini, N. Meyendorf, T.E. Matikas, R.S. Gilmore, 2000, ISBN: 1-55899-499-8

Volume 592— Structure and Electronic Properties of Ultrathin Dielectric Films on Silicon and Related Structures, D.A. Buchanan, A.H. Edwards, H.J. von Bardeleben, T. Hattori, 2000, ISBN: 1-55899-500-5

Volume 593— Amorphous and Nanostructured Carbon, J.P. Sullivan, J. Robertson, O. Zhou, T.B. Allen, B.F. Coll, 2000, ISBN: 1-55899-501-3

Volume 594— Thin Films—Stresses and Mechanical Properties VIII, R. Vinci, O. Kraft, N. Moody, P. Besser, E. Shaffer II, 2000, ISBN: 1-55899-502-1

Volume 595— GaN and Related Alloys—1999, T.H. Myers, R.M. Feenstra, M.S. Shur, H. Amano, 2000, ISBN: 1-55899-503-X

Volume 596— Ferroelectric Thin Films VIII, R.W. Schwartz, P.C. McIntyre, Y. Miyasaka, S.R. Summerfelt, D. Wouters, 2000, ISBN: 1-55899-504-8

Volume 597— Thin Films for Optical Waveguide Devices and Materials for Optical Limiting, K. Nashimoto, R. Pachter, B.W. Wessels, J. Shmulovich, A.K-Y. Jen, K. Lewis, R. Sutherland, J.W. Perry, 2000, ISBN: 1-55899-505-6

Volume 598— Electrical, Optical, and Magnetic Properties of Organic Solid-State Materials V, S. Ermer, J.R. Reynolds, J.W. Perry, A.K-Y. Jen, Z. Bao, 2000, ISBN: 1-55899-506-4

Volume 599— Mineralization in Natural and Synthetic Biomaterials, P. Li, P. Calvert, T. Kokubo, R.J. Levy, C. Scheid, 2000, ISBN: 1-55899-507-2

Volume 600— Electroactive Polymers (EAP), Q.M. Zhang, T. Furukawa, Y. Bar-Cohen, J. Scheinbeim, 2000, ISBN: 1-55899-508-0

Volume 601— Superplasticity—Current Status and Future Potential, P.B. Berbon, M.Z. Berbon, T. Sakuma, T.G. Langdon, 2000, ISBN: 1-55899-509-9

Volume 602— Magnetoresistive Oxides and Related Materials, M. Rzchowski, M. Kawasaki, A.J. Millis, M. Rajeswari, S. von Molnár, 2000, ISBN: 1-55899-510-2

Volume 603— Materials Issues for Tunable RF and Microwave Devices, Q. Jia, F.A. Miranda, D.E. Oates, X. Xi, 2000, ISBN: 1-55899-511-0

Volume 604— Materials for Smart Systems III, M. Wun-Fogle, K. Uchino, Y. Ito, R. Gotthardt, 2000, ISBN: 1-55899-512-9

MATERIALS RESEARCH SOCIETY SYMPOSIUM PROCEEDINGS

Volume 605— Materials Science of Microelectromechanical Systems (MEMS) Devices II, M.P. deBoer, A.H. Heuer, S.J. Jacobs, E. Peeters, 2000, ISBN: 1-55899-513-7
Volume 606— Chemical Processing of Dielectrics, Insulators and Electronic Ceramics, A.C. Jones, J. Veteran, D. Mullin, R. Cooper, S. Kaushal, 2000, ISBN: 1-55899-514-5
Volume 607— Infrared Applications of Semiconductors III, M.O. Manasreh, B.J.H. Stadler, I. Ferguson, Y-H. Zhang, 2000, ISBN: 1-55899-515-3
Volume 608— Scientific Basis for Nuclear Waste Management XXIII, R.W. Smith, D.W. Shoesmith, 2000, ISBN: 1-55899-516-1
Volume 609— Amorphous and Heterogeneous Silicon Thin Films—2000, R.W. Collins, H.M. Branz, S. Guha, H. Okamoto, M. Stutzmann, 2000, ISBN: 1-55899-517-X
Volume 610— Si Front-End Processing—Physics and Technology of Dopant-Defect Interactions II, A. Agarwal, L. Pelaz, H-H. Vuong, P. Packan, M. Kase, 2000, ISBN: 1-55899-518-8
Volume 611— Gate Stack and Silicide Issues in Silicon Processing, L. Clevenger, S.A. Campbell, B. Herner, J. Kittl, P.R. Besser, 2000, ISBN: 1-55899-519-6
Volume 612— Materials, Technology and Reliability for Advanced Interconnects and Low-k Dielectrics, K. Maex, Y-C. Joo, G.S. Oehrlein, S. Ogawa, J.T. Wetzel, 2000, ISBN: 1-55899-520-X
Volume 613— Chemical-Mechanical Polishing 2000—Fundamentals and Materials Issues, R.K. Singh, R. Bajaj, M. Meuris, M. Moinpour, 2000, ISBN: 1-55899-521-8
Volume 614— Magnetic Materials, Structures and Processing for Information Storage, B.J. Daniels, M.A. Seigler, T.P. Nolan, S.X. Wang, C.B. Murray, 2000, ISBN: 1-55899-522-6
Volume 615— Polycrystalline Metal and Magnetic Thin Films—2000, L. Gignac, O. Thomas, J. MacLaren, B. Clemens, 2000, ISBN: 1-55899-523-4
Volume 616— New Methods, Mechanisms and Models of Vapor Deposition, H.N.G. Wadley, G.H. Gilmer, W.G. Barker, 2000, ISBN: 1-55899-524-2
Volume 617— Laser-Solid Interactions for Materials Processing, D. Kumar, D.P. Norton, C.B. Lee, K. Ebihara, X. Xi, 2000, ISBN: 1-55899-525-0
Volume 618— Morphological and Compositional Evolution of Heteroepitaxial Semiconductor Thin Films, J.M. Millunchick, A-L. Barabasi, E.D. Jones, N. Modine, 2000, ISBN: 1-55899-526-9
Volume 619— Recent Developments in Oxide and Metal Epitaxy—Theory and Experiment, M. Yeadon, S. Chiang, R.F.C. Farrow, J.W. Evans, O. Auciello, 2000, ISBN: 1-55899-527-7
Volume 620— Morphology and Dynamics of Crystal Surfaces in Complex Molecular Systems, J. DeYoreo, W. Casey, A. Malkin, E. Vlieg, M. Ward, 2000, ISBN: 1-55899-528-5
Volume 621— Electron-Emissive Materials, Vacuum Microelectronics and Flat-Panel Displays, K.L. Jensen, W. Mackie, D. Temple, J. Itoh, R. Nemanich, T. Trottier, P. Holloway, 2000, ISBN: 1-55899-529-3
Volume 622— Wide-Bandgap Electronic Devices, R.J. Shul, F. Ren, M. Murakami, W. Pletschen, 2000, ISBN: 1-55899-530-7
Volume 623— Materials Science of Novel Oxide-Based Electronics, D.S. Ginley, D.M. Newns, H. Kawazoe, A.B. Kozyrev, J.D. Perkins, 2000, ISBN: 1-55899-531-5
Volume 624— Materials Development for Direct Write Technologies, D.B. Chrisey, D.R. Gamota, H. Helvajian, D.P. Taylor, 2000, ISBN: 1-55899-532-3
Volume 625— Solid Freeform and Additive Fabrication—2000, S.C. Danforth, D. Dimos, F.B. Prinz, 2000, ISBN: 1-55899-533-1
Volume 626— Thermoelectric Materials 2000—The Next Generation Materials for Small-Scale Refrigeration and Power Generation Applications, T.M. Tritt, G.S. Nolas, G. Mahan, M.G. Kanatzidis, D. Mandrus, 2000, ISBN: 1-55899-534-X
Volume 627— The Granular State, S. Sen, M. Hunt, 2000, ISBN: 1-55899-535-8
Volume 628— Organic/Inorganic Hybrid Materials—2000, R.M. Laine, C. Sanchez, E. Giannelis, C.J. Brinker, 2000, ISBN: 1-55899-536-6
Volume 629— Interfaces, Adhesion and Processing in Polymer Systems, S.H. Anastasiadis, A. Karim, G.S. Ferguson, 2000, ISBN: 1-55899-537-4
Volume 630— When Materials Matter—Analyzing, Predicting and Preventing Disasters, M. Ausloos, A.J. Hurd, M.P. Marder, 2000, ISBN: 1-55899-538-2

Prior Materials Research Society Symposium Proceedings available by contacting Materials Research Society

Relevance of Direct Write Processing

COMMERCIAL APPLICATIONS AND REVIEW FOR DIRECT WRITE TECHNOLOGIES

KENNETH H. CHURCH[1], CHARLOTTE FORE[1], TERRY FEELEY[2]
[1]CMS Technetronics, Inc., 5202-2 North Richmond Hill Road, Stillwater, OK, U.S.A. 74075.
[2]Laser Fare, 70 Dean Knauss Dr., Narragansett, RI, U.S.A. 02882

ABSTRACT

Direct write in the past has generated the excitement of possibly replacing photoresist for all electronic applications. Removing the mask would substantially reduce the number of steps required to produce electronic circuits. A reduction in steps represented time and dollar savings. The advantage of being able to direct write a manufacturable device would also save time and money in the design process as well. With all of the obvious advantages, it seemed inevitable that research dollars would continue to mount and thus overcome the obstacles preventing this technology from becoming more than a novel technique used in laboratories. As Moore's law began to settle in, so did photoresist and direct write was little more than a novelty.

That was then, and this is now. Developers have come to terms with the true value direct write can supply to the manufacturers and design engineers. Techniques such as Focused Ion Beam (FIB), Laser Chemical Vapor Deposition (LCVD), ink jetting and ink penning have found real applications that are making a difference in industry. A summary will be presented describing the various direct write techniques, their current applications and the possible or probable applications.

INTRODUCTION

Direct-write processes are fast, flexible, and forgiving. Masking and screen-printing processes take several steps to complete a circuit. Making a mask or a new screen can take days, even weeks. A direct-write process has now demonstrated the possibility of turning weeks of prototyping into hours. With this kind of improvement, it seemed inevitable that direct-write processes would take the electronics industry by storm. A masking process can take as many as 24 steps [1]. A direct-write process to create the same circuit could be reduced to 5 steps. If these are facts, the obvious question that must be asked is, "Why is direct-write such a novelty?"

As one electronics-company representative stated, "speed, speed, speed" is everything [2]. While this may not be the only issue in the electronics industry, it is the only one that matters if a new process cannot reach specified standards. To replace such masking techniques as screen-printing or photolithographic patterning, direct-write methods will have to complete these tasks in seconds, not hours.

Competing with masking production rates has proven to be an onerous task. Many direct-write believers found a niche that would keep these methods alive—rapid prototyping (RP). The biggest complaint from rapid-prototyping consumers is not speed; RP techniques are far superior in speed. It is also not necessarily poor performance; many rapid-prototyping techniques have demonstrated superior results. The main issue is repeatability on a manufacturing floor [3]. Many direct-write techniques have proven to be very effective for demonstrating a concept or device, but the prototyping process was not repeatable in a mass-production manner. This disadvantage would completely remove the benefits obtained from the fast fabrication and characterization results. Direct-write methods have made a difference in technology; they will make more of an impact in the next decade, due to the need for rapid changes in technology. Some of the well-known and obscure direct-write techniques are discussed below, with some of their possible applications.

THE BEGINNING

Rapid-prototyping began with the invention of stereolithography by Charles Hull [4]. It was based on an innovative approach to integrating CAD/CAM, lasers, and materials. It met a marketplace need, the desire to get conceptual models faster. Stereolithography is a direct-write process. What are the elements that drove RP from that first stereolithography machine to an industry that is now worth over $1.1 *billion*? The answer is successful commercial application

3

by appropriate use of the technology and a technology supplier base that consisted of more than one company. Companies are reluctant to try new technologies although they may seem faster, better, and more cost effective—because they are not proven. Companies are reluctant to buy expensive equipment when there is only one supplier because they do not want to be held hostage to a start-up company that may or may not succeed. Trying new technologies in their early stages can be a high-risk career move. If the implementers' efforts are successful, then further up the "food chain," other people take credit; if their efforts fail, they are blamed. The rule is straightforward—it helps to have more than one source for the technology. This reduces the risk for the innovator.

The rapid-prototyping industry started to grow when its customers began to see useful models being built by stereolithography. A critical mass of technology suppliers started to emerge. The first was enabled by "beta" programs and the advent of several service bureaus offering the technology on a per-part basis, because this enabled users to experiment with the technology without paying the high costs of equipment ownership. This reduced the risk for the consumer. The second was enabled by the emergence of competitors to the stereolithography approach, like selective laser sintering and fused deposition modeling. Competition gave the marketplace a sense of security that a lasting industry was emerging, not just a transient curiosity. The bottom line was that the growth of RP was driven by good applications in the marketplace, a critical mass of technology suppliers, and a critical mass of innovative service bureaus.

DIRECT-WRITE METHODS

Several approaches exist to direct-write methods, depending on function and device size, but all direct-write methods have three things in common: (1) materials; (2) delivery process; and (3) conversion process. In some approaches, the delivery process also serves as the conversion process. It is also important to note that direct-write systems use a computer aided design package to implement the desired pattern, shape, size and location of the deposits.

Physical Methods

Stereolithography (SLA) was developed by Charles Hull in 1986. His company, 3D Systems, is the trailblazer of the rapid-prototyping industry; it turned a niche demonstration into a multi-million-dollar business. SLA is capable of producing a vast number of unique objects in two or three dimensions. Its basic premise involves a platform of liquid polymer cured by an ultraviolet laser. The platform lowers as, layer by layer, the laser cures the desired pattern. Once the device is completed, it will need to be rinsed, and supports (built during the process) will be removed [5]. The practical resolution of SLA is on the order of 50 μm in the z direction and 75 μm in the xy plane [6]. The possibility of increasing resolution to less than 10 μm is being investigated; however, a significant thrust for this feature size has not yet emerged.

Carl Deckard patented *Selective Laser Sintering* (SLS) in 1989. This technique uses a CO_2 laser to fuse various materials. SLS is a bit more flexible than SLA in that it is not restricted to UV-curing polymers; it can be used on powders of nylon, elastomers, or metals. A new layer of powder must be applied to the part, then leveled for processing. Excess powder in each layer helps to support the part during the build. SLS machines are produced by DTM of Austin, TX.

The *Laser Engineered Net Shaping* (LENS™) process is a new solid freeform technology capable of producing metal parts with excellent materials properties directly from CAD files. The LENS™ process was initially developed at Sandia National Laboratories and is now commercially available through Optomec Design Co. This process goes beyond conventional rapid prototyping, which produces parts in plastic, to produce near-net-shape metal parts in a variety of materials, including stainless steel, tool steel, and titanium.

Fused Deposition Modeling (FDM) works by extruding molten material through an *xy*-plane-controlled nozzle onto a build platform. The build material is pushed through heated tips on the nozzle as it moves and extrudes. As each layer is completed, it hardens and "fuses" to the previous one. The platform is lowered and the process continues. Layer by layer, material is deposited until a 3D part or model is complete. Stratasys (Eden Prairie, MN) makes a variety of FDM machines. The first commercial systems were introduced in 1992.

Electrical Methods

The *Laser Particle Guidance* (LPG) process was developed by Michael Renn of the University of Colorado, who later joined Optomec [7]. A red, detuned laser beam is launched into the hollow region of a hollow-core fiber. Atoms in the guide propagate in a manner similar to that of light in a multimode fiber—axial motion is unconstrained and transverse motion consists of a series of lossless reflections from the potential established by the optical fields. Atoms exit the fiber through a numerical aperture that increases with increasing guiding-light intensity. Renn showed that atoms can be steered through the bends of the flexible fiber. This work has also involved transporting atoms through a portion of fiber exposed to atmosphere, demonstrating that the glass walls are sufficient to maintain vacuum in the guide.

The *Matrix Assisted Pulsed Laser Evaporation* (MAPLE) method was developed by scientists at the Naval Research Laboratory (NRL), this technology uses a laser to remove material from the backside of the disk and deposit on a substrate[8]. Resolution of this technique can reach sub 10 micron, depending on the spot size of the laser. A wide variety of materials can be directly deposited using this method.

Several groups have worked on the *Laser-Assisted Chemical Vapor Deposition* (LCVD) technique in hopes of reducing the number of steps required to produce integrated circuits. The original research work was done in the early 1970's; work continues today [9]. LCVD has been combine with other laser-assisted chemical processes and commercialized. Dan Ehrlich was one of the pioneers in this technology and started the company Revise, Inc., which manufactures etching/LCVD machines. Various chemical precursors can be introduced into a vacuum chamber to dissociate into desired products when elevated to a specified temperature, or dissociation can be photoinduced. In either case, a laser provides localized heating or photolysis leading to deposition.

Focused Ion Beam (FIB) systems use a gallium-ion beam to make precision modifications to wafer samples. A FIB machine can mill away material; alternatively, in the presence of an organometallic vapor, it can deposit metals and insulators in a process called ion-beam-assisted deposition (IBAD). The resolution of FIB devices can be less than 10 nm; therefore, their primary application is in the semiconductor industry. Uses include semiconductor sample preparation and analysis, modification of prototype circuits directly on the wafer, and chip repair.

Ink-jet and *nozzle/pen* techniques encompass a variety of machines that spray layers of material in droplet form onto a build platform. For example, 3D Systems uses "waxes" to create thermoplastic models with its ThermoJet Solid Object Printer. Other applications include MicroFab's MicroJet printing of solder. Droplet size is controlled by the print head aperture. Depending on the droplet material, feature resolution has reached 50 μm [10]. Ink-jet technology has been shown to reliably write a variety of materials very quickly. One of the main concerns of ink-jet technology is the viscosity of the material being dispensed—if it is too high, the piezoelectric pump cannot move it.

The highest value a nozzle or pen can bring to the direct-write arena is the variety of materials that can be dispensed through it. Pastes, which by implication have high viscosities, are normally used in screen-printing. Pastes can easily be transferred to a pen-dispensing system, thereby exploiting an established materials base. Pen-dispensing issues can be summed up as start/stop and agglomeration. Several companies have worked in this area, on a variety of projects, but one of the long established products in this area is the MicroPen™ from OhmCraft.

Optical Techniques

One of the newest members to join the direct-write family involves the laser/optical arena. The direct writing of optical waveguides has been demonstrated using a femtosecond-regime pulsed laser [11]. The team demonstrated the capability to write optical waveguides not just on the surface of a solid piece of glass, but internally as well by adjusting the focus depth. Interestingly, optical technology has lagged behind electronic technology in all areas, from demonstration to production to tools. The optics industry is now moving very quickly due to the advancements made in laser technology during the past decade, including ultrashort-pulse technology. Two years ago, "femtosecond" was a rare word in laser technology; today, it is possible to purchase a femtosecond-regime direct-write waveguide tool [12].

DIRECT-WRITE APPLICATIONS

Applications for direct-write methods have a vast array of possibilities, which include the mechanical/physical, electrical, and optical areas of engineering. The physical aspects of direct-write have an established market and products; however, ample room exists in which to grow and improve. The electrical engineering aspects have mostly been demonstration-oriented or commercially very small. The newest member to the direct-write family is in the optical engineering regime, but there it has tremendous promise. Materials with diverse properties have been produced with a broad range of tools and processes. Physical strength is the primary issue when dealing with such mechanical structures as tools. In direct-write electronics, a multitude of characteristics must be considered, including dielectric constant, conductivity, resistance, permeability, and physical strength. These constraints place rigid demands on materials and processing conditions to achieve specified goals.

The first and most obvious application is for rapid prototyping. Since the mid-1980's this area has grown significantly in dollars and more in interest. These machines can be very expensive ($500,000 to $2.2 million), but the return on the investment is great in terms of time, money, and flexibility. The SLA RP industry is well recognized; however, a very large market has opened in the semiconductor business for RP. FIB is a household name in the semiconductor business. FIB machines are used to cut traces and to add new ones. In the strictest since, this process should be labeled rapid alteration, but the "rapid" in the name carries the same implications. Semiconductor devices can have short lives; therefore, it is important to reach the market very fast. The masking process is slow to provide results; therefore, if several iterations through the design process are required, the turnaround time to get this done could extend beyond the need for the chip. The FIB is an expensive piece of equipment, but the savings it provides in time and reduced masking setup have made a major impact in this field.

LCVD is a competitor to FIB in some situations, but most of the time these processes complement each other very well. LCVD is fast compared to FIB. It also produces superior depositions with higher conductivity, and with a change of precursors it will etch without damaging the wafer. FIB has the required resolution to work with semiconductor circuits. The combination of these two can be advantageous to a vast array of semiconductor devices including "flip chips."

LCVD also has the ability to write in three dimensions by controlling the focus. This provides limitless possibilities for fractal antennas, vertical interconnects, bridges, and bumps for solder. The resolution of LCVD is superior to mechanical dispensing methods. Therefore, it fills a gap between the submicrometer and ten-micrometer regimes.

Significantly, each direct-write process can be different and still make an impact. SLA works well for RP models of a boat propeller, but would make a poor propeller. The materials in SLA would also make a poor conductive line, despite making a "nice"-looking line. These disadvantages do not make SLA useless; it has demonstrated its worth, so its disadvantages make it specialized. Plenty of such specialization exists. In fact, an overflow of specializations that currently exist right now must be filled. In direct-write electronics being developed today, each method has a different set of strengths. These strengths provide a large coverage area of the marketplace within that spectrum of opportunities. It is highly probable that each technique mentioned above will find a home.

Consider how powerful the potential for direct-write electronics is in the commercial sense, how pervasive this family of technologies may be in five years. One door opening is the field of hybrid and multi-chip module (MCM) circuitry. A direct-write tool capable of writing resistors, capacitors, and inductors would significantly impact this industry. These new hybrid and MCM chips would have less solder and therefore be more efficient and use less power. This would reduce battery size, thus making the device lighter or longer lasting. It would also make the circuit operate with less heat, thus reducing the cooling requirements.

Let us imagine the addition of a battery to a direct-write menu, then go one step further and add solar-cell capability. This is not a large stretch of the imagination because DARPA is

6

funding both of these projects. Another interesting fact is that DARPA is funding direct-write antennas [13]. The direct-write processes for batteries and antennas are providing unique opportunities in these areas which cannot reasonably be done using other techniques. Fractal antennas have tremendous potential to impact the Global Positioning System world. Current GPS antennas are large in comparison to PCS antennas. Incorporating GPS onto PCS telephones would be highly desirable; however, this cannot be reasonably obtained with current antenna technology. Direct-write will impact this area. It could be considered a niche market, since it is strongly bent toward 3-D fractal antennas, but a niche this big could change the definition of niche market.

One of the most promising opportunities for direct-write methods is sensors. Sensors that exist today range from sensors for heat to those that can detect parts-per-billion concentrations of toxic gases. The number of sensing devices continues to grow as research continues to explore new possibilities. Such a heat-sensing device as a thermocouple, which is as simple as two wires pressed together, has a market that exceeded $500 million in 1998 in the United States alone. The direct-write techniques mentioned above can do this today. BioMEMS is many times an extension of sensing, but also has extended possibilities that could revolutionize the ways that health-care delivery systems function. Most present health-care systems are financial nightmares. BioMEMS platforms offer the key to changing that fact. Successful direct-write commercialization will play a keynote in the success of BioMEMS by enabling the use of cost manufacturing techniques. The range of devices and capabilities needed to satisfy the BioMEMS opportunities are certainly covered, from SLA to FIB.

FROM NICHE TO MASS PRODUCTION

Success in this field will not come by accident. The best way to accomplish success is to look for a historical model from which to learn. Direct-write proponents are fortunate to have a very recent model in the birth and growth of the rapid-prototyping industry. Direct-write proponents as a group can learn a great deal from this experience. Jeffrey Moore's book *Crossing the Chasm* is a classic text describing the process through which a new technology goes in becoming a commercial success [14]. The growth of the rapid-prototyping industry is a classic example of what Moore teaches; direct-write electronics will be also. It is important to learn from experience; the existing RP industry has been a good teacher.

The existing RP industry did many things correctly, but its most significant contribution was to fill a need. Industry will not necessarily notice it *has* a need, but when the competition begins to move ahead, things begin to stir. Strategic partners in industry, who are willing to take risks, are very important. These groups can provide technical, real-life feedback. They are willing to accept "substandard" performance and high prices during the development and growth stages; however, they can grow impatient and demanding if improvements do not happen. The company Apple provides a good example of stirring industry in a positive fashion with less than ideal products. They did very well and made a huge impact upon the computer industry. However, when the time comes to change and raise standards, noncompliance can be costly [15].

Daily inroads are being made in the direct-write arena, providing new possibilities and changing industry. The performances of devices made via direct-write methods are in some cases superior to those made by conventional methods. Also, the speed at which these devices can be explored has been dramatically enhanced, and in some cases even enabled. Looking ahead, the compliance requirement for direct-write methods will be to make better-performing devices, but a major obstacle will be speed. If the devices made cannot be made fast enough to mass-produce, then that shortcoming will reduce the effectiveness of direct-write methods. It will not kill it; some applications are so important it will be accepted. This problem can be addressed in two ways: (1) better tools/processes; and (2) better materials.

CONCLUSIONS

Direct-write proponents are fortunate that DARPA has had the vision to fund several different technical approaches to address the issues associated with direct-write electronics. This vision helps ensure that its goal will be met and that the creation of a critical mass of tool suppliers is well on its way to becoming a reality. Proponents are also fortunate that they have learned to work together. They are building an industry together and they need each other for successful development of technologies and credible entry into the marketplace. Now what are needed are good applications of the technologies and continued production of superior materials, tools, and devices. Proponents need to match the marketplace expectations of what can be delivered to what they can actually deliver. This is not an easy process. As technology develops, each step forward will be heralded as it should be. Unfortunately, the amount of work required to go from one good deposition of the appropriate electronic materials to *ten million* good depositions is often underestimated. This is why it is so important that direct-write proponents have invited potential users of direct-write technologies into the program at the very beginning.

A direct-write industry is being created by researchers, developers, entrepreneurs, government agencies, and large corporations. Their work will lead to the development of a host of new, as yet unforeseen devices. Their work will also lead to improvements in quality and reductions in the costs of products people use every day. These everyday applications are the applications that will insure success.

ACKNOWLEDGMENTS

The author would like to thank Lowell Matthews of CMS for his assistance, and DARPA for funding support. He would also like to thank Dr. William Warren of DARPA specifically for his support and drive in the direct write community.

REFERENCES

1. R. Terrill, *IEEE Aerospace Applications Conference Proceedings* (Proceedings of the 1997 IEEE Aerospace Conference, 3, Snowmass Village, CO, 1997) pp. 481 - 488.
2. R. Terrill, K. Church, M. Moon, *IEEE Aerospace Applications Conference Proceedings* (Proceedings of the 1997 IEEE Aerospace Conference, 1, Snowmass Village, CO, 1997) pp. 377-382.
3. T. Plunkett, *Time-Compression Technologies,* 2000, *5(2)* 46
4. www.3dsystems.com, online reference, 2000.
5. M. Griffith and J. S. Lamancusa, "Rapid Prototyping Technologies," *Rapid Prototyping,* 1998, http://www.me.psu.edu/lamancusa/me415/rpintro2.pdf.
6. Ken Cornealy, Mark Moler, Tech Inc. (Stillwater, OK), private communication, 2000.
7. M. Renn, Phys. Rev. Lett. 75 (18), 3253 (1995).
8. Pique, A., (Mat. Res. Soc., San Francisco, CA) 1998.
9. D. Ehrlich, R. Osgood Jr., R. Deutsch, IEEE J. Quantum Electronics QE-16 (11), 1233-1243 (1980)
10. D. J. Hayes, D. B. Wallace, and W. R. Cox, "MicroJet Printing of Solder and Polymers for Multi-Chip Modules and Chip-Scale Packages," in *Proceedings of the IMAPS International Conference on High Density Packaging and MCMs, Denver, April 1999.*
11. K. Miura et al., Appl. Phys. Lett. 71, 3329 (1997).
12. P. Bado, Laser Focus World, April 2000, 73-78 (2000).
13. R. M. Taylor, K. H. Church, J. Culver, and S. Eason, "Direct-Write Techniques for Fabricating Unique Antennas," in *Materials Development for Direct Write Technologies,* edited by D. B. Chrisey, D. R. Gamota, and H. Helvajian (Mater. Res. Soc., Pittsburgh, PA, 2000), in press.
14. G. Moore, *Crossing the Chasm,* (Harperbusiness, New York, 1999).
15. G. Kawaski, Garage.com Bootcamp for Startups, Seattle, WA, 2000.

Real-World Applications of Laser Direct Writing

D. J. Ehrlich, Richard Aucoin, M. J. Burns, Kenneth Nill and Scott Silverman, Revise Inc.,
79 Second Ave., Burlington MA 01803, revise@revise.com

Abstract

Laser microchemical direct write deposition and etching methods have found an essential niche in debug and design for yield of wire-bonded and flip-chip integrated circuits. Future applications should develop in package-level system modification.

Introduction

In research dating back to the early 1980's laser and focused ion beam (FIB) direct write deposition and etching have been developed with an eye to a variety of microelectronic needs. The two methods referenced share much in the way of capabilities. Over time the two approaches have specialized and have been integrated into a powerful set of methods uniquely important to the microelectronics industry.

This paper will briefly summarize some of the established and emerging applications to more or less conventional circuit design debug; the elaboration of these methods to the particularly demanding testing and debug of packaged flip-chip parts, and the further evolution to package level debug. Surprisingly, the importance of this class of methods has been greatly intensified over the last several years. Leading microprocessor companies have begun to use FIB and laser direct writing methods to adjust designs to increase manufacturing yield and binning count at the factory as well as at the design center.

This paper will emphasize the practical applications of laser direct write methods and the integration of laser and FIB methods. The microchemical writing speed of the laser techniques are 2 to 6 orders of magnitude greater than they are for the ion beam analogs. Additionally, the electronic material quality of the laser deposited thin films are much higher, e.g. resistivity is typically 2 orders of magnitude lower than for the best focused ion beam deposited films. In the best cases the resolution of the laser techniques equal that of the dominant production technology for integrated circuits (ICs), which is optical lithography, but it cannot match the resolution of focused ion beams. As a result, users combine the virtues of both direct writing methods in actual practice. The diverse microchemical process technology used in laser direct writing is reviewed in Ref. 1.

Figure 1: Conventional debug of a wire-bonded front-surface part requires penetration of a passivation layer. In this case a compound silicon nitride/silicon dioxide layer was removed by a laser technique. Laser methods for chip depassivation are generally chosen for their process speed.

9

Design Modification of Conventional Wire-Bonded Circuits

The modification of conventionally bonded (circuit up) parts has been relatively straight forward as the electrical connection to individual transistors was possible through the passivation layer and the top of the part. Laser methods were chosen for long length discretionary interconnects over the passivation layer where the high conductivity and rate of laser deposited metal greatly exceeds FIB metal. Another application was removal of passivation, silicon dioxide, silicon nitride, polyimide, etc., where the rate of laser depassivation greatly exceeds the FIB (e.g., see Fig. 1). Two developments have complicated these applications; (1) the low accessibility of transistor connections due to multi-layer metallization, which now often covers >90% of the available surface, and (2) a strong trend toward nearly exclusive use of flip-chip circuits for high end systems. Both developments are driving design debug from the backside of the chip through the full thickness of the silicon wafer.

Flip-Chip Repair Process Flow

The debug/repair of flip chip parts imposes one new problem, the removal of the bulk silicon substrate in order to access the active device. Once this step is accomplished, probing and repair can proceed in analogy with more conventional front surface rework; examples being the use of a focused ion beam (FIB) to edit a circuit, or make probe points and a 3-beam prober for testing the circuit [3]. In fact access of active areas from the backside is often simpler and more easily interpreted than access through multi-level metallization. On integrated circuits in which the metallization completely covers the active device, diagnostic techniques such as photoemission must be completed from the backside.

Figure 2 depicts a flip-chip repair process flow. The first step is to globally thin the flipped part to a thickness that still maintains adequate mechanical strength and thermal dissipation capability. Typically a flip-chip may be globally thinned to 200µm using mechanical polishing without risking stress-induced fracture. This thickness is also sufficient for infrared through wafer viewing and global photoemission surveys.

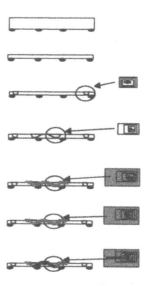

Figure 2: **Flip-Chip Repair Process Flow**

The second step is to reference front side fiducials in order to navigate to debug/repair sites. If accurate through wafer viewing is not available, trenches may be efficiently etched to the front surface. The laser microchemical process can etch completely to the field oxide, since oxide is an etch-stop layer, or leave a thin silicon layer to be processed by other tools. Other tools (e.g. a FIB) can locate the fiducials either optically or by etching through the remaining field oxide until the metal fiducial is exposed. Three fiducial trenches (200μm x 200μm x 200μm) can be laser processed in about thirty minutes.

The third step is to access circuit debug/repair sites by etching laser microchemical trenches over the regions of interest. Placement of these trenches may be aided by the use of computer aided navigation (CAV) e.g. Knight Navigation to correspond circuit to layout and navigate the laser microchemical etcher. Trenches need not be placed with extreme accuracy since they may be efficiently etched to hundreds of micrometers across; the trenches must, however, be flat, uniform, and stop close to the active device.

The fourth step is to deposit a dielectric isolation layer on the bottom and sidewalls of the trench. This can be accomplished globally as a separate step (typically time consuming), or locally using laser deposited oxide.

The fifth step is to make local repairs at the base of the isolated trenches. These repairs can be similar to those performed on the front side of the chip due to the proximity of the trench base to the active device. Typically a focused ion beam is used to create a high aspect ratio via to the node to be tested or rewired. For interconnects longer than a few tens of micrometers, laser deposition can be used to make interconnects with about 3μohm-cm conductivity.

The final step, if necessary, is to interconnect repairs between trenches. In this case a laser deposited interconnect is run up the sidewall of one trench over the top of the substrate and back down into another trench. Figure 3 shows a scanning electron micrograph of a typical laser etched trench.

Figure 3: A silicon trench etched using chlorine-assisted chemistry from the backside of a flip-chip integrated circuit. The terracing is more pronounced than typically necessary for the application. The above figure shows an excellent example of the process cleanliness and control possible with the laser method. (the typical depth is ten microns per small terrace in the scanning electron micrograph).

OBIC Endpointing Method

Laser microchemical processing enables efficient access to flip-chip devices for debug and repair. To be effective, the etched trenches must be flat, uniform, and stop close to the active device. The closer the base of the trench is to the active device the easier it will be to implement an edit. The required aspect ratio of a via made with a focused ion beam can double if the via must first be milled through tens of micrometers of silicon before accessing the diffusion regions.

The laser etch process is not typically the limiting factor in producing trenches close to the active device since the microchemical zone, the region melted by the laser and reacted with a gaseous ambient, is highly confined. Instead, knowing when to stop determines how close the base of the trench will be. If the initial surface is uniform and the thickness of the substrate is well known, the trench can be etched using dead reckoning, relying on the repeatability of the etch process to approach the active region. Typical long-term repeatability of trenches up to 200μm deep in homogeneous silicon is less than 2μm.

In general, the back surface of a flip-chip is not uniform after mechanical polishing to 200μm or less and the remaining thickness is not well known. An infrared confocal microscope can be used to measure the remaining thickness of silicon if the dielectric constant of the substrate is precisely known.

To eliminate the uncertainties of the substrate thickness and uniformity, the active device may be used as a reference. Laser microchemical etching of the backside of the flip-chip induces a current which may be measured between Vcc and GND. Figure 4 depicts schematically the fixture and electronics used to measure the induced current.

The fixture consists of a zero insertion force Socket 7 mounted on an aluminum stand inside the etch process chamber. Connections to the device are made via a vacuum feedthrough. Solder surfaces are coated so as not to react with the process gas and contaminate the chamber. A simple transimpedance amplifier is used to convert the current to a voltage that can be read by a control computer using an analog to digital converter.

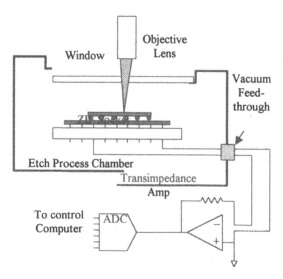

Figure 4: OBIC Endpoint Fixture and Electronics

A trench is etched by scanning an argon-ion laser (multi-line with primary wavelengths of 488nm, 514nm) across the silicon removing one layer at a time. Initially etching is done at 5 to 7μm per scan and then slowed to less than a micrometer per scan when approaching the active region. The induced current is measured during each scan line near the center of the trench. Figure 5 shows the current as a function of scans. Since the silicon strongly absorbs at visible wavelengths, the signal is sensitive to submicrometer thickness changes. By calibrating the signal to remaining thickness and using a threshold to complete the scan and then stop, the trench may be accurately etched to a known thickness from the active device.

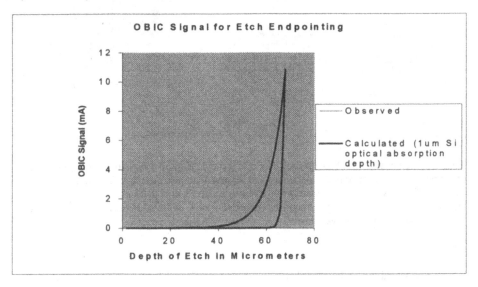

Figure 5: Measured OBIC current as a function of etch depth for a typical flip-chip part.

Package Level System Debug

As microprocessors migrate to multiple chip modules and as portable communication devices gather complexity and increasing integration, it is clear that direct writing methods will be required to debug systems at the package level. Methods for circuit adjustment on ceramic, laminate and flex-circuit packaging will be necessary. Many of the laser deposition chemistries developed for chip level will be equally applicable at package level. In addition there may be an opportunity of laser direct write techniques that will fabricate precision passive elements such as resistors, capacitors and inductors. These package level fabrication methods will need to be coupled closely to test apparatus to achieve highly precise passive tolerance. An example of a demonstration of this application is shown in Figure 6. The process challenge for capacitors and inductors is to achieve consistent, high quality dielectric deposition with a process step that remains compatible with a direct write electrode-patterning method.

13

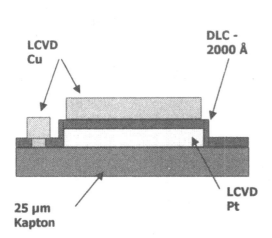

Figure 6. Demonstration of a capacitor on a free standing Kapton film, with laser deposited top (copper) and bottom (platinum) electrodes and plasma deposited dielectric (diamond like carbon). This device is approximately 500 microns square, with a measured value of 58 pF, low leakage and a Q of 30 at 1 MHz frequency. (Photo courtesy of Donald Foust, General Electric Corporate Research and Development).

Conclusion

We have reviewed the application of laser microchemical direct writing techniques as they are currently applied in silicon circuit debug and yield enhancement. These methods are now securely placed as part of the essential tool set for cost-effective manufacturing of microprocessors and other complicated silicon systems. Future technology trends will probably push applications of similar laser methods into use at the package level. These methods may involve laser fabrication of full discrete passive elements (resistors, capacitors, and inductors) as well as interconnect.

References

1. D. J. Ehrlich and J. Y. Tsao, eds., *Laser Microfabrication: Thin Film Processes and Lithography* (Academic, Boston, 1989).

2. S. Silverman, R. Aucoin, J. Mallatt, D. Ehrlich, Proceedings from the 23rd Annual International Symposium for Testing and Failure Analysis, 211-213 (1997)

3. R. H. Livengood, V. R. Rao, *FIB Techniques to Debug Flip-Chip Integrated Circuits*, Semiconductor International, March 1998.

4. S. M. Sze, *Physics of Semiconductor Devices*, Second Edition, p 42, John Wiley & Sons (1981)

Powder or Droplet Based
Direct Write Processing

CHARGED MOLTEN METAL DROPLET DEPOSITION
AS A DIRECT WRITE TECHNOLOGY

M. ORME, J. COURTER, Q. LIU, J. ZHU AND R. SMITH
Dept. of Mech. & Aerospace Eng., University of California, Irvine, Irvine CA 92697-3975
melissao@uci.edu

ABSTRACT

The formation of highly uniform charged molten metal droplets from capillary stream break-up has recently attracted significant industrial and academic interest for applications requiring high-speed and high-precision deposition of molten metal droplets such as direct write technologies. Exploitation of the high droplet production rates intrinsic to the phenomenon of capillary stream break-up and the unparalleled uniformity of droplet sizes and speeds attained with proper applied forcing to the capillary stream make many new applications related to the manufacture of electronic packages, circuit board printing and rapid prototyping of structural components feasible. Recent research results have increased the stream stability with novel acoustic excitation methods and enable ultra-precise charged droplet deflection. *Unlike other modes of droplet generation such as Drop-on-Demand, droplets can be generated at rates typically on the order of 10,000 to 20,000 droplets per second (depending on droplet diameter and stream speed) and can be electrostatically charged and deflected onto a substrate with a measured accuracy of ±12.5 μm.* Droplets are charged on a drop-to-drop basis, enabling the direct writing of fine details at high speed. New results are presented in which fine detailed patterns are "printed" with individual molten metal solder balls, and issues relevant to the attainment of high quality printed artifacts are investigated.

INTRODUCTION

The photograph in Figure 1 illustrates a pattern that was printed with charged and deflected solder balls onto a black card-stock substrate. The molten solder balls were generated from capillary stream break-up from a 100µm diameter orifice. Individual splat diameters are measured in the range of 375 – 400µm, and the scale shown with each image indicates 1.0mm lines. These particular realizations, which resemble "Moorish Arches", are included as a demonstration of the intricate detail possible with molten metal printing. The charge waveforms employed while making the prints were sine waves in which the time base of the charge waveform was not synchronized with that of the droplet generation waveform. Hence, the position of the droplets relative to the charge waveform was not fixed, allowing the droplets to essentially "walk" up and down the crests and valleys of the sine wave. When the droplets impinge upon the translating substrate, they sweep out the pattern shown.

Figure 2 illustrates a conceptual schematic of the charging and deflection technique for a simple sinusoidal pattern. In this case, the time base of the droplet generation waveform is synchronized with that of the charging waveform, and the frequency of the charging waveform is much lower than the droplet generation frequency allowing a dense sinusoidal

Figure 1: "Moorish Arches" printed with electrostatically charged and deflected solder balls. Individual splats are approximately 375µm and divisions shown are 1 mm.

Mat. Res. Soc. Symp. Proc. Vol. 624 © 2000 Materials Research Society

pattern to be printed. The substrate moves along one axis, and deflection occurs along the perpendicular axis. The time-varying charge signal is applied to the charge electrode, and a potential of ±3000V is applied to the deflection plates.

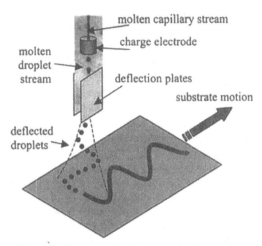

Figure 2: Conceptual schematic of molten metal droplet stream printing

Droplet Generation

The generation of highly stable droplet streams is a requirement for Direct Write applications. Any variations in droplet formation time will lead to significant errors in droplet placement via electrostatic charging and deflection is discussed in detail elsewhere [1]. Under carefully controlled conditions, droplet formation from capillary stream break-up provides droplet streams with speed dispersions as low as 1×10^{-6} times the average speed, and angular deviations as small as one micro-radian [2]. The molten metal capillary stream is excited with a periodic disturbance by means of a vibrating piezoelectric crystal whose frequency is chosen to be in the Rayleigh regime [3] for uniform droplet generation. The science of droplet generation from capillary stream break-up is well understood, and the interested reader is urged peruse the review articles by Bogy [4] and McCarthy & Molloy [5] for a more in-depth discussion. In brevity, Lord Rayleigh showed that if the applied disturbance is chosen such that it's wavelength is greater than the stream's circumference, the surface waves will become unstable and grow exponentially in time as $e^{\beta t}$, ultimately resulting in droplet formation, where β is the growth rate of the disturbance. Later, Orme [2] showed that the most uniform droplet stream is one that is perturbed with a disturbance whose frequency associated with the maximum β.

Droplet Charging and Deflection

Molten metal droplets are charged by electrostatic induction. The molten metal jet passes through the charge electrode that surrounds the jet at the point of droplet formation. The conductive molten liquid is grounded and a positive periodic potential, V_c, is applied to the charge electrode. As drops are formed from the continuous column of molten metal, a negative charge is induced on the drop that is that is proportional to the charge potential. In order to print the droplets to different locations, a distinct potential is applied to the charge electrode for every drop to be formed, and therefore, synchronization between the charge waveform frequency and the droplet production frequency must be maintained.

This method of droplet charging is similar to that employed in the technology of ink-jet printing. A few of the more important experimental works on ink-jet printing are given by Sweet [6, 7], Schneider et al. [8], Kamphoefner [9], and Fillmore et al. [10]. Analogous to the applications described in this work, an ink-jet jet printer produces characters on paper by deflecting charged droplets on one axis while the print head moves along the perpendicular axis. The droplet charge and the strength of the electric field through which the droplet moves determine the amount of deflection achieved by a droplet. The net charge on the droplet is acquired at the time of droplet formation.

In contrast to the ink-jet droplets, the molten metal droplets in this work attain significantly higher charges so that large lateral areas can be printed. The high charges cause significant inter-droplet mutual electrostatic interactions to occur which are not evident in the ink-jet printing technology. The importance of these interactions has been reported in the recent paper by Orme et al. [1]. The charge to mass ratio, Q/m, can be predicted by the relation given by Schneider et al. [8]:

$$\frac{Q}{m} = \frac{2\pi\varepsilon_o V_c}{\rho r_o^2 \ln(b/r_o)} \tag{1}$$

Here, m is the droplet mass, ε_o is the permittivity of free space, V_c is the charge potential, b is the radius of the charge tube, ρ is the molten metal density, and r_o is the radius of the unperturbed capillary stream. We have shown excellent agreement between measured droplet charge and the above equation in reference [1].

EXPERIMENT

The experimental setup consists of a droplet generator that is situated in a fume-hood, and injects droplets into a chamber that experiences a slow purge of an inert gas. We have found that it is necessary to eliminate traces of oxygen from the chamber in order to avoid the disruptive effects of oxidation, which act to impede molten metal droplet formation. Immediately below the droplet generator is the charging electrode and the parallel deflection plates as sketched in Figure 2. The separation between the orifice, which is contained at the lower end of the droplet generator, and the substrate, is 325 mm. The orifice diameter is 100 μm, and the mean droplet diameter is 189 μm. The substrate is mounted on an x-y table that moved at a speed of 5.2 cm/s. A microscope with long working distance optics and equipped with a camera is mounted outside the chamber. Properties used in the experiment are listed in Table 1 unless otherwise indicated in the text.

In previous work [11] we have found it necessary to actively control the position of the orifice in order to obtain a printing accuracy of ± 12.5 μm. To this end, the orifice is situated in a hemispherical seat assembly with the orifice at the center of curvature. Active feedback and control of the stream's position with information from two perpendicularly positioned CCD array cameras allows the adjustment of the hemisphere to compensate for any small deviations in the stream's trajectory. However, in this work the aforementioned feedback and control system was not utilized (though plans for it's implementation for future studies are underway), and therefore, the results presented here will possess dispersions in placement

droplet fluid	63% Sn, 37% Pb
specific gravity	8.420
surface tension	0.49 kg/s^2
viscosity	1.58x10^{-7}m^2/s
orifice diameter	100 microns
ambient & stagnation gas	Nitrogen
driving pressure	20 psi
solder reservoir temp.	200°C
disturbance frequency	12,000 Hz
charge electrode diameter	0.318 cm
deflection plate separation	1.27 cm
deflection plate length	5.08 cm

Table 1: Properties used in experiment

that are not representative of the "comprehensive" system. Nonetheless, the raw results below show excellent potential for the application of direct writing.

RESULTS

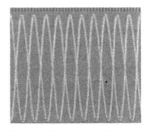

Figure 3: A double sine wave of solder onto a copper substrate. Grid lines are 1.0 mm apart.

Figure 3 illustrates a double sine wave that was printed with electrostatically charged and deflected solder droplets onto a copper substrate. In this case, the droplets were generated at a frequency of 20,000 Hz and the charge waveform was a sinusoid with charged with a frequency of 10,010 Hz. Had the charging frequency been exactly half of the droplet generation frequency (i.e., 10,000 Hz), the original stream would have been split in two angularly stable streams and the resulting printed pattern on the translating substrate would be two parallel lines. However, since the charging waveform was 10,010 Hz, it was intentionally slightly out of phase with the droplet generation frequency, causing the charge to drift between maximum and minimum values. Hence, the original droplet stream was split into two streams that were simultaneously scanning on the horizontal axis (i.e., perpendicularly to the translation axis) in mirror images. This example illustrates the stability of the droplet break-up and charge mechanism.

Effects of Droplet Stream Frequency

Figure 4 illustrates the effect of droplet generation frequency on the thickness and uniformity of the printed line. A "full" stream corresponds to a droplet stream generated at a frequency of 12,000 Hz as shown in the second trace in the figure. The fractions indicate the reduction in frequency accomplished by electrostatic charging and deflection. For example, the top line pictured was generated with a droplet production frequency of $1/24^{th}$ that of the full stream, i.e., every 500 Hz (12,000/24). Hence, in this case, every 24th droplet was deflected out of the stream by the same magnitude, thereby creating two angularly stable streams of different droplet densities, where in practice, either line may be collected in order to avoid impingement. The line width, the ball density, and the flow rate for each of the pictured cases are provided in Table 2 below. The droplet impingement temperature is constant for each line shown. The line width generally decreases with the droplet frequency. As the line becomes more dense, such as in the case with $1/4^{th}$ full or full frequency, the solidification time increases due to a larger accumulation of molten metal at one location, thereby providing the ability for the fluid to "roll" off the top of the printed line, leading to deviations in straightness. Alternatively, as the droplet

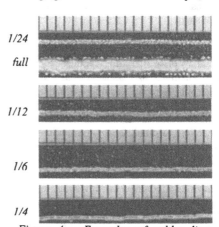

Figure 4: Examples of solder lines printed with different droplet frequencies. Droplets are deflected out of the stream in order to vary the frequency. Grid lines are 1.0 mm apart

20

Stream frequency	Full	1/4	1/6	1/12	1/24
mean line diam. (mm)	1.75 ± 0.2	.550 ± 0.05	.525 ± 0.05	.450 ± 0.05	.450 ± 0.05
number of balls/mm	230	47.6	38.4	19.2	9.6
flow rate (g/cm)	0.535	0.0134	0.00892	0.00446	0.00223

Table 2: properties of printed lines of varying frequency

frequency is reduced, the splats rapidly solidify, eliminating the possibility of relaxation into a smooth cord by the action of surface tension. Hence, for the production of straight and smooth lines, it is desired to simultaneously decrease the droplet frequency and increase the droplet temperature.

Line Separation and Thickness

Figures 5-7 illustrate magnified lines printed by three different means. In each case the substrate translation speed was 5.2 cm/s. The grid marks on each figure are 1.0 mm lines. In Figure 5, the lines were printed by electrostatically charging the solder stream generated at a frequency of 12,000 Hz with a 3,000 Hz sine wave. Synchronization between the charging waveform and the droplet generation waveform insured that each charge waveform cycle contained four droplets whose positions relative to the chrage waveform did not shift in time. Hence the effective droplet frequency of the each droplet stream impacting the substrate is $1/4^{th}$ that of the original uncharged droplet stream, i.e., 3,000 drops/second. Since the substrate speed was 5.2 cm/s, four 5.2 cm long lines were printed in 1.0 second. The separation between the inner-most printed lines is consistently of the order of 250 µm.

Figure 6 illustrates the printing of four lines by traversing the substrate four times under an angularly stable droplet stream generated with a frequency of $1/4^{th}$ that of the full stream, so that the line densities are the same as those shown in Figure 5. The printing speed of four lines 5.2 cm in length is 4.0 seconds in addition for the time required to re-position the substrate after each pass.

Figure 7 illustrates four lines printed by traversing the substrate under a stationary stream that was generated with a frequency of $1/24^{th}$ that of a full stream. Hence, the lines are narrower, but less smooth due to the rapid solidification. Nonetheless, it can be seen that four distinct lines can be printed within a 2.0 mm space. In order to abtain smoother lines, the droplets must impinge at a higher temperature in order to compensate for the reduction in mass delivery rate

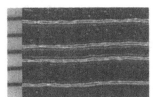
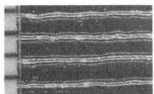

Figure 5: Four lines printed with a sinusoidal charging disturbance with four drops per cycle. Grid marks are 1.0 mm apart. Distance between center lines is approximately 250µm.

Figure 6: Four lines printed by traversing the substrate four times under an undeflected droplet stream. Droplet generation frequency is 1/4 full stream. Grid marks are 1.0 mm apart.

Figure 7: Four lines printed by traversing the substrate four times under an undeflected droplet stream. Droplet generation frequency is 1/24 full stream. Grid marks are 1.0 mm apart.

which reduces the heat flux to the specific location. Experiments to demonstrate this assertion are currently underway.

CONCLUSIONS

The "printing" of solder lines from discrete solder balls generated from capillary stream break-up was presented. It has been shown that fine detailed pictures (e.g., "Moorish Arches") can be printed at high speed with electrostatic charging and deflection of the molten metal balls. Studies showed that droplet streams generated at lower frequencies (by deflecting a desired fraction of droplets out of the main stream) result in finer pitch lines, though the use of such streams requires that additional heat must be supplied to the droplets if smooth lines are required. Our results show that four distinct parallel lines can be printed in a 2.0 mm space, providing excellent potential for fine-pitch printing.

ACKNOWLEDGMENTS

The authors gratefully acknowledge the generous support of the National Science Foundation; grant numbers DMI-9622400 and DMI-945720.

REFERENCES

1. Orme M., Liu Q., and Huang, C., "Electrostatic Interactions of Charged Micro-Liter Solder Droplets. *Journal of Atomization and Sprays*, in press

2. Orme, M., "On the Genesis of Droplet Stream Microspeed Dispersions", *The Physics of Fluids*, 3, (12), 1991

3. Lord Rayleigh, *Phil Mag.* 14, 184 (1882)

4. Bogy, D.B., "Drop Formation in a Circular Liquid Jet" *Ann. Rev. Fluid Mech.*, 1979, 11: 207-228

5. McCarthy and Molloy, "Review of Stability of Liquid Jets and the Influence of Nozzle Design' *Chem. Engineering*, 7, 1-20, 1974

6. Sweet R. G. "High-Frequency Oscillography with Electrostatically Deflected Ink Jets," *Stanford Electronics Laboratories Technical Report No. 1722-1,* Stanford University, CA, 1964

7. Sweet R. G. "High Frequency Recording with Electrostatically Deflected Ink Jets", *Rev. Sci. Instrum.* 36, 2, 131, 1965

8. Schneider J.M., N.R. Lindblad, and Hendricks C.D, "Stability of an Electrified Liquid Jet", *J Applied Physics.* 38, 6, 2599, 1967

9. Kamphoefner F.J. "Ink Jet Printing", *IEEE Trans. Electron Devices* ED-19, 584, 1972

10. Fillmore G.L., Buehner, W.L., West, D.L., "Drop Charging and Deflection in an Electrostatic Ink Jet Printer", *IBM J. Res. Develop.* Jan, 1977

11. Muntz EP, Orme M, Pham-Van-Diep G, Godin R, "An Analysis of Precision, Fly-Through Solder Jet Printing for DCA Components" presented at the 30th International Symposium on Microelectronics, Pennsylvania , October 1997

MANUFACTURE OF MICROELECTRONIC CIRCUITRY BY DROP-ON-DEMAND DISPENSING OF NANO-PARTICLE LIQUID SUSPENSIONS

J.B. SZCZECH *, C.M. MEGARIDIS *, D.R. GAMOTA **, J. ZHANG **
*ME Department, University of Illinois at Chicago, Chicago, IL 60607, jszcze1@uic.edu
**Motorola Labs, Inc., 1301 E. Algonquin Rd., Schaumburg, IL 60196

ABSTRACT

An emerging selective metallization process utilizes Drop-On-Demand (DOD) inkjet printing, and recent developments in nano-particle fluid suspensions to fabricate fine-line circuit interconnects. The suspensions consist of silver or gold particulates of 1-10 nm in size that are homogeneously suspended in an organic carrier solvent. A piezo-electric droplet generator driven by a bipolar voltage signal is used to dispense 50-70 µm diameter droplets traveling at 1-3 m/s before impacting a compliant substrate. The deposit/substrate composite is subsequently processed at 300°C for 15 minutes to allow for evaporation of the solvent carrier and sintering of the nano-particles, thereby yielding a finished circuit product. Test vehicles created using this technique exhibited features as fine as 120-200 µm wide and 1-3 µm thick. The circuitry performed well during environmental conditioning studies. However, repeatability of the results showed sensitivity to the generation of steady, satellite-free droplets. In an effort to generate droplets consistently, it is essential to develop a strong fundamental understanding of the correlation between device excitation parameters and fluid properties, and resolve the microrheological behavior of the conductive ink as it flows through the droplet generator.

I. INTRODUCTION

Over the past 15 years integrated circuit feature dimensions and pitches have seen a dramatic reduction in size. An excellent indicator of this trend is the wireless communications market. Cellular manufacturers are continuously striving to improve capital utilization, reduce downtime associated with changeover and obtain higher yields and output of their products. Recently, interest has increased in the rapid prototyping capabilities of Direct Write Technologies (DWT), propelling the development of a number of candidate deposition systems for the fabrication of micro-electronic components. For example, Bharathan and Yang [1] have successfully created an electro-luminescent device using inkjet technology and light-emitting polymers. In response to this growing interest in DWT, a preliminary investigation has been conducted to determine whether it is possible to use DOD inkjet technology and recent developments in nano-particle suspension technology as a means to fabricate horizontal, fine line circuitry for electronic assemblies.

A common method of fabricating fine-line conductor formations or printed wiring boards (PWB) involves the selective metallization of non-conductive substrates using a subtractive etching process [2]. Traditional selective metallization processes use non-conducting substrates that have one or both surfaces electroless-deposited with a metal coat only to have the majority of the deposited metal etched away. Such practice not only wastes material but also generates substantial amounts of chemical pollutants. Furthermore, this method of producing a PWB does not lend itself to rapid translation due to the need of masks and rather lengthy substrate processing steps.

An advantage of using DOD inkjet technology in circuit interconnect manufacturing is that the process is additive. Material is only deposited in desired locations, thereby reducing the amount of chemical and material waste. In addition, the ability of DOD to produce entire circuit patterns that are simply drawn using computer graphics software makes it a very powerful rapid prototyping technology. To demonstrate the feasibility of DOD to fabricate fine-line conductors on conformal substrates, two test vehicles having different pattern geometries were created.

23

II. EXPERIMENT

A. Material System

One of the major setbacks to using DOD inkjet technology for circuit interconnect fabrication is the stringent requirement of the conductive inks' physicochemical properties [3]. Most commercially available conductive inks possess too high of a viscosity and surface tension to allow droplet formations of micron size. However, recent developments in nano-particle suspension technology have created a new category of low viscosity conductive inks suitable for use with DOD. The particular material used in this study, shown in **Figure 1**, consists of silver (Ag) nano-particles on the order of 10 nm in diameter. Suspensions containing gold nano-particles have also been used.

The nano-particles are suspended in a toluene-based solvent that acts to separate the particles, thereby creating a near homogeneous dispersion. In addition to toluene, other metals are included in the suspension to improve adherence to the substrate. Typical physicochemical properties of the nano-suspension are listed in **Table I**. The properties may be altered by the addition or removal of toluene. Unfortunately, due to the rapid evaporation rate of toluene the properties of the nano-suspension vary dramatically over time and can change from a low viscosity liquid to a viscoelastic body in a matter of a few seconds upon exposure to ambient conditions. This phenomenon adversely affects jetting performance, since it can create a membrane at the exit orifice of droplet ejection. After using this material in preliminary tests, it was observed that the membrane creation due to evaporation of the carrier solvent is the primary obstacle preventing repeatable and reliable jetting.

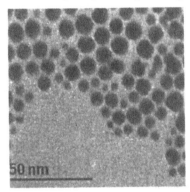

Figure 1. TEM image of the silver nano-particle suspension. The silver particles shown in black are supported by a carbon substrate on a copper electron microscope grid. The particle diameters vary from 5 to 15 nm and appear well dispersed.

Table I
Physicochemical properties of silver nano-particle suspension.

property	value	comments
viscosity [mN s m^{-2}]	1-2*	at 25 °C
density [g cm^{-3}]	1.32	at 25 °C
surface tension [mN m^{-1}]	27-29	at 25 °C
shelf life [months]	6*	when stored at 10 °C
sinter time [minutes]	15	at 300 °C
resistivity [Ω cm]	4.2x10^{-6}*	for suspension containing 30% wt Ag cured at 300 °C of film thickness 250 nm

* Values obtained from material vendor.

B. Dispensing System

To dispense the nano-particle liquid suspension, a solder jet system was modified to enable dispensing of materials that exhibit low viscosity behavior at ambient conditions. The use of commercially available piezoelectric printers, such as those used to print on cellulose based substrates, is not possible with the nano-suspension due to the plastic construction of their print heads and the aggressive nature of the toluene solvent. As indicated in **Figure 2**, the jetting system is comprised of various components, all of which must be optimized to ensure reliable jetting performance.

The droplet generator used in the set-up consists of a glass capillary tube surrounded by a radially polarized piezoelectric ceramic crystal (PZT). The crystal is affixed to the glass capillary by an epoxy.

Droplets are ejected from a converging nozzle ending at a single 60 μm circular orifice. Extensive research and modeling of the droplet formation process of this device has been done by [4], [5], and [6], therefore its jetting principles are well documented. The typical jetting conditions employed in this study are given in **Table II**.

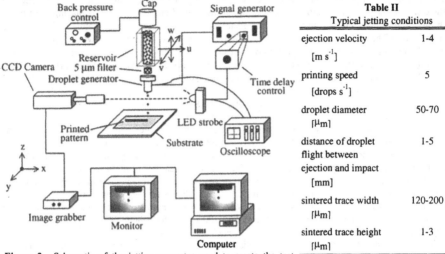

Table II Typical jetting conditions	
ejection velocity [m s^{-1}]	1-4
printing speed [drops s^{-1}]	5
droplet diameter [μm]	50-70
distance of droplet flight between ejection and impact [mm]	1-5
sintered trace width [μm]	120-200
sintered trace height [μm]	1-3

Figure 2. Schematic of the jetting apparatus used to create the test vehicles. The primary components are the droplet generator, fluid reservoir, signal generator, strobe and a CCD camera/microscope. Using the strobe and CCD camera it is possible to observe and characterize the droplets during their formation process, provided the ejection process is highly repetitive.

A bipolar voltage excitation signal, shown in **Figure 3**, was incorporated to drive the PZT crystal. The signal parameters v_0, v_1, and v_2 are the baseline, rise and fall voltage, respectively. They can vary from 0-100 volts and are used to control the radial displacement of the PZT. The movement of the PZT, upon activation, generates pressure waves inside the capillary tube. The parameters f, t_1, and t_2, are the signal frequency, rise and fall dwell periods, respectively. The rise and fall dwell time can range from 0-800 μs and govern the duration for which the PZT is in a radially expanded or contracted state, respectively. This allows the pressure waves to travel throughout the capillary tube. Optimization of the signal parameters is crucial for reliable jetting.

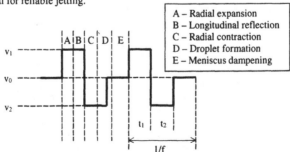

A – Radial expansion
B – Longitudinal reflection
C – Radial contraction
D – Droplet formation
E – Meniscus dampening

Figure 3. Shown is the waveform of the excitation signal used to excite the droplet generator and its effects. The signal causes the PZT to expand and contract, which in turn generates pressure waves that propagate through the nano-suspension and eventually eject a droplet.

C. Substrate Processing

Once the nano-suspension has been deposited onto the substrate, it undergoes a sintering process to drive off the carrier solvent and to allow the nano-particles to melt and sinter. Studies conducted by Buffat and Borel [7] provide an excellent guideline for the processing temperatures as a function of particle size of the gold nano-particles. Processing of the printed patterns is done by placing them into a preheated, forced convection oven at 300°C for 15 minutes. The processing temperature was determined as a result of a Thermal Gravimetric Analysis (TGA) which revealed a substantial weight change occurs at 300°C, thus suggesting that most if not all of the carrier solvent has evaporated. According to the nano-suspension manufacturer, improved continuity can be obtained if the processing temperature is increased to 600°C, but unfortunately most organic substrates cannot handle such high levels of temperature.

III. RESULTS

Two test vehicles shown in **Figures 4** and **5** were fabricated on 0.6 mm thick polyimide substrates. Printed on each test vehicle were two simple, type I and II. Polyimide was selected as the substrate of choice for it could withstand the 300°C sintering process. Other substrates such as glass or ceramic can withstand this process temperature, but one of the features of the study was to demonstrate the ability of DOD to print on conformal substrates. Resistance measurements were recorded for each test vehicle along various paths between the dispensed pads. The path lengths were recorded along with the average path width and heights. A video caliper was used to measure the path widths. The widths of the interconnects from both samples varied from 120-200 μm. The heights of the paths were obtained using cross-sectional techniques. This was done by encasing a portion of the path in an epoxy and micro-polishing the plane perpendicular to the direction of the path. The polished cross-section was then examined under a Scanning Electron Microscope (SEM), see **Figure 6**. Observed interconnect heights in both test subjects ranged from 1-3 μm. With the resistance measurements and profile dimensions known an average resistivity value of approximately 2.7×10^{-5} Ω cm was calculated. In comparison, industrial standard materials such as polymer thick films and copper have resistivity values on the order of 10^{-3} and 10^{-6} Ω cm, respectively.

Test vehicles printed using pattern type I were subjected to environmental conditioning and reliability studies. The conditioning involved storage at 85% humidity at 85°C. The test conditions were selected to simulate the harsh conditions that PWBs typically encounter. Between timed intervals resistance measurements were taken along various interconnect paths and were compared with initial resistance values to yield a percent change in resistance. There were no signs of dramatic change in

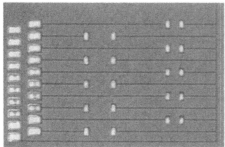

Figure 4. Photograph of test pattern I. The pattern was created using DOD and a silver nano-suspension. The nominal trace width and thickness is 200 x 3 μm. The distance from the pads (left) to the rightmost ends of the pattern is 64 mm. The square pads at the far left measure 2 x 1 mm.

Figure 5. Photograph of test pattern II. The pattern was created using DOD and a gold nano-suspension. The nominal trace width and thickness is 120 x 1 μm. The distance from the pads (left) to the rightmost ends of the trace is 62 mm. The square pads measure 1.5 x 1.5 mm.

resistance of the processed interconnects during the study, see **Table III**. The total average of the change in resistance was approximately 4%.

Figure 6. Partial cross-section of a pad created with the silver nano-suspension observed under SEM. The pad was processed at 300°C for 15 minutes. The average height of the pad was approximately 3 μm. The porous structure suggests that the sintering process is not complete. To further process the pad, a much higher temperature would be required.

Table III

Percent variance in resistance after environmental conditioning at 85% humidity and 85°C (test vehicle printed with pattern type I).

Cycle time [hrs]	Path Length [mm]						
	71.5	59.0	64.1	13.7	31.7	13.7	31.7
31	0.0%	5.4%	1.7%	1.1%	3.9%	2.4%	3.4%
55	2.3%	2.8%	4.6%	10.6%	3.8%	3.4%	3.9%
172	0.5%	2.0%	2.1%	12.2%	4.2%	10.2%	4.2%
Total average	4%						

IV. CONCLUSION

The use of DOD inkjet printing for the deposition of horizontal circuit interconnects was demonstrated. Two types of test vehicles each containing different pattern geometries were successfully created through the use of a single 60 μm orifice piezoelectric droplet generator and ink consisting of 1-10 nm silver or gold particles suspended in an organic solvent (toluene). The techniques used to manufacture the test vehicles demonstrate the capability of DOD and nano-suspension technology to generate uniform and functional fine-line conductors. The interconnect profiles of both test vehicles used in the study ranged from 120-200 μm in width, 1-3 μm in height and possessed an average resistivity value of 2.7×10^{-5} Ω cm. Variance in the interconnect profile is directly attributed to the droplet generator performance and excitation. For instance, increasing the excitation voltage amplitudes by 1 volt can cause an 0.5 m/s increase in droplet velocity. When the droplet impacts the substrate at this increased velocity, it further spreads radially onto the substrate. The result is then an interconnect with increased width and decreased height dimensions. Test vehicles subjected to 85% humidity at 85°C environmental conditioning for up to 172 hours showed on average a 4% variance in resistance.

V. ACKNOWLEDGMENTS

The authors would like to thank Motorola Labs (Advanced Technology Center) for the generous use of their facility, deposition and rheology instrumentation.

VI. REFERENCES

[1] J. Bharathan and Y. Yang, "Polymer Electroluminescent Devices Processed by InkJet Printing: I. Polymer Light-Emitting Logo," *Applied Physics Letters*, vol. 72, no. 21, pp. 2660-2662, May 1994.

[2] G. Leonida, *Handbook of Printed Circuit Design, Manufacture, Components, and Assembly*. (Electrochemical Publications Limited, Scotland, 1981), pp. 198-203.

[3] M. D. Croucher and M. L. Hair, "Design Criteria and Future Directions in InkJet Ink Technology," *Ind. Eng. Chem. Res.*, vol. 28, no. 11, pp. 1712-1218, 1989.

[4] J. F. Dijksman, "Hydrodynamics of Small Tubular Pumps," *Journal of Fluid Mechanics*, vol. 139, pp. 173-191, 1984.

[5] D. B. Wallace, "A Method of Characteristics Model of a Drop-On-Demand Ink-Jet Device Using an Integral Method Drop Formation Model," *Proc. ASME Winter Annual Meeting.*, San Fransico, CA, Dec. 1989.

[6] D. B. Bogy, and F. E. Talke, "Experimental and Theoretical Study of Wave Propagation Phenomena in Drop-On-Demand Ink Jet Devices," *IBM Journal of Research and Development*, vol. 28, no. 3, pp. 314-321, May 1984.

[7] Buffat and J. P. Borel, "Size Effect on the Melting Temperature of Gold Particles." *Physical Review A*, vol. 13, no. 6, pp. 2287-2298, Dec. 1975.

DEPOSITION OF CERAMIC MATERIALS USING POWDER AND PRECURSOR VEHICLES VIA DIRECT WRITE PROCESSING

P.D. Rack[1], J.M. Fitz-Gerald, A.C. Geiculescu[2], H.J. Rack[2], A. Piqué, R.C.Y. Auyeung, and D.B. Chrisey
Naval Research Laboratory, Washington, D.C.
[1]Rochester Institute of Technology, Dept. of Microelectronics, Rochester, NY
[2]Clemson University, Department of Ceramic and Materials Engineering, Clemson, SC

ABSTRACT

Dry powder and sol gel ceramic films were deposited using a Matrix Assisted Pulsed Laser Evaporation Direct Write (MAPLE-DW) technique developed at NRL for optical and electrical device applications. The MAPLE-DW technique uses a high-energy focussed photon source in combination with a "ribbon" to fabricate materials onto a range of substrates at room temperature without material degradation. Two different classes of materials were processed in this research, (1) ceramic dry powder materials and (2) sol gel precursor materials. Cathodoluminescent measurements demonstrated that the efficiencies of the transferred phosphor materials were not degraded during the laser transfer process. Scanning electron microscopy and 3-D surface profilometry of the ribbon after the MAPLE-DW process revealed a 90-95% transfer efficiency for the dry powders. Scanning electron microscopy and energy dispersive spectroscopy revealed that the sol gel materials also transferred with an efficiency in the 90-95% range. Three distinct regions were identified on the sol gel ribbon after the laser transfer process. The regions suggest that the matrix absorbing layer and the sol gel materials are completely removed for the areas irradiated by the laser pulse, and in a second so-called "heat affected region," only the sol gel material is ejected from the ribbon.

INTRODUCTION

Many direct write techniques based on laser-induced processes have been developed for depositing materials for a variety of applications. Among these techniques, laser induced forward transfer (LIFT) has demonstrated the ability to direct write metals for interconnects and mask repair and also simple dielectric materials such as metal oxides [1]. LIFT was first demonstrated using metals such as Cu and Ag over substrates such as silicon and fused silica utilizing excimer or Nd:YAG lasers [1-2]. It is a simple technique that uses laser radiation to physically vaporize a thin film ($\approx$ 100 nm) from the far side of a laser transparent support into a roughly similar pattern on a substrate placed in close proximity ($\leq$ 100 μm) to it. In order to utilize the process, the laser fluence should be adjusted so that the process is carried out near the energy threshold to remove only the film material and not to damage the support. Target films should not exceed an experimentally determined thickness, generally less than a few 100 nm. Overall, the laser induced forward transfer process has proven to be a technique that can be used on a wide variety of target films of metals, and simple oxides. However, it is not suitable for complex multi-component materials such as high dielectric constant ceramics or phosphors.

29

A new, vacuum deposition technique, known as matrix assisted pulsed laser evaporation (MAPLE) has been developed at the Naval Research Laboratory (NRL) for depositing thin, uniform layers of chemoselective polymers [3-5] as well as other organic materials, such as carbohydrates [6]. MAPLE is a variation of the conventional pulsed laser evaporation process with respect to the laser interactions and the spatial dynamics of the thin film growth. The "soft" transfer mechanism associated with the matrix assisted pulsed laser deposition process enables the deposition of complex and fragile organic molecules into thin films without denaturing. Irreversible structural degradation is typically observed with conventional pulsed laser deposition even at low fluences.

MAPLE direct write is a new process that combines aspects of the transfer mechanism of matrix assisted pulsed laser evaporation with the direct write resolution of laser induced forward transfer [7]. This enables the transfer of polymer, metallic, ceramic and electronic materials without degradation in performance. MAPLE direct write can be utilized for micromachining, drilling and trimming applications, by simply removing the transfer material support from the laser path, and as such it is both an additive as well as subtractive direct write process. In this paper, we demonstrate the application of the MAPLE direct write process to both thick film powder phosphors and sol gel precursors.

EXPERIMENTAL PROCEDURES

The thick film phosphor ribbons were fabricated by initially sputtering a thin (100nm) gold film onto a 5cm diameter x 2mm thick quartz wafer. The phosphor powders ($ZnSi_2O_4$:Mn – green, Y_2O_3:Eu – red, and $BaMg_2Al_{16}O_{27}$:Eu – blue) were suspended in a glycerin/isopropanol solution with $LaNO_3$ and $Mg_3(NO_3)_2$ salts and electrophoretically deposited onto the gold coated quartz wafers to form thick dense ribbons. The output from a KrF excimer laser (λ = 248 nm, 25 ns pulse) was directed through a variable circular aperture and then through a 10x ultraviolet grade objective lens. By changing the aperture size, beam spots from 10 to 300 microns were generated. The laser fluence (0.100-2.5 J/cm^2) was estimated by averaging the total energy of the incident beam over the irradiated area. All laser transfers were performed at room temperature and atmospheric pressure.

Two different solutions were examined for the MAPLE-DW process of sol gel materials: 1) a ZrO_2 solution comprised of zirconium acetate + DI-water and 2) a PZT solution comprised of lead acetate trihydrate + titanium isopropoxide + zirconium acetylacetonate + DI-water + acetic acid. For the sol gel materials investigated, a ND:YAG laser (frequency tripled λ = 355 nm, 5-10 ns pulse) was used with a constant spot size of ~40 μm and a fluence of ~1 J/cm^2. Optical absorption measurements showed both solutions are transparent at the laser wavelength of 355 nm, therefore a thin absorbing layer of zirconium metal was used to transfer the sol gel materials. These ribbons were fabricated by sputter depositing a thin (50 nm) zirconium layer onto the quartz wafer. The sol gel solutions were coated onto the zirconium absorbing layer by dispensing 0.5ml of the solution and spinning at 2000 rpm for 45 seconds. The spin coating procedure was performed four times for each ribbon. Again, all laser transfers were performed at room temperature and atmospheric pressure

Each of the ribbons and transferred materials were characterized by scanning electro microscopy, energy dispersive x-ray spectroscopy (EDS), and 3-D surface profilometry. The cathodoluminescent properties of the thick film phosphor powders were characterized with a custom built electron flood gun. Low voltage cathodoluminescence (CL) spectra were measured with a Plasma Scan PSS-2 spectrometer, and CL efficiency measurements (1000V 22μA/cm^2) were taken with an International Light 1700 photometer equipped with a photopic filter.

RESULTS AND DISCUSSION

Figure 1 shows scanning electron micrographs and 3-D surface profilometry measurements of the ZnSi$_2$O$_4$:Mn ribbon after the MAPLE-DW process. Very efficient transfers of the ~15 μm thick film was achieved, as very little phosphor powder remained after the single 25 ns pulse. Close inspection of of the quartz ribbon after the laser transfer revealed tiny gold droplets on the quartz wafer. In addition, high resolution microscopy of the transferred material also showed gold droplets. These observations suggest that the ~100 nm gold layer is absorbing the laser radiation and ejecting the thick phosphor powders toward the close-proximity receiving substrate. At 248 nm, gold has a low reflectivity (~30 %) and high absorption. Calculations show that ~99.5% of the unreflected radiation is absorbed in the 100 nm of gold. Further studies on the transfer mechanisms and particle velocities are currently under investigation.

Figure 2 shows a scanning electron micrographs and a 3-D surface profilometry measurements of the transferred ZnSi$_2$O$_4$:Mn powders deposited onto glass substrates. Very good line width control and very dense thick film transfers are observed. Y$_2$O$_3$:Eu and BaMg$_2$Al$_{16}$O$_{27}$:Eu phosphors were also processed with the MAPLE-DW technique with similar transfer characteristics. To examine the effect that the MAPLE-DW process has on the phosphor materials, cathodoluminescence emission spectra and efficiency measurements were performed on the phosphor coated ribbons prior to the MAPLE-DW process and on the transferred materials. Cathodoluminescent (CL) measurements confirm that the laser transfer process does not affect the phosphor performance. The emission spectra were measured from Zn$_2$SiO$_4$:Mn and Y$_2$O$_3$:Eu ribbons prior to the transfer and on the transferred substrates. The emission spectra of the transferred Zn$_2$SiO$_4$:Mn and Y$_2$O$_3$:Eu were identical to the CL spectra of the phosphor ribbons. The CL efficiencies before and after the MAPLE-DWprocess were 1.3 lumens/Watt for the Zn$_2$SiO$_4$:Mn materials and 1.1 lumens/Watt for the Y$_2$O$_3$:Eu materials (measured at 1kV and 22.5μA/cm^2). The consistent CL spectra and efficiencies before and after the laser processing confirm that the MAPLE-DW process did not deleteriously affect the phosphor material.

Figure 3 shows scanning electron micrographs of the PZT sol gel ribbon after the MAPLE-DW process. Again, very efficient transfers of the ~1.2 μm thick film was achieved, as very little gel remained after the single 15 ns pulse. Inspection of the ribbon at high magnification, reveals 3 distinct regions on the ribbon after the MAPLE-DW process. To understand these three regions, energy dispersive x-ray spectroscopy (EDS) was performed on each of the regions. Region 1 (SiO$_2$) is the area irradiated by the laser pulse and EDS analyis revealed that in this region the zirconium matrix absorbing layer and the sol gel material are both ejected from the ribbon material. This transfer

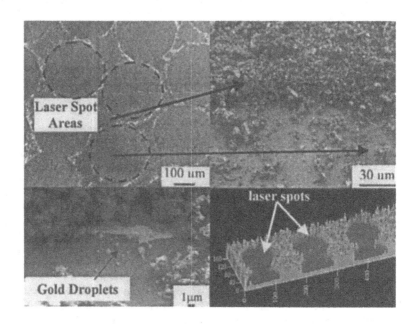

Figure 1. SEM micrographs and 3-D surface profilometry of a $ZnSi_2O_4$:Mn ribbon after the MAPLE-DW process.

Figure 2. SEM micrograph and 3-D surface profilometry of MAPLE-DW processed $ZnSi_2O_4$:Mn lines.

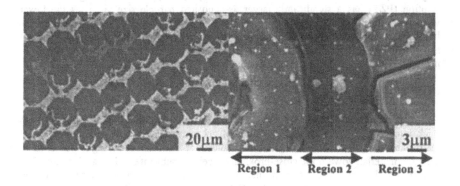

Region 1   Region 2   Region 3

Figure 3. SEM micrographs of a PZT sol gel ribbon after the MAPLE-DW process.

Figure 4. SEM micrographs of the MAPLE-DW deposited PZT (left) and ZrO2 (right) sol gel materials.

mechanism is believed to be similar to the thick powder because at 355 nm zirconium has a very low reflectivity (~1.3 %) and reasonable absorption (ie. 50 nm of zirconium ~26 % is not absorbed). Region 2 (Zr + SiO_2) extends ~ 5μm radially from the laser irradiated spot, and in this region only the sol gel has been ejected. Region 2 appears to be a so-called "heat affected region." The thermal energy generated by the absorbed laser pulse is conducted radially (~5 μm) and raises the temperature high enough to to volatalize the sol gel material, however, is not high enough to evaporate the zirconium metal. Finally, region 3 is simply the unaffected ribbon material which is representative of the bulk sol gel ribbon.

Figure 4 illustrates that both the PZT solution (a) and the ZrO_2 solution (b) deposit as a dense film. Inspection of the micrographs suggest a correlation of the deposited particle size and the initial viscosity of the solution. Though the viscocities were not quantitatively measured, the PZT solution was more viscous as evidenced by its larger as-spun thickness (~1.3 μm versus ~ 0.5 μm). Apparently the higher viscosity solution (PZT) has a significantly larger as-deposited particle size vis a vis the lower viscocity solution (ZrO₂). This is believed to be due to larger cohesive forces in the higher viscocity solution. EDS and x-ray diffraction analysis of the sol gel solutions show that the chemistry and amorphous structure of the as the as-spun ribbon is maintained after the MAPLE DW process.

CONCLUSIONS

A new direct write process, MAPLE-DW, was successfully used to deposit both thick film powder materials and thin film sol gel materials. In both cases a thin metal matrix absorbing layer with a high optical density at the laser wavelength was used to eject the materials toward the close-proximity receiving substrate. SEM analysis and 3-D surface profilometry demonstrated that in both cases the transfers were dense and maintained good line width control. Cathodoluminscence analysis showed that the MAPLE-DW process did not degrade the performance of the phosphor powders. EDS analysis of the sol gel ribbon revealed 3 specific regions; 1) the laser irradiated region, 2) a heat affected region, and 3) the unaffected ribbon. Finally, the chemistry and amorphous structure of the sol gel material is maintained during the MAPLE-DW process.

REFERENCES

1. J. Bohandy, B.F. Kim, and F.J. Adrian, J. Appl. Phys. **60**, 1538 (1986).
2. J. Bohandy, B.F. Kim, F.J. Adrian and A.N. Jette, J. Appl. Phys. **63**, 1558 (1988).
3. R.A. McGill, R. Chung, D.B Chrisey, P.C. Dorsey, P. Matthews, A. Piqué, T.E. Mlsna, and J.L Stepnowski, IEEE Trans. On Ultrasonics, Ferroelectrics and Frequency Control, **45**, 1370 (1998).
4. R. A. McGill, D. B. Chrisey, A. Piqué, T. E. Mlsna, SPIE Proceedings, **3274**, 255-266, (1998).
5. Piqué, R.C. Aeuyung, R.A. McGill, D.B. Chrisey, J.H. Callahan, and T.E. Mlsna, Mat. Res. Soc. Proc. **526**, 375, (1998).
6. A. Piqué, D.B. Chrisey, B.J. Spargo, M.A. Bucaro, R.W. Vachet, J.H. Callahan, R.A. McGill, D. Leonhardt, and T.E. Mlsna, Mat. Res. Soc. Proc. **526**, 421, (1998).
7. D.B. Chrisey, A. Pique, J.M Fitz-Gerald, R.C.Y. Auyeung, R.A. McGill, H.D. Wu, M. Duignan, Applied Surface Science, 154-155, (2000) p.593-600.

ROLE OF POWDER PRODUCTION ROUTE IN DIRECT WRITE APPLICATIONS

M. B.(Arun) Ranade*, Z. Serpil Gonen** and Bryan W. Eichhorn**
*Particle Technology, Inc., P. O. Box 712, College Park, MD 20741
**Chemistry Department, University of Maryland, College Park, MD 20742

ABSTRACT

Various production techniques are available for many materials for the direct write (DW) applications. Single metals and alloys are prepared from vapor and spray reactors as well as by precipitation techniques. Single and multicomponent oxides such as ferrites are also possible from these techniques in the nanometer to micrometer size ranges. Examples with the use of these techniques for production of metals and ferrites for DW are described with examination of crystallinity and morphology.

INTRODUCTION

Direct write transfer for fabrication of electronic circuits involves making an ink or a paste with powders of a variety of functional materials along with additives and a vehicle. The mixture is then transferred onto a substrate using one of several direct write techniques currently available or in development. Depending on their desired function these powders may vary from metals to oxides and other. The transferred material must consolidate on the substrate to maximize their functional characteristics and must adhere to well to the substrate. Feature Characteristic feature dimensions may vary from submicron to several hundred microns.

The functions required from powders may range from dielectric properties, resistors, inductors, and capacitors to solid electrolytes for active energy supply. Most of the materials to satisfy these needs are single and multicomponent oxides, metals and alloys. Because of the small feature size, particle size is limited to less than 2-3 μm.

The powder has to be well dispersed and incorporated with various additives, binders and a vehicle to make ink or a paste. Once the material is transferred on the substrate the vehicle and solvents have to be removed by evaporation and other chemical transformation. To achieve a dense layer of the functional material in the final feature predictable high loading of the powder in the paste or ink is necessary. A spherical particle shape is desirable to permit high loading at a given viscosity. Multimodal or broad size distributions are best attributes of the powder to achieve high loading by distribution of smaller particles between the interstices formed by larger particles.

POWDER PRODUCTION ROUTES

Table I is a summary of powder processes available to produce the desired oxide and metal powders. Three major routes, solid state; aerosol; and liquid precipitation are represented.

Table I: Powder Production Routes

Production Process	Applicability	Particle Size, μm.	Shape	Post Production Treatment	Comparative Cost
Solid State Reaction	Single, multicomponent oxides	> 5	Irregular	Calcination, size reduction	Primarily Inexpensive, post production processing may be expensive
Spray Pyrolysis	Single, multicomponent oxides, metals,alloys	0.5 – 15	Spherical	Calcination may be needed	Value added offsets higher cost
Vapor Phase Reactor	Single oxides, metals, nitrides,carbides	10 nm to 1 μm	Equi – axial, lacy agglomerates	Calcination may be needed	Inexpensive in high volumes
Solution Precipitation	Single, multicomponent oxides, hydroxides	10 nm to several μm	Varity of shapes possible	Calcination required	Contamination freedom and size control may be expensive
Hydrothermal	Single, multicomponent oxides	10 nm to several μm	Generally equi-axial	Calcination may not be required	Moderately expensive – high pressure equipment, batch processing

SOLID STATE ROUTE

Solid state reactions represent the conventional technology used in producing many single and multicomponent oxides such as dielectrics and ferrites. As the name implies solid reactants are mixed and heat-treated to achieve the desired crystallinity. Grinding of the product is necessary to obtain fine particles. The lower size is generally 5 μm, and the particles are very irregular in shape. The process is capable of low powder cost but the grinding costs, the possibility of contamination during grinding and the irregular shape make it less attractive for the direct write applications.

AEROSOL ROUTE

Spray Pyrolysis (SP) and Vapor Phase (VP) reactor technologies are based on processing while suspended in a gaseous medium capable of producing spherical or equi-axial powders in the desirable size range.

A mixture of soluble metal compounds such as nitrates and acetates or chlorides are dissolved in water to from a feed solution which is then atomized into a mist using one of several types of atomizing devices which include single fluid, two fluid, rotary or acoustic atomizers. The droplets are conveyed through a hot zone where the mist drops are evaporated and the remaining precursor compound is decomposed or converted into the final product.

The principal advantage offered by SP is that the powder particles are spherical and the size can be selected by choice of the solution concentration and the atomizing device. Broad size distributions may be achieved by simultaneously feeding mists of different droplet size and solution concentrations. The process is continuous and scalable to high production rates.

The residence time in the reactor hot zone where the conversion takes place is limited to several tens of seconds and post treatment such as calcination may be required to achieve the desirable crystalline phase. Some of the advantages of shape and size may be lost to sintering or crystal facet development if too high a calcination temperature is necessary. The process is especially attractive for multicomponent oxides and alloys.

The cost of powder is primarily determined by the particle size as the ratio of the product mass to carrier gas has determines the thermal energy costs for the energy lost in heating the carries gas. The ability to tailor size distribution and spherical shape often offset the extra processing costs and makes the SP process competitive with the solid state route.

Example of SP Powder – We have used the SP process to produce many materials suitable for DW applications. Yttrium Iron Garnet (YIG) powder was made following a paper by Matsumoto et al.[1] with minor modifications. Nitrate solutions in appropriate stoichiometric proportion were atomized using a 1.65 MHz. acoustic transducer. The mist was carried through a furnace at 1000°C. And the product was collected on a filter bag. The powder examination indicated a mass median size of 1.6 μm and undeveloped crystallinity.

Two powder samples were calcined in air at 950°C and 1000°C for 3-4 hours in air. The sample calcined at both conditions showed good crystallinity. However, the sample treated at higher temperature showed considerable deformation from sphericity and aggregation while the other sample was mostly spherical with some new features on surface as seen in Figure 1.

Figure 2 shows an example of nickel conductor powder with good crystallinity made using nickel nitrate precursor solution[2]. We have made several other mixed oxide materials such as MnZn ferrite and barium alumino boro silicate glasses.

In VP, the particles are condensed from the vapor phase. Vapor is produced either by evaporation of the desired material or by chemical reaction between precursor chemicals. An example of the technology on a large scale is titania pigment production by gas phase reaction between titanium tetrachloride and oxygen. The process is especially suitable for nanometer size powders[3].

The main problem with VP technology at high production rates is coagulation between the primary particles forming agglomerates which become hard aggregates if the temperature is high enough for sintering to occur. Rapid quenching and dilution minimizes the aggregation formation. The VP process is continuous and can be very cost effective as seen in the commodity level production of titania pigments.

Since particle formation is by nucleation from vapor phase, materials of different vapor pressure condense at different times making the process suitable primarily for single component materials.

Example of VP Powder – We have used the VP to produce nanoscale silver powder by evaporating silver in a carrier nitrogen gas and rapidly quenching ånd dilution in a continuous process. The micrograph in Figure 3 indicates primary particles in 20-50 nm range.

LIQUID PRECIPITATION ROUTE

Precipitation from liquid phase precursor is widely used for single and multicomponent oxides and metals. Very uniform particles are possible in the 0.5 to 3 μm size range. Generally these powders are have poorer crystallinity and are amenable to contamination as the liquid is in contact with the apparatus. Hydrothermal processing using high pressures has the potential of preparing desirable powders in nano to sub micrometer size range at costs comparable to spray pyrolysis.

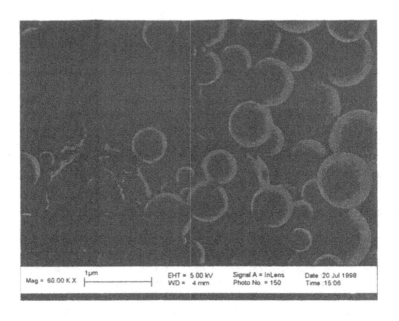

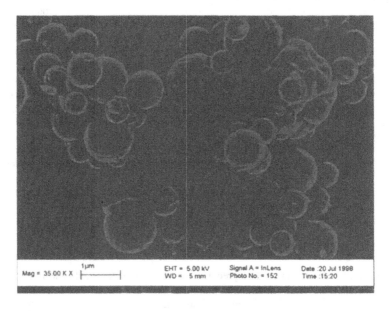

**Figure 1: Scanning Electron Micrograph of YIG Ferrite Powder -
Top: Uncalcined Powder; Bottom: Powder Calcined at 950°C for 4
Hours**

38

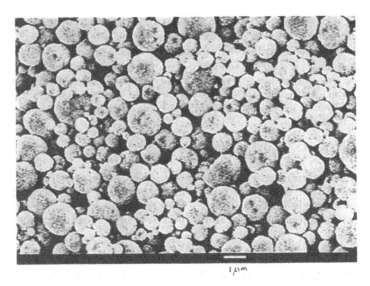

$1 \mu m$

Figure 2: PTI Nickel Powder for Multilayer Capacitors

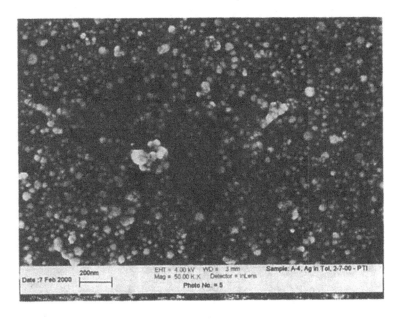

Date :7 Feb 2000   200nm   EHT = 4.00 kV   WD = 3 mm   Sample: A-4, Ag in Tol, 2-7-00 - PTI
Mag = 50.00 K.X   Detector = InLens
Photo No. = 5

Figure 3: Nanoscale Silver Powder Produced by Vapor Phase Route

Example of Hydrothermal Powder: The manganese zinc ferrite $Mn_{0.5}Zn_{0.5}Fe_2O_4$ was prepared hydrothermally from nitrate precursors according to the method of Lin et al[4]. Nitrates in stoichiometric proportion were dissolved in a minimum amount of water. Aqueous NH_4OH was added as prescribed. The mixture was sealed in a long quartz tube and loaded into a LECO hydrothermal reactor. The mixture was heated for 5 hrs. at 150 °C (autogeneous pressure) and cooled back to room temperature. The product was then filtered and washed with ethanol and acetone. XRD analysis of the product showed it to be a microcrystalline single phase product without further calcination.

SUMMARY

Amongst the various route available for powder production, the aerosol route with spray pyrolysis and vapor phase process offer several distinct advantage by producing particle of tailored size and spherical or equi-axial shape. Hydrothermal process offers a potential for mixed oxides, especially in nanometer size scales where the other processes do not work well.

REFERENCES

1. K. Matsumoto, Y. Yamanabe, S. Sasaki, T. Fuji, K. Honda, T. Miyamoto, J. Appl. Phys. **70,** pp. 5912-5914 (1991)
2. M. B. Ranade, D. V. Goia, G. J. Varga, B. W. Gamson, J. Bara, USP #5,928,405(1999)
3. H. Hahn, Nanostructured Materials, 9, pp. 3-12 (1997)
4. W. H. Lin, S.-K. J. Jean, C.-S. Hwang, J. Mater. Res., 14, 204-208, (1999).

MATERIAL SYSTEMS USED BY MICRO DISPENSING AND INK JETTING TECHNOLOGIES

Jie Zhang*, Irina Shmagin, James Skinner, John Szczech** and Daniel Gamota

Motorola Labs - Advanced Technology Center
Schaumburg, IL
*Email: ajz016@email.mot.com
**University of Illinois at Chicago

ABSTRACT

In today's electronic industry, manufacturers are continuously improving capital utilization, developing flexible manufacturing processes, reduce changeover time and improving yield and throughput. Interest in rapid prototyping and 3-D fabrication capabilities are rapidly increasing, and a number of candidate direct writing technologies are in development to meet these demands.

This work studies material systems used by data driven materials deposition (DDMD) technologies for potential low temperature reel-to-reel high volume manufacturing on low cost substrates. Characterization results of fabricated discrete and RF devices using commercially available micro dispensing and ink jet systems will be discussed. Material rheological properties, deposition process characterization, deposition repeatability, fabricated device reliability and electrical performance will be presented. The test vehicles contain resistors and capacitors, transmission lines, open and short series stub filters, and half-wavelength resonators. The material/substrate compatibility will be demonstrated through environmental conditioning of the test vehicles. In addition, a cost estimate for using micro dispensing technologies was conducted to compare current manufacturing technologies to DDMD.

INTRODUCTION

Product miniaturization and functionality enhancement have demanded the development of advanced manufacturing technologies and processes. DDMD technologies contain many attributes: low cost, flexible patterning, fast change over, and reel-to-reel processes. Micro dispensing and ink jetting may replace traditional manufacturing methods to meet market demands [1].

Micro dispensing is the most familiar DDMD technology to high volume manufacturing. It has been used for chip encapsulation, molding, conductive adhesive deposition and solder joint repair. The micro-dispensing tools from the different suppliers are based on different pump technologies: positive displacement pumps, auger screw driven pumps, and various valve pump systems. The deposition methods also include contact and non-contact processes. The micro dispensing vendors for electronic applications are Asymtek, Camalot, Nordson and MRSI, Ohmcraft, Vacuum Metallurgical Corp. Lt. (VMC). Micropen^{TM} from Ohmcraft uses contact dispensing while the others are non-contact systems. The aforementioned suppliers sell production ready systems for integration into electronic product manufacturing lines. These systems are capable of dispensing materials having a broad range of rheological properties.

Ink jet technology is a material deposition method used in the printing industry. The first attempt to use this technology in electronic manufacturing was the solder jet system – jetting molten Sn/Pb solder on IC chips to form IC interconnects [2]. There are two types of jetting systems: drop-on-demand and continuous jetting. The driving force for ejecting droplets can be electro-mechanical (PZT), electro-magnetic, thermo-fluid-dynamic, or hydrodynamic. Several drop-on-demand PZT ink jet device suppliers, InkJet Technologies and Leader Corporation, have offered jetting systems for pharmaceutical, electronic and automotive industries in an effort to develop the processes for jetting unique materials. MicroFab has demonstrated jetting DNA solutions, thermoplastic polymers, and eutectic Pb/Sn solder. The jetting systems have restricted material properties, e.g. viscosity of 3- 100cps.

In this study, the feasibility of fabricating passive and RF devices using micro dispensing and jetting technologies was evaluated. The materials system characterization included: rheological properties, deposition characteristics, and electrical properties of fabricated structures. The characteristics of depositing polymer thick film (PTF) pastes (micro dispensing) and conductive inks (jetting) were benchmarked. The electrical and reliability performances of DDMD structures were studied and compared to conventional electronic devices.

MATERIALS SYSTEMS FOR DDMD

Material systems for DDMD are technology dependent. For micro dispensing and ink jetting, the materials rheological properties are important. Polymer pastes have adequate electronic properties for forming PCB components, are compatible with substrate materials, and are easy to dispense. Low viscosity and low surface tension of organo-metallic solutions and nanoparticle suspensions are suitable for ink jet technologies.

Micro dispensing systems are capable of delivering pastes having viscosities up to 300K Pa-sec. PTF systems consist of a polymer binder in solvent and solid flake(s). The solvent/resin determines the thermal processing conditions, physical property of the paste upon curing and adhesion to the substrate are determined by the resin. The type of solid flake determines the electrical property requirements, i.e. metals for conductors, carbon for resistors, and Barium-titanate for dielectrics. For a given polymer and solvent system, the physical properties of the flakes - size, size distribution, shape, and strength - strongly influence the paste rheological properties, e.g. static viscosity. The particle size and size distribution are also important parameters for choosing dispensing needle dimensions to avoid needle clogging.

Viscosity studies for selected Silver and Carbon pastes were conducted using a Bohlin viscometer fitted with a conicylinder (couette) geometry-testing fixture. The tests were conducted at constant ambient conditions (25°C). PTF pastes selected for DDMD processing studies were subjected to a range of shear rates to examine the rheological properties of the materials (Table 1). The two C pastes represent different sheet resistance. All materials displayed shear-thinning behavior.

Table 1. Viscosity responses for PTF pastes

Shear Rate (/sec)	Ag (Pa-sec)	*C1 (Pa-sec)	*C2 (Pa-sec)
0.0465	401	706	1750
0.117	180	436	1176

0.463	59	230	694
1.17	31	151	354
4.65	14	112	238

*The sheet resistance per mil thickness of C1 and C2 are 1kΩ/sq and 10kΩ/sq respectively

Materials suitable for ink jetting require an optimal apparent viscosity, surface tension and density. Material properties affect the flow through capillary nozzle, pressure wave propagation, and subsequent droplet formation. There were two materials evaluated using piezo ink jet technology: organo-metallic ink and nanoparticle suspensions.

An experimental palladium precursor solution was used to evaluate the jetting ability to fabricate circuitry on FR4, ceramic, polyimide and PEN substrates. For these tests, the Pd precursor ink had 14wt% to 17wt% solids. The Pd ink requires rapid pyrolysis for thin film formation. Commercially available Gold (Au) and Silver (Ag) nanomaterial suspensions have been jetted. The nanosized particles were suspended in an organic solvent. The weight percentage of solids in the suspensions varies from 30% to 40%. The organic solvent volatilized during heating and the high surface energy nanoparticles fused together to form high purity and low resistivity thin metal film traces on polyimide and ceramic substrates [3].

The rheological properties of the conductive inks are listed in Table 2. The properties for Isopropyl Alcohol (IPA), Pd, and nano inks are optimal for the jetting system. A contact angle goniometer (Kruss) and DMA 4500/5000 U-tube oscillation density meter (Anton Paar) were used for surface tension and density measurements, respectively. The viscosity was measured using a Bohlin viscometer.

Table 2. Rheological properties of jetted material

Material	Surface Tension (mN/m)	Viscosity (cps)	Density (g/cm^3)
IPA	27.29	2	0.98
Pd	27.81	2	1.13
Ag	28.90	1	1.4

In addition to rheological property requirements, the material processing conditions and electrical properties are also critical for low cost and conformal electronic design and manufacturing. These properties are listed in Table 3. The processing conditions for selected PTF pastes and organo-metallic inks are compatible to low Tg organic PCB and flex, while the thin film inks are compatible to higher Tg flex.

Table 3. Material system process conditions and electrical properties

Materials	Process Conditions	Environment	Electrical Property
PTF Ag	< 150 C, <20 min	Ambient	7-10 mΩ/squa
PTF C	<170 C, < 20 min	Ambient	50 –100K Ω/squa
Nano inks	300 C, 30 min	Ambient	0.1 mΩ-cm
Organo-Metallic inks	130 C, 3 min	Ambint	--

DDMD FABRICATION, DEVICE RELIABILITY AND ELECTRICAL PERFORMANCE

1. Thick Film Structures

A Camalot micro dispensing system was used to fabricate thick film structures. The system uses a non-contact dispensing scheme, which is sensitive to the rheological properties of the paste and surface energy of the substrate. Due to constant volume flow rate during micro dispensing, the dispensed structure dimensions, width and thickness, are controlled by adjusting the dispensing speed. In addition to dispensing speed, the dispensing nozzle to substrate gap, dispensing pump initial and ending scheme also have to be optimized according to material properties and writing speed.

Test vehicles for passives and coplanar wave-guides (CPW) RF devices have been designed for dispensing process optimization and material electrical property characterization. The passives included resistors and capacitors. The RF devices were CPW transmission lines, half wavelength resonators, and open and short series stub filters.

As an example, Figure 1 shows the fabricated resistor test vehicle on a flexible polyester film (Tg= 150°C). The circuits and the resistor terminations were fabricated using 7 mΩ/sq PTF Ag paste. The resistors were made from 400 Ω/sq carbon paste. The critical dimensions of this test vehicle are the distances (gaps) between two termination pads, and width and thickness of resistor patterns. The cross section shown in Fig. 1b illustrates these critical dimensions. There are four different aspect ratio resistors in the test vehicle, 20, 4, 2, and 1 referred to as R1, R2, R3, and R4, respectively. As it was expected, the low aspect ratio resistor had the greatest variation (up to 8%). Studies showed that the resistance variation was introduced by the Ag termination gap and the resistor cross-section area. The test vehicles fabricated with PTF carbon resistor was subjected to 85% humidity and 85 °C temperature (85%/85°C) test for 168 hours. The test results showed that the resistance changes were less than 10%, as indicated in Fig. 1c. The lower aspect ratio resistors had the greatest changes due to the contact resistance effects.

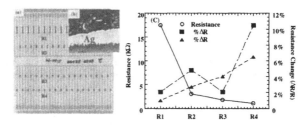

Figure 1. DDMD resistor test vehicle

The test vehicle for evaluating DDMD transmission lines is shown in Fig. 2, and consisted of CPW transmission lines. The transmission lines were fabricated on a 32 mils thick FR-4 substrate (ε = 4.2). The transmission line dimensions are: signal line width 80 mils, ground line width 100 mils, spacing between ground and signal line 10 mils, line length 2.5 inches. The test vehicle included a copper transmission line, structure A in Fig. 2a. The DDMD transmission lines were fabricated by dispensing a silver conductor line on the copper pads, structure B in Fig. 2a. Similar RF responses of these two CPW structures were resulted, shown in Fig. 2b.

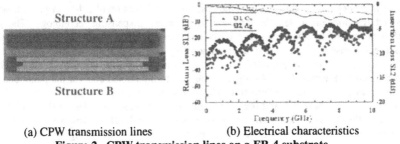

Structure A

Structure B

(a) CPW transmission lines   (b) Electrical characteristics

Figure 2. CPW transmission lines on a FR-4 substrate

2. Thin Film Structures

The PZT ink jetting device operation uses fluidynamic principles for discrete droplet formation from a liquid capillary flow. The frequency and size of the droplets are controlled by the electrical signal: voltage amplitude, frequency, and dwell time. The jetting system used in this study was a MicroFab jetting device integrated on a Universal Corporation 3D control platform (X-Y placement accuracy of $\pm 10\mu m$).

The ink jet system was used to deposit organo-metallic and nano particle inks to form thin film structures on rigid and/or flexible substrates. Detailed deposition process development has been reported in [3]. The thickness of the ink jet deposited thin film was 0.4 to 3 micron. The minimum width of the thin film line was defined by the individual droplet size and the surface energy of the substrates. To form a continuous line similar to those shown in Fig. 3a, the droplet deposition pitch had to be slightly smaller than the collapsed droplet radius on the substrate, which formed the minimum line width of the droplet diameter on the substrate. The surface morphology and microstructure of thin film lines formed with nano Ag particles are shown in Figs 3b and 3c. The conductivity of the Ag traces after sintering at $300°C$ for 30 min was 10^{-4} Ω-cm. The thin Ag film test vehicles fabricated by jetting were subjected to 85%RH/85°C conditioning. Less than 1% resistance change for all 20 samples was observed after 168 hr.

Figure 3. Ag thin film, a) 200 μm wide Ag conductor on a ceramic substrate, b) surface morphology of the Ag conductor, and c) vertical microstructure of 3 μm thick Ag line

COST ESTIMATION FOR USING DDMD TECHNOLOGIES

Cost is one important measure of feasibility for using DDMD technologies when compared to conventional manufacturing methods. A DDMD cost model was developed in an effort to compare DDMD systems to conventional manufacturing processes. This model is sensitive to several categories of manufacturing parameters: 1) general factory data; 2) equipment; 3) material costs; and 4) processing parameters.

The model was used to evaluate the cost for the fabrication of a 4-layer pager board with embedded resistors on a flexible substrate using micro dispensing technology. The general factory data, including labor and utility costs, etc., were industry standards. The equipment and processing parameters were based on commercially available systems. The cost of DDMD materials and substrates were quoted for medium volume manufacturing by vendors. The cost data generated by the model was 4x higher compared to the quoted cost for the pager board with SMT resistors in the same medium throughput. The distribution of DDMD cost in which the material cost is less than 2%, and the equipment and labor costs are 81% of the total cost. It is obvious that the low throughput and low system automation levels contributed to the high cost. Acceptance of DDMD will require multiple print head micro dispensing systems. The model showed that a 4-head system was cost competitive with current manufacturing process. This advanced system can provide parallel printing operations thereby increasing manufacturing throughput.

CONCLUSIONS

PTF conductive and resistive materials were characterized and used to fabricate passives and RF devices on organic and flex substrates. Studies have shown that PTF system dispensing can be utilized in applications where electrical properties of PTF material are adequate.

Jetting conductive inks to form thin film, ≤3 μm thick, is feasible. The resistivity of the thin film is an order of magnitude lower than PTF systems. The rheological properties of conductive inks to enable the fabrication of these structures have been identified.

Reliability studies of DDMD materials and structures were conducted. The results demonstrated good electrical performance of rigid and flexible test vehicles under various accelerated testing conditions. All reliability testing results showed adequate performances for consumer electronics.

The cost for using DDMD technologies in manufacturing depends on the system throughput. The major cost for micro dispensing technology was contributed to capital and labor. For DDMD technology to be cost competitive, the system throughput and automation level must be enhanced.

ACKNOWLEDGEMENTS

The authors would like to thank to Jocilin Odulio, Judy Liu, Charley Ding and Dave Patton for their many contributions for the DDMD studies and RF testing. We would like to thank Mark Cholewczynski for conducting reliability testing.

REFERENCES

1. J. Zhang et al., "Evaluation of Advanced Materials Systems and Data Driven Materials Deposition Technologies" internal document

2. D. Hayes et al., Int. J. Microcircuits & Electronic Packaging, Vol. 16, p174, 1993

3. J. Szczech et al., "Manufacture of Microelectronic Circuitry by Drop-on-Demand Dispensing of nano-Particle Liquid Suspensions", MRS2000, Symposium V, V.1.6.

OFFSET PRINTING OF LIQUID MICROSTRUCTURES FOR HIGH RESOLUTION LITHOGRAPHY

SCOTT M. MILLER*, ANTON A. DARHUBER*, SANDRA M. TROIAN* and SIGURD WAGNER**
*Interfacial Science Laboratory, Dept. of Chemical Engineering, Princeton University
**Dept. of Electrical Engineering, Princeton University, Princeton, N.J. 08544

ABSTRACT

We have investigated the direct printing of polymer solutions from a chemically patterned stamp onto a hydrophilic target substrate as a new high-throughput alternative to optical lithography. The patterns on the stamp, which are typically in the micron size range, define regions of alternating wettability. They are produced by patterning a hydrophobic self-assembled monolayer previously deposited onto a hydrophilic surface, typically a glass slide or silicon wafer with a natural oxide coating. Polar liquids or aqueous polymeric solutions are then deposited only onto the hydrophilic surface patterns by dip-coating the stamp in a liquid reservoir. The deposited film thickness depends critically on the speed of withdrawal and the feature size and shape. For vertically oriented hydrophilic stripes dipped in a reservoir containing a polar liquid, we have developed a theoretical model whose prediction for the maximum deposited film thickness agrees exceptionally well with experimental measurements. After deposition, the wetted stamp is pressed against a target substrate by means of a motion controlled press. In this way we have so far printed 5μm wide polyethylene oxide lines onto a silicon wafer.

INTRODUCTION

The production of low–cost large–area electronics requires the development of new fabrication techniques that are considerably less expensive and allow higher throughput than conventional photolithography. The offset printing of materials for electronic device fabrication is one such approach under investigation. This type of contact printing, also known as lithographic printing, is heavily used in the graphic arts industry for the fast reproduction of newspapers, magazines, and other high–volume products. Ink is confined in regions of alternating wettability on the plate roller, which is then pressed onto a blanket roller. The desired ink patterns are then transferred from this soft roller surface onto the desired medium, such as paper. Since the layers that define the wetting and non–wetting regions are microscopically thin, the entire surface can be regarded as truly planar and only the surface chemical heterogeneities determine the ink placement. Pattern fidelity upon transfer to the target surface is achieved even at high roller speeds. Alternative techniques for ink confinement which also allow for high resolution reproduction, like gravure printing, depend instead on geometric constraints such as etched grooves. Extending the conventional offset process for use in microelectronics fabrication requires advances and optimization of three key steps. The resolution of the stamp pattern must be scaled down to the micron range. This first step is rather easy to accomplish given the number of self assembling monolayer treatments that have been introduced to produce substrates of tunable wettability at length scales down to the nanometer range [1]. Secondly, a simple but rapid deposition process is required so that the volume of liquid deposited and printed can be accurately controlled. Finally, it may be necessary to develop new inks that are compatible with the first two steps and useful for device fabrication, such as etch resists or insulating and conducting materials. In this paper, we focus on preparation of offset printing plates with micron–size features and the inking and printing of liquids with such plates.

Mat. Res. Soc. Symp. Proc. Vol. 624 © 2000 Materials Research Society

EXPERIMENT

Printing Plate Fabrication

The stamps used in our study of offset printing are fabricated by patterning the surface of a hydrophobic self–assembled monolayer which resides on the surface of a hydrophilic substrate. Since the monolayer is molecularly thin, on the order of 3 nm, while the feature sizes of the imprinted pattern are on the micron scale, the stamp surface can be regarded as flat. The only deviations from the flat profile are the liquid microstructures deposited during the printing process.

We have prepared two types of printing plates as illustrated in Fig. 1, each with distinct advantages. The first type of substrate consists of a patterned monolayer of octadecyl-trichlorosilane (OTS) on a silicon wafer or hexadecane thiol (HTD) on a gold film deposited against a chromium adhesion layer on a silicon wafer. In each case, areas coated with the monolayer are strongly hydrophobic while areas which expose the silicon oxide or gold surface are completely hydrophilic. These surfaces can be rapidly patterned by microcontact printing [2] or deep UV photo–cleavage [3]. The second type of printing plate we have used is prepared by first depositing a thin film of gold, approximately 50 nm, above a 5 nm adhesion layer of chromium on an oxidized silicon wafer and then patterning the gold and chromium using photolithography and wet chemical etching. After patterning the metal layers, an HDT monolayer is deposited selectively on the gold surface from a 1 mM ethanolic HDT solution. This renders the gold coated areas hydrophobic and leaves the exposed silicon dioxide surfaces hydrophilic. Because the metal layer is much thinner than the pattern feature size, this structure is functionally equivalent to the first type of printing plate. This design has two advantages. Not only is the wettability pattern easily refreshable by monolayer removal and redeposition but it is optically visible because of the difference in reflectivity between gold and silicon surface. This allows direct observation of the degree of pattern fidelity between the imposed surface pattern and the areas of liquid pickup and ink transfer. Silicon wafers ([100]-oriented p-type doped) were used as the base substrate because they are very smooth and flat which allows good contact between the two surfaces for printing macroscopically large areas.

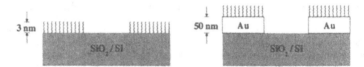

Figure 1: Two printing plate structures with surface wettability patterns defined by hydrophobic self–assembled monolayers. (a) A patterned monolayer formed on an unpatterned substrate. (b) A monolayer formed selectively on a patterned gold film on a substrate. In both case, the thickness of the monolayer or gold and monolayer combination is negligible in comparison to the patterned microscale features.

Liquid Deposition and Printing

The first step of the printing process is the deposition of liquid "ink" onto the printing plate. We accomplish this by selective dip coating, in which the patterned printing plate is withdrawn at constant velocity from a bath of liquid which preferentially wets the hydrophilic areas of the plate, generating liquid microstructures. We have studied the stability of liquid microstructures on such patterned surfaces in Ref. [4]. Inks used in this study were glycerol, tetraethylene glycol (TEG), and aqueous solutions of 15 wt% poly(ethylene oxide)

(PEO, mol. wt. 100 kg/mol). The first two liquids have no significant evaporation on the time–scale of the experiments because of their low vapor pressure. Water does evaporate from the PEO solution but the residual polymer is left only on the hydrophilic portions of the dip-coated sample. The viscosity, η, of glycerol and TEG, both Newtonian liquids, was measured with a capillary viscometer which gave 975 mPa·s and 51 mPa·s, respectively. Glycerol is a hygroscopic liquid and this value indicates the presence of absorbed water. Measurement of the viscosity versus shear rate behavior of the PEO solution using a cone–and–plate rheometer confirmed Newtonian behavior up to a shear rate of approximately $30s^{-1}$, with a limiting viscosity of 1970 mPa·s. The liquid surface tensions, measured with a platinum Wilhemy plate, were determined to be $\gamma_{\text{glycerol}} = 63.0$ mN/m, $\gamma_{\text{TEG}} = 41.4$ mN/m, and $\gamma_{\text{PEO}} = 53.5$ mN/m. The height of each liquid structure formed by dip-coating was measured by interferometry. Microstructures were visualized using a reflected light microscope equipped with a green bandpass filter whose transmittance peak was centered about 550 nm. The inter–fringe spacing was of the order of 200 nm. The exact value depends on the liquid index of refraction.

The second step in the printing process is the transfer of ink from the printing plate to an unpatterned (hydrophilic) substrate. A motion controller was used to contact the inked printing plate with the target substrate (an oxidized silicon wafer) and to separate the plates.

RESULTS

Dip Coating

Landau and Levich analyzed the dip coating of a Newtonian liquid onto a homogeneous wetting plate withdrawn normal to the surface of a liquid reservoir [5]. Their analysis is valid for Newtonian liquids in the lubrication regime at low capillary number; the film thickness is small relative to the plate dimensions, and surface and viscous forces are dominant over inertial forces. The analysis relies on matching the curvature of the deposited film far above the liquid bath to the curvature in the meniscus region, which for small capillary numbers assumes its static shape. The final film thickness, h_∞, is determined to be

$$h_\infty = 0.946 \left(\frac{\gamma}{\rho g}\right)^{1/2} \left(\frac{\eta V}{\gamma}\right)^{2/3} \tag{1}$$

where ρ is the liquid density, g is the gravitational constant, and V is the plate withdrawal speed. The capillary number $\text{Ca} = \eta V/\gamma$.

In Ref. [6] this analysis is extended to a plate patterned with a vertical wetting stripe. The curvature perpendicular to the vertical stripe modifies the static meniscus profile changing the final exponent in Eq. (1). The maximum deposited film thickness is instead given by

$$h_\infty \propto W \left(\frac{\eta V}{\gamma}\right)^{1/3} , \tag{2}$$

where W is the (hydrophilic) stripe width.

Figure 2(a) presents experimental data for the height of a liquid line deposited on a hydrophilic stripe versus velocity of withdrawal for a stripe 40 μm wide withdrawn from a bath of glycerol. The data shows excellent agreement with the theoretical prediction of a 1/3 exponent. Further study has also confirmed the linear dependence on the width W [6]. Figure 2(b) shows the effect of varying the azimuthal angle of the hydrophilic stripe on the deposited liquid height. The data points represent experimental measurements at 40 μm/s withdrawal velocity for a line width of 40 μm. The deposited film thickness changes little for angles less than approximately 45° but rises steeply by a factor of 2.3 for larger angles. For the larger angles, the effective line width increases, which modifies the meniscus shape

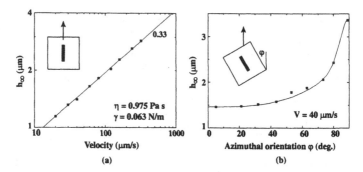

Figure 2: The thickness of a liquid film deposited on a wetting stripe on a hydrophobic surface by dip coating. Measurements were conducted using a 40 μm wide wetting line. (a) The thickness as a function of velocity of withdrawal. Points are experimental measurements, the line is the analytical 1/3 power law prediction. (b) Film thickness as a function of azimuthal angle of the wetting line. Points are experimental measurements made at 40 μm/s withdrawal velocity, the curve is a guide to the eye.

in such a way as to increase liquid pickup. The effective withdrawal velocity along the stripe direction also increases.

Figure 3 shows typical liquid structures obtained by dip coating. Figure 3(a) shows three lines of glycerol 18 mm long, and 16 μm wide, separated by 7 μm. The length to width ratio exceeds 1000. Figure 3(b) shows a 3.5 μm wide line of PEO. The structures in Fig. 3(a) and (b) were both formed on the first type of printing plate (OTS/SiO₂). Liquid structures other than straight lines can also be produced by dip-coating, as illustrated in Figure 3(c), which shows a half–loop of glycerol formed on the second type of printing plate (HDT/Au/SiO₂). The line width of this half–loop is 100 μm not because of any fundamental limitations related to the shape, but because the printing plate was patterned from an inexpensive photolithographic mask made from a commercial image–setter [7]. Much smaller pattern features require the use of a more expensive chromium mask, as was used for other printing plates in our study.

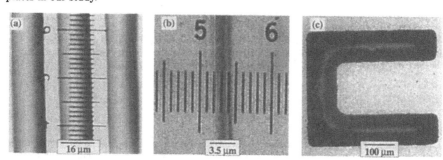

Figure 3: Materials deposited onto patterned printing plates via dip coating. (a) Three parallel lines of glycerol, each 16 μm wide with 7 μm interline separation. (b) A 3.5 μm wide PEO line. Both (a) and (b) were formed on an OTS/SiO₂ surface. (c) A half loop of glycerol with line width 100 μm formed on an HDT/Au/SiO₂ surface.

Liquid Printing

Polymer structures were printed onto unpatterned silicon wafer target substrates as illustrated in Fig. 4. The left panel shows a printed line 3 μm wide, while the right panel shows two printed droplets, each 4 μm wide. The height of the droplets was estimated from the interference fringes to be 190 nm, which corresponds to a volume less than 2 femtoliters.

Both the printed line and droplets display an asymmetry to the left. This effect is likely caused by a non–uniformity in the printing process. When the printing plate and the target substrate were drawn apart, one side of the substrates separated first, causing the polymer stripe which bridged the two substrates to undergo an instability similar to viscous fingering. For the case of a liquid sheet between separating solid substrates, the fingering instability can be analyzed in the Newtonian or perfectly–plastic limit [8]. The fingering wavelength, λ, is given by:

$$\lambda_{Newtonian} = \pi\sqrt{\frac{\gamma h^2}{\eta V}} \approx 26\mu m \tag{3}$$

$$\lambda_{plastic} = \pi\sqrt{\frac{2\gamma}{\tau_0}h} \approx 0.21\mu m \ , \tag{4}$$

where τ_0 is the yield stress for the plastic liquid. (Between the dip–coating and printing processes, much of the water evaporates from the PEO solution, leaving behind a semi-solid substance.) We estimated λ using $\gamma_{PEO} = 53.5$ mN/m (which assumes that the semi-solid PEO has a surface tension equal to that of the aqueous solution), $h = 0.5\mu m$, $V = 100\mu m/s$, and $\tau_0 = 11.7$ MPa [9]. The values obtained represent two possible limiting behaviors. The experimentally observed wavelength [Figure 4 (a)] is about 2μm, which lies inbetween the two estimates. Since the fingering process cannot be eliminated altogether, one possible solution is to increase the value of λ beyond the typical feature size, by increasing the film thickness or decreasing the plate separation velocity for Newtonian liquids. Alternatively, a post–processing step like annealing can eliminate some asymmetries or defects introduced during printing. Control of ink rheology is equally important in preventing filament formation (Fig. 4(b)), as well as ribbing and misting during the printing process [10].

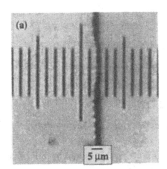 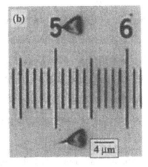

Figure 4: Polymer structures printed onto an unpatterned (hydrophilic) substrate. PEO was deposited from aqueous solution by dip-coating and then printed onto an SiO$_2$ surface. (a) A stripe of PEO 3.5 μm wide. (b) Two droplets of PEO printed onto a SiO$_2$ surface. The droplet volume is $\sim$ 2 femtoliters.

SUMMARY

Stable liquid and polymer structures can be formed by dip-coating chemically patterned substrates into a liquid reservoir and withdrawing at constant speed. The desired liquid shapes form on substrates coated with a pre-patterned self-assembled monolayer of OTS or HDT. Microscale feature sizes are easily obtained. For a dip-coated sample consisting of well spaced vertical hydrophilic stripes surrounded by hydrophobic regions, the maximum deposited film thickness scales as W $Ca^{1/3}$, where W is the stripe width and Ca the capillary number based on the plate withdrawal speed, a result well confirmed by experiment. This correlation is markedly different from the classic Landau-Levich result for the dip-coating of chemically homogeneous surfaces. The dependence on the stripe width, W, permits the deposition of very thin liquid layers. The liquid microstructures formed by dip-coating are then brought into contact with a hydrophilic surface to mimic an offset printing press. This techique has so far printed $5\mu m$ wide PEO stripes. Optimization of this process should lend itself to the rapid reproduction of liquid and semi-solid microstructure elements.

ACKNOWLEDGMENTS

The authors gratefully acknowledge support from the Electronic Technology Office of the Defense Advanced Research Projects Agency, the Eastman Kodak Co. graduate fellowship program (SMM), and an Austrian Fonds zur Förderung der wissenschaftlichen Forschung postdoctoral fellowship award (AAD).

REFERENCES

1. R. Moaz and J. Sagiv, J. Coll. Interface Sci. **100**, 465 (1984).

2. A. Kumar, H.A. Biebuyck, and G.M. Whitesides, Langmuir. **10**, 1498 (1994).

3. C.S. Dulcey, J.H. Georger Jr., V. Krauthamer, D.A. Stenger, T.L. Fare, and J.M. Calvert, Science. **252**, 551 (1991).

4. A.A. Darhuber, S.M. Troian, S.M. Miller, and S. Wagner, J. Appl. Phys. **87**, 7768 (2000).

5. L. Landau, and B. Levich, Acta Physicochimica U.R.S.S. **17**. 42 (1942).

6. A.A. Darhuber, S.M. Troian, J.M. Davis, S.M. Miller, and S. Wagner, J. Appl. Phys. *submitted.*

7. D. Qin and Y. Xia, and G.M. Whitesides, Adv. Mater. **8**. 917 (1996).

8. R.J. Fields, and M.F. Ashby, Phil. Mag. **33**, 33 (1976).

9. F.E. Bailey, Jr., and J.V. Koleske, ed. *Poly(ethylene oxide)*, (Academic Press, New York, 1976), p. 132.

10. J.E. Glass, and R.K. Prud'homme, in *Liquid Film Coating*, edited by S.F. Kistler, and P.M. Schweizer (Chapman Hall, London, 1997), pp. 137-182.

SOL–GEL-DERIVED 0–3 COMPOSITE MATERIALS
FOR DIRECT-WRITE ELECTRONICS APPLICATIONS

STEVEN M. COLEMAN,[1-2] ROBERT L. PARKHILL,[2] ROBERT M. TAYLOR,[2] AND
EDWARD T. KNOBBE[1]

[1]Thin Film and Materials Processing Group, Department of Chemistry, Oklahoma State
University, Stillwater, OK, U.S.A. 74078.
[2]CMS Technetronics, Inc., 5202-2 North Richmond Hill Road, Stillwater, OK, U.S.A. 74075.

ABSTRACT

The use of sol–gel-derived 0–3 composite ceramics for low-temperature direct-write
electronics applications was investigated. The 0–3 composite paste materials were prepared
using selected metal alkoxides and commercial low- and high-κ' dielectric powders. The
composite pastes were deposited onto alumina and polyimide substrates using conventional
screening and micro-dispensing techniques. The deposited films were oven-dried at or below
200°C and thermally densified using a CO_2 laser. The 0–3 composites exhibited good
adhesion and structural density. Electrical characterization of the laser-processed dielectrics
revealed κ' values as high as 295 and tan δ as low as 0.02 on polyimide substrates.

INTRODUCTION

High-technology activities aimed at the rapid prototyping of such advanced devices as
miniaturized Global Positioning System transceivers have initiated a large amount of research
in the area of low-temperature thick-film microelectronics. Fundamental to many such
devices are components comprised of high-dielectric-constant media, usually in the form of
thin and/or thick films. Such films function in a variety of roles, including substrates for
printed antennas and capacitor components. Currently, dielectric pastes used in the
microelectronics industry require processing temperatures in excess of 800 °C, which tends to
limit their use to ceramic substrates. Thus, in order to expand the diversity of potential
substrates for thick-film microelectronics, such as lightweight systems utilizing polymeric
substrates, low-temperature dielectric pastes must be developed.

In recent years, interest has increased in using sol–gel-based materials processing
methods to deposit thin films of such dielectric materials as silica, alumina, zirconia, and
various titanates. The precursors involved in sol–gel chemistry, usually metal alkoxides, are
typically hydrolyzed at temperatures at or below room temperature. The resulting
macroscopically liquid sol can be deposited by various solution-based processing methods,
including dip-coating, spin-coating, and MicroPen™ dispensing techniques. The advantages
of the sol–gel deposition approach include: (1) extremely good film-thickness uniformity; (2)
homogeneous compositions of very high chemical purity; (3) tenacious adhesion to substrate
materials of many types, including metals, semiconductors, glasses, ceramics, and selected
plastics; (4) a wide range in available active film areas (*e.g.*, μm^2 to cm^2); and (5) low
consolidation and annealing temperatures compared to those of conventional powder
processing and melt methods.

Historically, sol–gel-based application methods have not enjoyed widespread use in
capacitor film deposition because of difficulties in fabricating dense, crack-free films of
sufficient thickness for device applications (*e.g.*, 250 nm to 10 μm). Typically, sol–gel
processing is limited to film thicknesses on the order of 100 nm per coating to achieve crack-

53

free layers. Thicker, crack free films are only obtainable, by traditional sol–gel methods, by depositing multiple thin-film depositions and/or through the use of drying control chemical additives (DCCA's). Recent developments in such new sol–gel-based methods as "0–3 composites" have demonstrated that thickness constraints are substantially alleviated. Sol–gel derived 0-3 composite systems are presently being investigated for use in thick-film microelectronics applications.

In 1997, Barrow and coworkers reported a new method, based on the earlier work of Yi *et al.*, for the deposition of lead zirconate titanate (PZT) capacitor films using a hybrid crystallite particle dispersion/sol–gel approach [1–4]. In this work, the authors reported the deposition of so-called "0–3 ceramic/ceramic composites" derived by suspending colloidal PZT particles into a PZT sol–gel precursor solution. The term "0–3" composite indicates the connectivity of the two phases present in the solid state (*i.e.*, the crystallite powder phase is assumed to be discontinuous in all directions, while the sol–gel matrix is assumed to be continuous in all three dimensions). Films prepared using the 0–3 composite approach have been reported to have substantially improved substrate adhesion and film integrity when compared to powder processing or conventional sol–gel-based coating methods. The lower total volume fraction of volatile component and added particle density in the deposited solution lowers relative solvent loss, which results in less shrinkage and a substantial reduction in drying stresses. Barrow *et al.* reported the preparation of fired and annealed, crack-free 0–3 composite films having single deposited-layer thicknesses of up to 10 μm [2–3]. Thickness control was achieved by changing the total volume fraction of suspended crystallites in the sol. Fully densified layers deposited onto Pt–Ti–SiO$_2$–Si substrates were obtained by firing at 500 °C and annealing at 650 °C. Thus, the 0–3 composite method provides an approach to the deposition of thick ceramic films deposited in a single run, densified at comparatively low temperatures, and produce a wide range in available thicknesses and compositions.

EXPERIMENT

Dielectric pastes were prepared using commercially available powders and selected metal alkoxide sol–gel precursors. Powders of low-κ' (e.g., TiO$_2$) and high-κ' (e.g., BaTiO$_3$) dielectric materials were purchased from Atlantic Equipment Engineers and Superior MicroPowders, respectively. Particle sizes ranged from less than 1 μm to 5 μm. Metal alkoxides were purchased from Gelest Chemicals, and used in the as-received condition. The primary considerations for the selection of metal alkoxides were (1) compositional compatibility with the powder used, (2) reactivity, (3) physical properties (i.e. viscosity and phase), and (4) thermal decomposition characteristics. Furthermore, the metal alkoxides were chosen so that the molecular precursors would also act as solvents, thereby reducing the need for organic solvents. The dry powders were blended with the metal alkoxides using a high-speed rotary tool. Compatible solvents were added to the final compositions to obtain pastes with a viscosity suitable for thick-film deposition. Alumina and polyimide substrates, each having pre-deposited silver pads (Ferro 3309 silver paste), were used for these tests. Sol–gel derived pastes were deposited onto the conductive pads using either simple masking techniques or an OhmCraft MicroPen™ dispenser. The green dielectric films were dried at room temperature, followed by solvent removal at or below 200°C for 2 hours. Densification of the dielectric pastes was achieved using a CO$_2$ laser operating at 10.6 μm. Laser irradiation tests were performed with both CW (continuous wave) and pulsed operational modes. Local power density was also varied. After processing top electrodes (area = 2mm^2) of silver were sputtered onto the densified capacitor layer.

Structural and compositional characterizations were performed using a JEOL Model 3400 scanning electron microscope (SEM) and a Phillips XL30 environmental SEM equipped with an EDAX (energy dispersive X-ray) system. The X-ray system was equipped with a Si(Li) detector and a Be window, enabling the detection of light elements ($Z > 11$).

Electrical characterization was performed using a Hewlett–Packard 4284A precision *LCR* meter. Test frequencies ranged from 100 Hz to 300 kHz, with oscillation levels of 100 mV. Additional instrumentation was used to characterize the films at a frequency of 5 MHz.

RESULTSM

A variety of 0–3 composite dielectric film compositions was deposited onto silver/alumina and silver/polyimide substrates, then characterized. Single-layer film deposits were found to range from 5 μm to 30 μm, with little or no cracking and good adhesion. These films also exhibited desirable electrical characteristics, including a reasonable tan δ for a wide range of κ′ values. The results and limitations have been identified and are summarized below.

Physical Film Properties

Figure 1 illustrates a film of barium titanate particles bound together in a titania matrix. The micrograph revealed that the composite exhibits a microporous character. It also revealed that the film was free of major structural defects, *i.e.* cracks or continuous pinholes. At high magnification, inter-particle necking was observed and identified with hydrolysis and condensation of the titanium alkoxide precursor. The observed microporosity and loss of packing density were attributed to a combination of solvent loss and particle reactivity. Such effects are common in paste materials and can be expected when processing temperatures are low and/or the processing time at elevated temperatures is short.

The integrity and adhesion of the films to the substrate and silver pad were observed to be good for polyimide, and very tenacious for alumina substrates. Films applied to alumina were scratch-resistant and easily withstood repeated adhesion-tape tests. Films applied to polyimide were not as durable and often failed under adhesion-tape tests. The exceptional durability observed in the films was attributed to the continuous connectivity between the particles, as illustrated in Figure 1. The variation in substrate adhesion was postulated to occur as a result of mismatches in coefficient of thermal expansion (CTE) between the substrate and films. The Δ(CTE) between ceramic films and polyimide substrate creates substantial problems for integrated device applications even over small processing temperature ranges.

A number of process-induced defects were observed and attributed to physical effects associated with rapid solvent loss, high Δ(CTE), and rapid thermal processing (RTP) effects. Major defects, such as cracking and pinholes (see Figure 2) were observed for selected compositions in the initial stages of solvent loss and during oven-drying. Film cracking due to rapid solvent loss has been reported in the literature extensively for both sol–gel-derived materials and 0–3 composites. Extensive cracking observed primarily on the polyimide substrates after oven drying was attributed to a large Δ(CTE) between the film and substrate. Pinhole defects were attributed to a combination of solvent loss, poor particle packing, and insufficient molecular-precursor volume. In both cases, defects appear to arise as the matrix phase continues to shrink after the drying film reaches critical packing concentration.

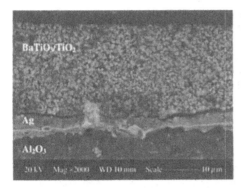

Figure 1: Scanning electron micrographs of a barium titanate capacitor deposited on an alumina substrate with a silver electrode.

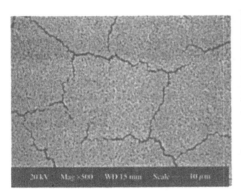

Figure 2: Micrographs of surface defects observed in 0–3 composites.

Dielectrics on Polyimide Substrates

Laser processing has the advantage of generating localized heating in the deposited film up to several hundred degrees Celsius, while leaving the substrate relatively cool. This approach can be especially advantageous for densifying high-temperature films on polymeric substrates, a case where there is a large Δ(TE). When fired in a furnace, ceramic films cracked and delaminated from the electrode, even when very low temperature-ramp rates of 0.1 °C/minute were used. However, functional dielectric capacitors are readily processed on polyimide substrates, using laser processing techniques. Optical pyrometer data indicated that surface temperatures greater than 345 °C could be reached without delamination or cracking.

Electrical Properties of Processed Films on Polyimide Substrates

A wide range of dielectric constants (κ') and loss factors (tan δ) were measured for the 0–3 composite capacitors on polyimide substrates. Low-frequency measurements taken from 100 Hz to 300 kHz resulted in a compositionally dependent range of κ' values (14 to 295) while tan δ was found to vary from 0.02 to 0.18. Data for these films was also taken at a frequency of 5 MHz with a measured range of κ' from 14 to 295 and tan δ varying from 0.022 to 0.042. Table I summarizes the electrical properties of laser-processed films on polyimide substrates for both low- and high-κ' films.

The frequency-dependent loss characteristic of the capacitor layers shown in Figure 3 implies the presence of residual organic species and/or non-terminated hydroxyl end-groups in the processed film. This is partially attributed to insufficient heating throughout the thickness of the film. EDAX analysis indicates a small amount of residual carbon is present in the matrix after laser sintering. Further investigation into this phenomenon should allow the adjustment of laser parameters resulting in the formation of lower-loss capacitive films without damaging the polyimide substrate.

CONCLUSIONS

Analysis of the described sol–gel-derived 0–3 composite ceramics implies that low-temperature direct-write electronics materials can be applied to alternative low-temperature substrates using laser processing techniques. The laser processed 0–3 composites were identified to have good adhesion, microporous networks, and inter-particle necking. Structural defects observed in the films were attributed to solvent loss and poor consolidation between particles. Electrical characterization of the laser-processed dielectrics revealed κ' values as high as 295 and tan δ as low as 0.02 on polyimide substrates. Variations in κ' were attributed to microporosity, while increases in tan δ were attributed to residual organics and unterminated hydroxyl end groups.

**Table I: Frequency Response of Laser Processed
Low- and High-κ' 0–3 Composite Dielectrics on Polyimide**

Frequency	Parameter	High-κ' Dielectric Films	Low-κ' Dielectric Films
	κ'	67–295	14–16
100 Hz–300 kHz	tan δ	0.03–0.18	0.020–0.022
5 MHz	tan δ	0.022–0.042	0.025–0.029

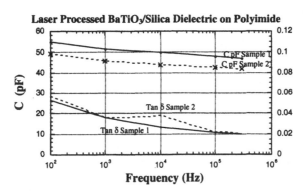

Figure 3: Frequency response of a barium titanate/silicate capacitor deposited on a polyimide substrate with a silver electrode.

ACKNOWLEDGMENTS

The authors would like to thank Paul Clem and Tim Schaefer for the electrical characterization of the films presented in this work and Lowell Matthews for assistance in manuscript preparation. This work was supported by DARPA.

REFERENCES

1. D. A. Barrow, T. E. Petroff, R. P. Tandon, and M. Sayer, J. Appl. Phys. **81**, p. 876 (1997).

2. D. A. Barrow, T. E. Petroff, and M. Sayer, J. Surf. Coat. Technol. **76–77**, p. 113 (1995).

3. D. A. Barrow, T. E. Petroff, and M. Sayer, Mat. Res. Soc. Symp. Proc. **60**, p. 103 (1995).

4. G. Yi and M. Sayer, Ceramic Bulletin, **70**, p. 1173 (1991).

Direct Write Metallizations for Ag and Al

C. J. Curtis, A. Miedaner, T. Rivkin, J. Alleman, D. L. Schulz[†] and D. S. Ginley
National Renewable Energy Laboratory, Golden, CO USA 80401
[†]Present address: CeraMem Corporation, 12 Clematis Ave., Waltham, MA 02453

Abstract

We have employed inks containing nanometer-sized particles of Ag and Al (nano-Ag and nano-Al, respectively) as precursor inks for the formation of contacts to n- and p-type Si, respectively. The particles as formed by the electroexplosion process were dispersed in toluene, applied to Si and annealed above the respective eutectic temperatures. In the case of nano-Ag, this directly yields an ohmic contact. However, the nano-Al was found to be coated with an oxide layer that impairs the formation of an ohmic contact. A chelating chemical etch involving treatment with hexafluoroacetylacetone was developed to remove this oxide coat. This treated nano-Al produced a good ohmic contact. Smooth, pure Ag films have also been deposited by spray printing organometallic inks prepared from Ag(hfa)(SEt$_2$) and Ag(hfa)(COD). These films are deposited in one step onto heated glass and Si substrates at one atmosphere pressure. The films show resistivities of ~2 $\mu\Omega$·cm. These inks appear to be amenable to ink-jet printing of Ag lines and as a low temperature glue for the Ag nanoparticles for thicker metallizations.

Introduction

There is an increasing drive to replace current metallization approaches for microelectronics, photovoltaics, toys and a variety of low resolution packaging applications with lower cost simpler approaches. For example in photovoltaics, metallizations involve the deposition of metal films and lines onto the junction part of the device to extract current. At the present, metallizations are carried out by high vacuum approaches (sputtering or evaporation) or screen printing using micron-sized particle inks. Both approaches are mask-based technologies. The ability to directly write conducting metal layers and grids using non-vacuum, spray and ink-jet deposition techniques would greatly simplify the metallization process, lowering both the capital and material costs. Ink-jet printing can also give narrower grid lines than screen printing, decreasing shading losses and therefore increasing cell efficiency[1-3]. Figure 1 shows a pictorial representation of a Si solar cell. Typically it consists of a thin crystalline p-type wafer with a diffused front n-type layer to make a homojunction. Carriers are collected at the front and back contacts after they are generated in the wafer and then separated by the junction field. Contact is normally made to the p-type and n-type sides by alloying Al and Ag metals with the Si to form p+ and n+ contacts, respectively.[4-7]

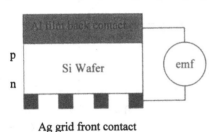

Ag grid front contact

Figure 1. Pictorial representation of a Si solar cell.

The Al layer and Ag lines act both as contacts to the Si and as electrodes for the cell, with sunlight passing between the Ag grid lines.

To allow for direct write metallizations, we are investigating the use of nanoparticle-based inks. In the simplest case, an ink could be formulated using only nanoparticles, however

59

in some cases, to get good particle-particle and particle-substrate interconnections requires addition of an active agent to promote interconnection. This agent may act to enhance the surface reactivity or serve to actively join particles together. We are investigating both approaches. We have investigated chelating etches and organometallic complexes of the same metal to enhance film and contact formation. We discuss inks based on metal nanoparticles, organometallic complexes, and combinations of both, that can be used in direct write metallization processes to print Ag and Al lines and films on Si and other substrates using spray and ink-jet printing. We report here use of nano-Ag and nano-Al to form ohmic contacts to n- and p-type Si, a chemical etching process to remove the oxide coating from Al particles, and spray printing of Ag films on glass and Si using organometallic inks.

Experimental and Results

Nano-Ag Contacts to Silicon

Nano- Ag, produced by electroexplosion of Ag wire, was obtained from Argonide Corporation (Sanford, FL). TEM characterization of this material shows agglomerated, crystalline particles 100-300 nm in diameter. The agglomerated clusters are held together by small necks that connect particles and probably form during the electroexplosion process. TEM-EDS shows the particles to be Ag with no other contaminants. The particles were slurried in toluene and dropped on to n-type Si substrates in air at room temperature. After the toluene evaporated, the samples were annealed at 882°C for 1 hr under Ar. After annealing, the surface was coated with a gray powder residue. This was removed by wiping with a cotton swab, electrodes were attached with silver paint and the contact was characterized by measuring the current-voltage (I-V) curve. A typical I-V characterization result is shown in Figure 2. The linear relationship between current and voltage observed indicates that a good ohmic contact is formed.

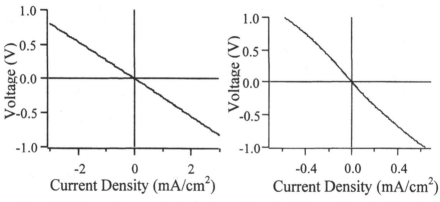

Figure 2. I-V curve for nano-Ag derived contact on n-type Si

Figure 3. I-V curve for contact to p-type Si derived from oxide-coated nano-Al

Nano-Al Contacts to Silicon

Nano-Al produced by the electroexplosion process was also obtained from Argonide Corp. TEM characterization of this material shows the majority to be spherical particles with diameters of 50-300 nm, with small irregular chunks as small as 10 nm interspersed. TEM-EDS shows Al with a significant amount of O present. This is undoubtedly due to surface oxidation of the particles resulting in an aluminum oxide surface layer. These particles were slurried in toluene, applied to p-Si substrates and annealed at 650°C for 1 hr under Ar. The I-V curve of the resulting contact is shown in Figure 3. The curvature observed at higher current densities indicates that the contact formed is not ohmic.

Chemical etches to remove the oxide coating on nano-Al particles were then examined. Etches described in the literature for cleaning Al [8] completely dissolved the nanoparticles. Treatment with a variety of organic chelating agents also proved unsuccessful. However, treatment of oxide-coated nano-Al with neat hexafluoroacetylacetone (Hhfa) for 16 hr followed by decanting the liquid phase, washing with toluene and drying was found to remove the oxide coat. Figure 4 shows TEM-EDS characterization of nano-Al before and after this Hhfa treatment. After the etch, the surface of the cleaned nano-Al particles is coordinated with hexafluoroacetylacetonate (hfa) ligands. Figure 5 shows an FT-IR spectrum obtained from a nujol mull of treated nano-Al powder. The peaks at 1675, 1600 and 1536 cm^{-1} are C-O stretches and the peaks at 1636, 1261 and 1212 cm^{-1} are C=C and C-F stretches of the coordinated hfa ligand [9].

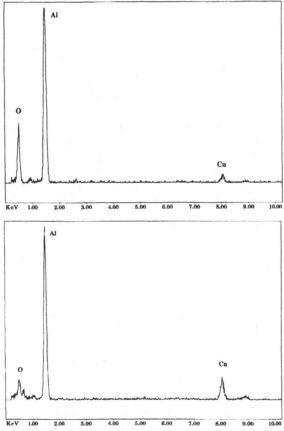

Figure 4. TEM-EDS characterization of nano-Al before (top)
and after (bottom) Hhfa treatment.

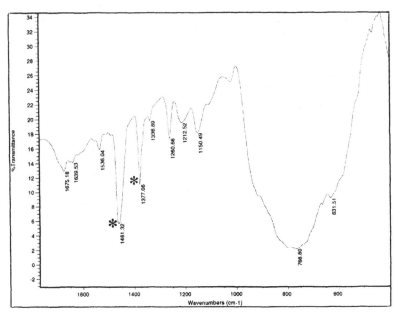

Figure 5. IR spectrum of Hhfa treated
nano-Al. Nujol peaks marked *.

When Hhfa –treated nano-Al was used to form a contact to p-Si in the manner described above, the linear I-V curve shown in Figure 6 was obtained. This indicates the viability of the approach of creating more reactive particle surfaces through controlling the surface chemistry. Not only does the Hhfa treatment displace the aluminum oxide layer, but the ligand is capable of stabilizing the surface and preventing further oxidation. The approach may be general and coupled with the enhanced reactivity of nanoparticles, could lead to low temperature contacts.

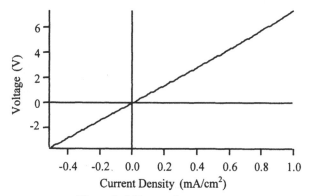

Figure 6. I-V curve for contact to p-type
Si derived from Hhfa-treated nano-Al

Spray Printed Ag Films on Glass and Silicon

We have been investigating the application of organometallic inks directly or in conjunction with nanoparticles to act as a bonding agent. We have demonstrated the viability of the concept with Cu. Here we report on the efficacy of using organometallic silver inks in a direct write approach. Organometallic inks derived from Ag(hfa)(L) complexes, where L= SEt$_2$[10-12], 1,5-cyclooctdiene (COD) [13], have been used to deposit Ag films on glass and Si substrates using a spray printing process. A typical ink was prepared by dissolving Ag(hfa)(SEt$_2$) (2.0 g) in toluene (2.3 g) and forcing the resulting solution through a syringe filter. This ink was sprayed in several coats onto heated substrates using a hand-held sprayer (Paasche Type VL Airbrush) in a N$_2$-filled glove box. Figure 7 shows SEM images of Ag films sprayed on glass and Si substrates at 400°C. The films are seen to be smooth and dense with a thickness ranging from 0.8 to 1.5 μm (Dektak3 profilometer). Surface resistance measurements performed by 4-point probe give a resistivity of ~2 μΩ·cm for both of these films. XPS data as a function of depth for a Ag film spray printed with Ag(hfa)(COD) ink shows C, O and F present in substantial amounts at the surface, but these levels fall rapidly within the top 30 Å. No O or F is found at 50 Å and the C concentration falls from 2 to 1 atom % between 100 and 300 Å deep. Smooth Ag films with a thickness of ~1 μm and excellent resistivities have thus been deposited in a single step at atmospheric pressure using these organometallic inks. We are currently developing the apparatus necessary to print these inks onto hot substrates using an ink-jet print head.

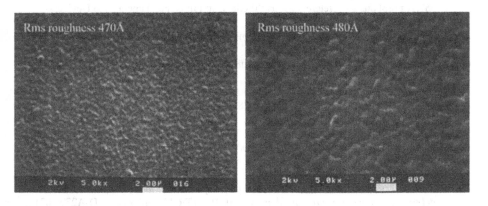

Figure 7. Ag film spray printed on glass (left) and Si (right) using ink prepared from Ag(hfa)(COD).

Conclusions

Nano-Ag and nano-Al were used to form ohmic contacts to n- and p-type Si respectively. It was necessary to remove the oxide coat from the Al particles before an ohmic contact could be achieved. A chelating etch process using hexafluoroacetylacetone was developed for this purpose. Spray printing using organometallic inks based on Ag(hfa)(SEt$_2$) and Ag(hfa)(COD) has been used to deposit smooth, pure Ag films with resistivities that approach that of the bulk metal. These films are deposited in one step onto heated substrates at one atmosphere pressure. These inks appear to be amenable to ink-jet printing. They will serve with the Ag nanopowders as a versatile ink system for lines of a wide variety of widths and thicknesses.

Acknowledgments
The authors gratefully acknowledge the gift of nano-Ag and nano-Al from Dr. Fred Tepper of Argonide Corporation. The NREL Silicon Team graciously provided the Si substrates. We also thank Kim Jones (TEM) and Rick Matson (SEM) for technical assistance. This work was funded by the U.S. Department of Energy, National Photovoltaic Program and IPP Program.

References

1.   Nade, K., K. Kozuka, and T. Isogai, *Inks for printed circuits.* Jpn. Kokai Tokkyo Koho, 1979: p. 3 p.
2.   Schulz, D.L., et al., *Particulate contacts to Si and CdTe: Al, Ag, Hg-Cu-Te, and Sb-Te.* AIP Conf. Proc., 1999. **462**(CPV Photovoltaics Program Review): p. 206-211.
3.   Stark, J.V., et al., *Nanoscale Metal Oxide Particles/Clusters as Chemical Reagents. Unique Surface Chemistry on Magnesium Oxide As Shown by Enhanced Adsorption of Acid Gases (Sulfur Dioxide and Carbon Dioxide) and Pressure Dependence.* Chem. Mater., 1996. **8**(8): p. 1904-1912.
4.   Von Roedern, B. and G.H. Bauer, *Material requirements for buffer layers used to obtain solar cells with high open circuit voltages.* Mater. Res. Soc. Symp. Proc., 1999. **557**(Amorphous and Heterogeneous Silicon Thin Films: Fundamentals to Devices--1999): p. 761-766.
5.   Rohatgi, A., et al., *Rapid processing of low-cost, high-efficiency silicon solar cells.* Bull. Mater. Sci., 1999. **22**(3): p. 383-390.
6.   Haoto, D. and Y. Iwamoto, *Thin film solar cells and their manufacture.* Jpn. Kokai Tokkyo Koho, 1999: p. 8 p.
7.   Miyoshi, K., *Solar cells.* Jpn. Kokai Tokkyo Koho, 1999: p. 4 p.
8.   Petzow, G., *Metallographic Etching.* 1978, Metals Park, Ohio: American Society of Metals. 39-43.
9.   Farkas, J., M.J. Hampden-Smith, and T.T. Kodas, *FTIR Studies of the Adsorption/Desorption Behavior of Copper Chemical Vapor Deposition Precursors on Silica. 2. (1,1,1,5,5,5-Hexafluoroacetylacetonato)(2-butyne)copper(I).* J. Phys. Chem., 1994. **98**(27): p. 6763-70.
10. Corbitt, T.S., et al., *Aerosol-assisted chemical vapor deposition of copper: a liquid delivery approach to metal thin films.* Report, 1994(TR-12; Order No. AD-A279 702): p. 14 p.
11. Jain, A., T.T. Kodas, and M.J. Hampden-Smith, *Thermal dry-etching of copper using hydrogen peroxide and hexafluoroacetylacetone.* Thin Solid Films, 1995. **269**(1-2): p. 51-6.
12. Jain, A., et al., *Chemical vapor deposition of copper from (hfac)CuL (L = VTMS and 2-Butyne) in the presence of water, methanol, and dimethylether.* Chem. Mater., 1996. **8**(5): p. 1119-27.
13. Doyle, G., K.A. Eriksen, and D. VanEngen, *Alkene and carbon monoxide derivatives of copper(I) and silver(I) .beta.-diketonates.* Organometallics, 1985. **4**(5): p. 830-5.

INK JET DEPOSITION OF CERAMIC SUSPENSIONS: MODELLING AND EXPERIMENTS OF DROPLET FORMATION

N. REIS AND B. DERBY
Manchester Materials Science Centre, UMIST, Grosvenor St., Manchester, M1 7HS, U.K.

ABSTRACT

We have successfully printed green ceramic objects from slurries of Al_2O_3 dispersed in paraffin wax using a commercial ink-jet printer developed for pattern making (Sanders Prototype MM6PRO). Concentrated suspensions are generally more viscous than the fluids normally passed through ink jet heads. This may alter the response of the printing system to its process parameters, e.g. driving voltage and frequency. We have explored the influence of fluid properties on the ink jet behaviour using Computational Fluid Dynamics (CFD) modelling and a parallel experimental study to determine the optimum printing conditions for particulate suspensions.

INTRODUCTION

Ink jet printing is an attractive option for direct write technologies and for the micro-manufacture of parts. Drop-on-demand printers provide a relatively inexpensive means of accurately delivering small volumes of material to precise locations enabling the reproduction of predetermined patterns stored in computer files. By overprinting, three dimensional objects can be constructed. Multi-material or composite structures can be readily fabricated if more than one ink jet droplet formation device is used. If the building droplets are sufficiently small, graded or functionally gradient structures can also be deposited.

We have recently demonstrated that ink jet printing, using a piezoelectric drop-on-demand printer, can be used to successfully deposit ceramic suspensions with a very high volume loading of ceramic particles [1, 2]. This paper reports on our current progress on printing concentrated ceramic suspensions. In it we present the results of a simple computational fluid mechanics model of the ink jet printer, assuming Newtonian flow. This can be used to successfully predict the jetting parameters of a number of fluids and thus allow the identification of the optimum printing parameters.

EXPERIMENTAL

Suspension preparation

Alumina suspensions were prepared by dispersing a fine, sub-micron powder (α-Al_2O_3 RA45E, Alcan Chemicals Ltd., U.K.) in paraffin wax (Mobilwax135", Mobil Special Products, U.K.) containing variable additions of sterylamine (1-Octadecylamine, Lancaster Synthesis, U.K.) and a proprietary dispersant (Hypermer LP1, ICI Surfactants, U.K.). All formulations were mixed by conventional ball milling at 100...C for periods of 10 hours. By controlling the ratio between surfactants molecular weight, and terminal group functionality (i.e., acidic or basic), suspension shear viscosities as low as 40 mPa.s were obtained for apparent particulate volume fractions of 0.4 (steady shear viscosity was measured in a concentric cylinder Brookfield Viscometer at $100s^{-1}$ and 100...C). Detailed preparation procedures and findings have been reported previously [1].

65

In situ monitoring of drop formation

In this study, phase change drop on demand print heads were used (Sanders Design Inc., Wilton, NH). These are tubular piezoelectric transducers surrounded with a temperature controlled heating tape to ensure temperature stability. Monitoring and recording of drop formation at the jet nozzle outlet were performed using a conventional CCD camera and a PC frame grabber card. Since CCD frame rates are not compatible with typical jet repeat rates, a sub-microsecond light emitting diode strobe was used to back-lit the forming droplets, at controllable delays from the transducer excitation pulse. The set-up allows freezing the droplet motion at constant delays, and also following their perceived motion using linear voltage vs. time ramps. Details on this set-up can be found elsewhere [3].

CFD modelling

To model the jet dynamics, a commercial CFD package (FLOW-3D$^\Box$) was used. It is based on a refined Volume of Fluid (VOF) method coupled with Fractional Area-Volume Obstacle Representation (FAVOR") algorithms to define obstacle s geometry within an Eulerian grid. This was primarily chosen because of its simplicity and robustness to model the free surface flow characteristic of drop formation in ink jet heads.

For this study, all models of drop ejection use a pressure pulse condition at the bottom boundary of the computational domain, as introduced by Fromm [4]. For the evolving free surface, an outflow continuative boundary condition was used. This consists of zero normal derivatives at the boundary for all quantities to ensure smooth continuation of the flow through the boundary. Although convenient for sub-sonic incompressible flows, this condition must be regarded with caution since it has no physical foundation.

The flow was assumed to be laminar and axisymmetric, which are valid assumptions for ink jet printing given the steadiness and repeatability of the process. Other assumptions included isothermal conditions, no wall slip and Newtonian viscosity. Although the first two are fair assumptions, the latter only applies to simple fluids and dilute suspensions, and hence is not applicable to concentrated suspensions. An effort is currently being made to include shear and temperature dependence of the viscosity and also introduce mass particles in the flow. Due to the flow symmetry, only a two dimensional representation of half of the jet was used in the computations. Grids were refined until no significant changes were observed for the description of flow. All computations used cylindrical coordinates, with the mesh becoming finer towards the nozzle axis.

RESULTS AND DISCUSSION

Experimental observations

Our experiments revealed three key aspects regarding drop ejection and stability in piezoelectric ink jets, operating in air at atmospheric pressure:
1. The acoustic pressure at the nozzle inlet, which depends on the fluid acoustic properties, chamber (including the piezoelectric) properties, geometry and dimensions of the fluid filled cavity, and the voltage pulse train used to excite the transducer. This pressure is responsible for the momentum imparted to the fluid at the nozzle and also the deceleration forces causing constriction of the jet [5].

2. The viscous loss at the nozzle due to the constriction of flow, dependent on the nozzle configuration and dimensions, and fluid viscosity. For a given acoustic pressure wave at the nozzle, the lower the viscosity the greater are both velocity and the amount of fluid propelled forwards, which usually lead to the formation of long tails behind the head of the drop.
3. The surface tension forces acting at the free surface. These depend not only on the fluid surface tension but also on the wetting characteristics between the fluid and nozzle material. The former is responsible for the spheroidisation (tail recoil or satellite disruption) of the liquid thread emerging from the nozzle. The latter is important in controlling the wetting of the nozzle outlet face, which inevitably results in spray formation. Its effect on drop volume and velocity due to meniscus reverberations has been recently discussed [6].

All the above considerations are still valid at high frequencies. Operating in the high frequency regime has a number of advantages. First it is easier to tune variables such as drop velocity and mass deposition rate, which are crucial for the splat formation mechanisms, [3] and reduce building times. Secondly, it is possible to manipulate the acoustic pressure wave by tuning the pulse train applied to the piezoelectric transducer for jetting higher viscosity fluids. Figure 1 shows that above a certain frequency, sub-harmonic effects can be used to increase the acoustic pressure amplitude and hence drop velocity

Jetting at high frequencies however, brings additional considerations into play, since the rate of droplet generation is usually faster than the time required to form a single drop. This results in the formation of two or more connected droplets near the nozzle, which detach during flight. As a consequence the tail behind a forming droplet is now merged with the droplet immediately behind and thus regular droplets can be obtained at distances acceptable for printing — figure 2a.

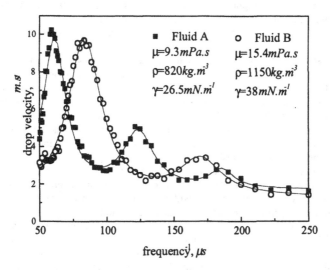

Figure 1- Effects of driving pulse frequency and fluid properties on drop velocity for the same jet head construction . The peaks and valleys correspond to sub-harmonic resonance and anti-resonance for each fluid.

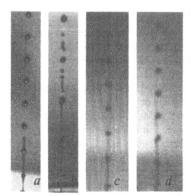

Figure 2 — Drop formation at high frequencies for different fluids. (a) is Protosupport"; (b) is an eutetic mixture of camphor and naphthalene; (c) and (d) are 30 and 40 vol.% Al$_2$O$_3$ suspended in paraffin, respectively.

	ρ [kg.m^{-3}]	μ [mPa.s]	γ [mN.m^{-1}]	f / ΔV / T [kHz / V / ...C]
(a)	820	9.3	26.5	10 / 70 / 110
(b)	970	1.3	46	10 / 70 / 110
(b)	1800	14.5	25	10 / 70 / 100
(d)	2100	38	25	13 / 80 / 110

Although higher acoustic pressure amplitudes characteristic of high operating frequencies can be used to compensate for viscous losses, care must be taken while jetting low viscosity fluids. The inevitable long threads of fluid expelled from the nozzle may cause the formation of satellites for slightly higher values in surface tension— figure 2b. Figures 2c and 2d show the jetting behaviour of 30 and 40 vol% Al$_2$O$_3$ suspended in paraffin, respectively.

For the 40 vol% suspension, a small pressure had to be applied to force the flow of the fluid through the lines and replenish of the fluid in the jet head chamber. Since the tested suspensions are shear thinning, this is assumed to be due to the highly viscous flow of concentrated suspensions under low shear.

CFD calculations

In order to assess the assumptions and numerical schemes used to track the free surface, candelilla wax was used as model fluid. This wax chosen since it has viscosity similar to other fluids of interest (steady shear viscosity of 17 mPa.s at 100...C) and exhibits Newtonian behaviour.

Figure 3 shows a comparison between numerical calculations and in situ monitoring of the forming droplets, for which a good agreement was found. It can also be observed that for the same velocity, the model has predicted shorter detaching times and hence smaller tails following the head of the drop. This was found to depend not only on the fluid properties but also, and in accordance with earlier studies using different numerical schemes [4,7], on the shape and amplitude of the pressure pulse applied as bottom boundary condition. Another factor that has not been considered is the extensional character of the flow and hence the introduction of elongational viscosity as opposed to shear viscosity.

We are currently developing the capability to model the acoustics of the ink jet chamber in order to, optimally synchronise driving parameters and obtain higher pressures needed for jetting higher viscosity fluids. These results will also be used to improve our free surface modelling assumptions, primarily regarding boundary conditions.

Numerical simulations can also be explored to provide useful information regarding the mechanisms of drop formation, and determine the window of fluid properties needed for jetting. Figure 4 shows how surface tension and viscosity can affect drop volume and velocity. These numerical predictions confirm our experimental findings regarding drop formation. It can be seen that viscosity has a great effect in the velocity and amount of fluid ejected from the nozzle, whereas surface tension is responsible for the minimising the surface area by spheroidising the fluid columns. High values of both properties can restrain jetting by retracting the fluid column

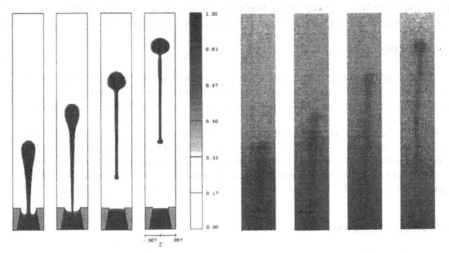

Figure 3 CFD calculations (left) and experimental observations (right) of candelilla wax being jetted at 100...C. Frames are separated by 50 µintervals, and are 1 mm in height. Length units (r and z) are displayed in cm.

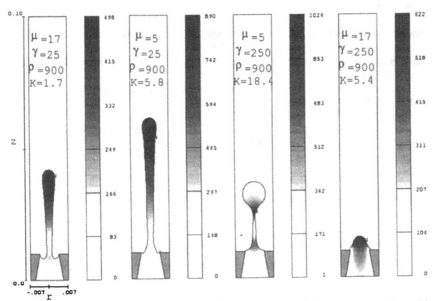

Figure 4 — Velocity magnitude contours after 80µs, for the fluid properties indicated in the frames and using the same numerical schemes and boundary conditions as in figure 3 (velocity in cm.s^{-1}). The parameter K indicates the ratio between Reynolds and Weber numbers ($K=(\rho L\gamma)^-/\mu$).

back into the nozzle before conditions for droplet detachment are achieved. It was also numerically predicted the formation of satellite droplets as illustrated in figure 2b.

CONCLUSIONS

The feasibility of direct ink jet deposition of concentrated suspensions up to 40% solids by volume has been proved, by using drop on demand piezoelectric print heads. Viscous losses due to the particulate filling can be compensated controlling the process acoustics through excitation pulse adjustments.

Simple CFD calculations were developed to simulate the jetting process with considerable success when compared to experimental observations. These can be used to narrow the number of experiments needed to determine optimum jetting conditions for each fluid. These codes are being improved to include more realistic assumptions concerning boundary conditions, introduce elongational viscosity and also to consider the non-Newtonian behaviour of concentrated suspensions.

ACKNOWLEDGEMENTS

We would like to acknowledge the support of the EPSRC and the Ministry of Defence through grant GR/L42537. We would also like to thank Julian Evans, John Halloran and Kitty Seerden for helpful comments and discussion.

REFERENCES

1. N. Reis, K.A.M. Seerden, P.S. Grant, B. Derby, and J.R.G. Evans, presented at the IoM Annual Ceramics Convention, 10-11 April 2000, Cirencester, U.K.
2. B. Derby, N. Reis, K.A.M. Seerden, P.S. Grant, and J.R.G. Evans, in this issue, MRS Spring Meeting 2000, San Francisco, U.S.A.
3. N. Reis, K.A.M. Seerden, B. Derby, J. W. Halloran and J.R.G. Evans, Mater. Res. Soc. Symp. Proc., **542**, p. 147-152 (1999).
4. J. E. Fromm, IBM J. Res. Develop., **28**, p. 322-333 (1984).
5. R. Badie and D. F. de Lange, Proc. R. Soc. Lond. A, **453**, p. 2573-2581 (1997).
6. J. F. Dijksman, Flow Turbul. Combust. **61**, p. 211-237 (1998).
7. R. L. Adams and J. Roy, ASME J. Appl. Mech. **53** [1], p. 193-197 (1986).

ELECTROSTATIC PRINTING, A VERSATILE MANUFACTURING PROCESS FOR THE ELECTRONICS INDUSTRIES

Robert H. Detig, Ph.D.
Electrox Corporation, Denville, NJ 07834

ABSTRACT

Functional materials configured as liquid toners are printed on a variety of substrates for various manufacturing processes. The materials include metal toners, resistor toners, high k dielectric toners, phosphors and glass. The substrates printed upon include glass, bare and coated metal, polymeric films and even paper. A fixed configuration electrostatic printing plate is used in most manufacturing applications though traditional photo receptor plates can be used if electronic addressability is desired.

Applications of electrostatic printing for electronic packaging products (printed wiring boards and flex circuits) and of passive electronic components themselves will be shown. Results with a pure silver toner printed on both glass and paper will be reported. Examples of passive electronic components like resistors, capacitors, and even inductors that have been electrostatically printed with liquid toners will be shown. Possible applications of toners to the manufacture of flat panel displays will be discussed.

EXPERIMENT

There are three key elements to the electrostatic printing process:
1. The Electrox electrostatic plate[1,2]
2. Functional materials configured as liquid toners
3. Non-contact or gap transfer in which toner particles are transferred by an electric field across a gap of the order of 50 to 150μ to the receiving surface[3]

The steps of the process are best illustrated in the following figures. Figure 1 shows the plate-making step. A photopolymer layer is coated on an electrically grounded substrate. The substrate can be metal, metallized polyester or polyimide film, or even glass made conductive with an indium tin oxide coating (ITO). The photopolymers typically vary in thickness from 10μ to 50μ. The photopolymer is exposed to UV radiation in the 300 to 400nm region which causes exposed areas to undergo a chemical change. This raises the electrical resistivity of these regions significantly so that they can store electrostatic charge for a useful period of time. The plate-making step is now complete; there is no chemical or aqueous processing of the plate.

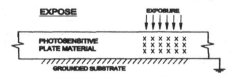

Figure 1. Plate Making

The plate is sensitized by corona charging it. The resulting surface potential for a typical 37μ thick plate is from 500 to 1000 volts. After a short period of time the unexposed regions of the plate self discharge due to their relatively low electrical resistivity. We now have a

Mat. Res. Soc. Symp. Proc. Vol. 624 © 2000 Materials Research Society

traditional latent electrostatic image. The latent image is processed by development with an electrophoretic liquid toner

Figure 2 shows the transfer step wherein the toner is transferred across a finite mechanical gap to a receiving glass plate by an electrical field created by the electric field plate

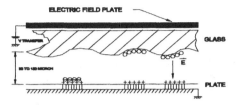

Figure 2: Toner Transfer Across a Gap

on the other side of the glass driven to a suitable potential. An alternate scheme not shown in Figure 3 is to corona charge the other side of the glass with a charge of polarity opposite that of the toner. Figure 3 shows a highly irregular surface, not an exaggeration. This is a particular advantage of electrostatic printing over other printing or deposition techniques. The toner travels across the gap following the parallel electric field lines and it does not disperse as a function of distance between glass and plate. Therefore high resolution images, true to their design, can be deposited on the glass substrate, even if it has imperfections or a relief structure already on it. We have also printed on metal plates, polymeric films and paper. An example of printing on steeply relieved metal surfaces is the printing of images on U.S. coins.

The Plate

While a photo addressable drum or plate could be used. Electrox chose an electrostatic plate for the following reasons:

1. The electrostatic plate offers a superior latent image over that of the photo receptor plate.

2. Small features like 10 micronss or even smaller are possible.

3. The electrostatic plate offers reasonable process speeds of 250 mm/sec for high through put.

The Toner

In this area, our toner technology, is far removed from traditional liquid electrographic toners. The following table shows these differences

	Materials	Particle
Metals	Aluminum	30 micron
	Silver	0.2 micron
Glasses, Ceramics	Glass frits	0.5 to 7 micron
	Phosphors	0.6 to 80 microns
Catalysts	Palladium	0.33 micron
	Tin	0.8 mircons
Composites	Conductor-Silver filled resin	
	Resistors-Carbon filled resin	
	Capacitor Barium Titanate filled resin	

Capabilities of Functional liquid Toners:

High densities, up to 10 gm/cm^3
Board range of Electrical resistivity
     Silver 1.63x10^{-6} ohm cm
     Glass 10^{+15}
Board range of particle sizes 0.05 micron to 100 micron
Broad range of mechanical properties, soft resins to very hard materials
Resinless or very little unwanted materials
Virtually any material not swelled nor dissolved by the Isopar diluent liquid can be made into a liquid toner. There are few "process burden" materials required to make the liquid toner. Our solid silver particle toner is 96% silver with no resin to interfere with electrical conductivity.

Non-contact or Gap Transfer

One of the principle advantages of liquid toner systems is their ability to be transferred across a significant mechanical gap between image plate and receiving plate (3). This is very important in some manufacturing applications where the receiving surface is either metal or irregular (like glass); or where high overlay accuracy is needed thereby eliminating elastometric roller transfer. We will show samples of US coins printed with toners where the edge of the coins were spaced 125° C above the printing plate. The image show fine features with good edge acuity.

Fusing or Sintering

Necessarily the toner image is particulate and must be fused to result in a useful structure. In some cases, like metal, heating to near their melting point is not allowable. A significant enhancement of toner technology is the silver Parmod Toner of Parelec LLC of Rocky Hill, NJ. Here we have a solid silver particle coated with a MOD coating (metallic organo decomposition product). Which when heated to a modest temperature (230°C-2min) reduces itself to pure silver thereby chemically sintering the Ag particles together.

Printed on paper and processed at 230°C for 2 min yields a bulk resistivity of about 5 micron ohm. cm, 10 to 20 times better than the best silver filled resins. Printed on glass with processing at 400 to 430°C range gives bulk resistivity values close to that of pure silver.

RESULTS

Recently we have found that certain thin coatings of resins on PET will chemically enhance the MOD decomposition process allowing it to function in the 125°C to 150°C range, compatible with PET and PEN films and useful for many commercially attractive applications.

At these processing conditions we get bulk resistivities of 15 micro ohm cm, 3 to 6 times better that the best composite inks and useful, as is, for many applications were low cost of production is essential.

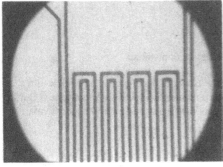 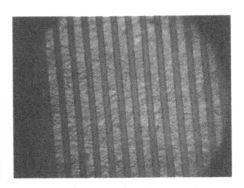

Figure 3. Silver Toner Printed on Soda Lime Glass

Figure 4. Silver Toner Printed on Paper and Sintered at 230°C

Figure 3 show the silver toner printed on glass. The conductor patterns are 40 microns wide with 60 micron spaces; the metal is 1 to 2 microns thick. Figure 4 shows that same silver toner printed on paper with thermal processing at 230°C-2mins. Even though it is a smooth coated sheet of food packaging paper, the discrete fibers shown in the photomicrograph are of the order of 12 microns. Again the silver patterns are about 1 micron thick

Figure 5 shows the silver toner printed on coated 75 micron thick PET, heat processed at 125°C for 30 min. Bulk resistivity of the silver material with this low temperature processing is 15 micro ohm cm. Adhesion of the silver metal traces to the coated PET is excellent.

Figure 5. Silver Toner Printed on Special Resin Coated PET, Sintered at 125°C

Figure 6. A Nest of Electronic Components Printed on Glass

Figure 6 shows a nest of electronic components printed on glass. On the bottom are 200 micron wide silver conductor patterns. Upper left shows a 9mm x 9mm and a 5mm x 5mm capacitor. Upper center shows a 5 turn inductor and the upper right shows six resistors.

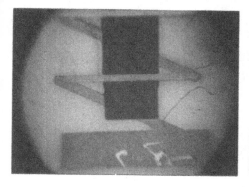

Figure 7. A Detail of the Inductor, Four layer part   Figure 8. Phosphor Toner Printed at the Bottom of Plasma Panel Trenches

Figure 7 shows a detail of the inductor. This is a four layer part. The bottom terminal and the silver lines at 30 degrees comprise the first layer; then two layers of ferrite toner and finally the fourth layer which is the silver metal printed horizontally. This five turn coil has a dc resistance of 4 ohms.

In another venue figure 8 shows a phosphor toner printed in the trenches of an ac plasma display back plate. The dark vertical line are the barrier ribs, approx 120 microns high and 40 micron across. The width if the trench is 120 microns. Every third trench is one of the primary colors; red, green, and blue.

CONCLUSIONS

We have assembled the elements of a generic manufacturing technology. A wide variety of materials can be formulated into liquid toners. They can be imaged by a suitable plate or drum then transferred in a non-contact mode to a wide range of substrates. In this program we have primarily printed on glass plates, usually soda lime class 2.25 mm thick. But printing on copper, brass, aluminum, stainless steel, and US coins is easily done.

Electrostatic printing of functional materials configured as liquid toners is a versatile manufacturing process. A solid, resinless Ag toner allows one to print pure metal in an additive process. Furthermore it can be printed on low temperature substrates like PET films and paper as well as glass. With resistive barium and ferrite toners one can print resistors, capacitors and inductors. Finally phosphors were printed in 120 micron deep trenches for ac plasma display manufacture.

REFERENCES

1. J Reisenfeld, US Patent No 4,732,831 (22 March 1988)
2. R Detig and D Bujese, US Patent No 4,859,557 (22 August 1989) and 5,011,758 (30 April 1991)
3. D Bujese, US Patent No 4,786,576 (22 November 1988) and 4,879,184 (7 November 1989)

Laser Direct Write
Techniques

A UV DIRECT-WRITE APPROACH FOR FORMATION OF EMBEDDED STRUCTURES IN PHOTOSTRUCTURABLE GLASS-CERAMICS

P.D. FUQUA*, D.P. TAYLOR*, H. HELVAJIAN**, W.W. HANSEN**, M.H. ABRAHAM**
*Materials Processing and Evaluation Department, Space Materials Laboratory **Center for Microtechnology; The Aerospace Corporation, Los Angeles, CA 90009-2957

ABSTRACT

Photostructurable glass-ceramics are a promising class of materials for MEMS devices. Previous work micromachining these materials used conventional photolithography equipment and masking techniques; however, we use direct-write CAM tools and a pulsed UV laser micromachining station for rapid prototyping and enhanced depth control. We have already used this class of materials to build components for MEMS thrusters, including fuel tanks and nozzles: structures that would prove difficult to build by standard microfabrication techniques.

A series of experiments was performed to characterize process parameters and establish the processing trade-offs in the laser exposure step. The hypothesis that there exists a critical dose of UV light for the growth of an etchable crystalline phase was tested by exposing the material to a fluence gradient for a variety of pulse train lengths, and then processing as usual. By measuring the dimensions of the etched region, we were able to determine the dose. We found that the dose is proportional to the *square* of the per-pulse fluence. This has allowed us to create not only embedded structures, but also stacked embedded structures. This also implies that we can embed tubes and tunnels with a single exposure inside a monolithic glass sample. We feel that this technique has promise for a number of applications, including microfluidics.

INTRODUCTION

One of the many factors that drives progress in Microelectromechanical Systems (MEMS) is the improvement in materials and materials processing. Much of the work is based on silicon; however, semiconductors are not optimal for some MEMS applications. One particularly interesting alternative class of materials is photostructurable glass ceramics. Stookey and others did the earliest work at Corning in the 1950s. [1,2] That work nucleated a number of wildly successful commercial product lines including Corningware. These materials are typically exposed by ultraviolet lamps and patterns are created using shadowmasks. Recent work in this laboratory has involved the use of ultraviolet lasers to create MEMS components for space applications. [3,4]

The use of laser light in the exposure process has several advantages to conventional techniques. Our laser micromachining station moves the work piece relative to the laser beam according to a predefined computer program. The exposure process is maskless and thus amenable to rapid prototyping. Depth control is achieved by the proper choice of exposure wavelength and dosing. As one tunes the laser further into the material's UV absorption band, the absorption of light increases and the penetration depth decreases resulting in shallower features. Furthermore, the beam can be shaped and the resulting structures will maintain the shape. For example, a collimated beam can result in a cylindrical hole. A focused beam results in either a cone section, or a hyperboloid-like structure. Such structures may be useful for microthruster expansion nozzles.

The primary material used in this work is a photostructurable glass ceramic available in wafer form from Schott glass under the trade name Foturan [5]. This material is composed of a lithium aluminosilicate glass doped with trace amounts of silver, cerium and antimony [6].

79

Mat. Res. Soc. Symp. Proc. Vol. 624 © 2000 Materials Research Society

According to the literature, cerium is a photosensitizer.[1,2,5] In the presence of ultraviolet light, Ce^{3+} gives up an electron to become Ce^{4+}. A fraction of the free electrons find Ag^+ which reduces to Ag^0. In subsequent thermal processing, silver atoms diffuse together to form clusters. If a cluster is larger than 80Å, it can provide the nucleus for the growth of a crystalline phase into the amorphous phase. The crystalline phase is composed largely of lithium metasilicate, which is preferentially soluble in hydrofluoric acid. Soaking the sample in HF solution results in dissolution of the exposed regions.

One might describe the processing of this material as direct-write three dimensional volumetric lithography. Other laser direct-write techniques structure material by selective deposition, as in laser CVD, or by selective removal of material, as in laser ablation. This technique involves selective removal, but is more photon efficient because the laser doesn't have to provide the energy to remove material. Here developing the pattern and etching the wafer is done as batch processes. In this way, it combines the selectivity of direct-write with the efficiency of batch processing.

While the processing parameters for broadband lamp exposure have been established, we found little published on laser exposure of photostructurable glass-ceramics [7]. This work was performed to establish laser process parameters including wavelength and fluence. We began with a hypothesis that there exists a critical dose, D_c, above which photostructurable glass forms a latent image, and below which, no image is formed. If this is true, then the boundary of the region formed after etching the sample would define a "surface of critical dose." It can be argued that the volume that was etched away was subjected to a dose higher than D_c and the glass that remains did not reach D_c. Conceptually, the critical dose is the dose required to create a density of nuclei large enough to result in an interconnected network of crystallites. Of course it is these crystallites that are etched away in subsequent processing. Certainly one would expect that the density of nucleation sites would be proportional to the dose and that the critical dose will be a function of the composition and process parameters. Therefore, the critical density of sites should be fluence dependent, which could be described mathematically as follows:

$$\rho = KF^m N \tag{1}$$

Where ρ is the density of nuclei, F is the per-pulse fluence, m is the power dependence, N is the number of pulses, and K is a proportionality constant. Per-pulse fluence is defined as the local irradiance integrated over time over the length of a single pulse. We could further refine this equation by arbitrarily defining a dose, D, that is equal to ρ/K or

$$D = F^m N \tag{2}$$

For a given number of pulses, D_c will correspond to a critical fluence F_c.

$$D_c = F_c^m N \tag{3}$$

EXPERIMENT

Coupons of Foturan were exposed to a laser beam with a Gaussian spatial profile for a predetermined number of pulses. The coupons were then baked and etched to reveal a "surface of critical dose." Since the spatial profile of the laser was known, the fluence at the edge of the etched surface could be found. Given our hypothesis above, one expects that as the number of pulses increases, the critical dose will be found further out in the wings of the fluence profile.

The illumination source for these experiments was a Continuum HP-1000 diode pumped Nd:YAG laser that was either frequency tripled to 355nm or quadrupled to 266nm. The pulse length was given as 8 nsec. The beam passed through a quartz lens with a 500 mm focal length. The coupons were exposed at three different points along the focusing beam, effectively changing the exposure area. At each spot size, the sample was illuminated with bursts of 30,

80

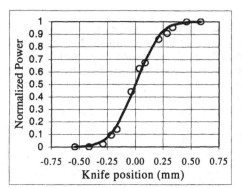

Figure 1. A typical plot of normalized power vs. knife edge position. The solid line is fit to equation 4.

100, 300, 1000, 3000, and 10000 pulses and then heat treated equivalently. The spot size of the laser beam was measured using the knife-edge method, which involves measuring power in the beam as a function of the distance that a blade encroaches into the beam. By fitting the data to the following equation, which assumes a Gaussian profile, the spot size can be determined.

$$\frac{P}{P_o} = \frac{1}{2} erfc\left(\frac{\sqrt{2}x}{\omega_o}\right) \qquad (4)$$

Here P/P_o is normalized power, ω_o is the $1/e^2$ spot radius, and x is the knife edge position relative to the center of the beam.

Since knowledge of the spatial illumination profile is essential to interpreting the data, a typical beam profile is presented in Figure 1. The solid line is a fit of experimental data points assuming a $1/e^2$ spot radius of 0.35 mm. For the purposes of this work, the assumption of a Gaussian beam is clearly reasonable.

The exposed coupons were developed in a programmable furnace. The temperature was ramped at 5°C/min to 500°C and held for an hour. At this temperature, neutral silver atom diffusion permits the formation of silver clusters in the UV exposed regions. The temperature was then raised again to 605°C at 3°C/min and held for another hour. At this higher temperature, the glass devitrifies, nucleating a crystalline phase at the silver clusters. It is important to note that the crystalline phase (also called ceramic phase) is slightly less dense than the amorphous phase, so there is a small expansion associated with the devitrified regions. The temperature is sufficient for softening of the amorphous phase, so the glass flows slightly to accommodate the expanded volume. After completing the nucleation and growth steps, the sample has an image consisting of brown crystalline regions in an clear amorphous matrix. It is a property of the crystalline phase that it is more rapidly soluble in hydrofluoric acid than the surrounding phase. In a 5% solution of hydrofluoric acid at room temperature, the etch ratio is 50.

Figure 2. A Cross-sectional view of a crystallized region that was exposed to 300 pulses of 266 nm light.

Figure 3. A Cross-sectional view of a crystallized region that was exposed to 10,000 pulses of 266 nm light.

RESULTS

Figure 2 is an optical micrograph of a cross section of Foturan exposed with a 266nm Gaussian beam. The dark region is crystalline and colored by silver clusters. Notice that after 300 pulses, the feature is shallow. As more pulses illuminate the sample, the crystalline volume expands into the wings of the Gaussian spatial distribution, and it extends further into the material. The absorption coefficient of unexposed Foturan is 3.4 mm^{-1} at 266 nm, so individual laser pulses are attenuated with penetration into the material. Figure 4 is a scanning electron micrograph that shows a similar sample that has been baked, cross-sectioned and then etched.

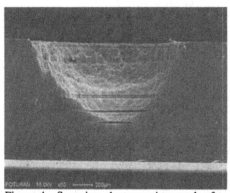

Figure 4. Scanning electron micrograph of an etched volume. The bar represents 200 microns.

Figure 5. SEM close up of a planar feature at the etch boundary. The bar represents 50 microns.

The planar features are thought to be regions of a slightly different composition, which are more sensitive to UV light. Under an optical microscope, striae were also visible in cross section. It is unlikely that these features are due to polishing defects since they penetrate into the material away from the polished surface, where they can be seen after etching. Increased cerium or silver concentrations is a likely cause. We tried to obtain a compositional map using SIMS and EDX, but the signals were too small to be conclusive.

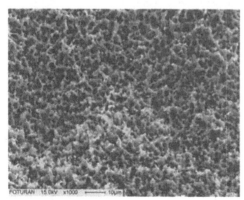

Figure 6. Coral texture at the etch interface.

Figure 5 is a SEM image that shows the interface between the etched and nonetched regions. This sample was etched after cross sectioning, so one can see etch pits that correspond to crystallites that were formed, but were not part of an inter-connected network. Clearly, as one crosses the etch boundary, the crystallite density drops. Figure 6 shows a normal view of the etched surface. This "coral-like" texture is typical of these processing conditions, and is a result of the nucleation density being insufficient for complete crystallization. The cavities in the image are a few microns in diameter. A smooth texture can be achieved by other means including exposure with a beam that has a clipped spatial profile.

After the coupons were exposed, baked, and etched, the size of the etched opening was determined by optical microscopy. Because the absorption coefficient at 355 nm is small, samples exposed at that wavelength resulted in a via that penetrated the 1 mm sample. Samples that were exposed to 266 nm light had openings that generally didn't pass through the sample. The depth of those openings was measured by tracking the change of focus on an optical microscope. With the assumption of a Gaussian fluence profile, and by knowing the spot size, one can calculate the per-pulse fluence at the etch boundary.

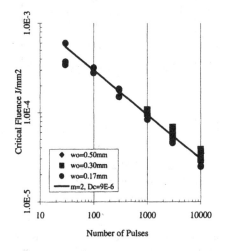

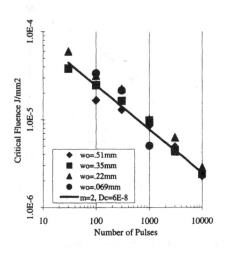

Figure 7. Critical Fluence vs. Number of Pulses of 355 nm light.

Figure 8. Critical Fluence vs. Number of Pulses of 266 nm light.

Figure 7 is a log-log plot of the critical fluence, F_c, calculated at the etch boundary, versus the number of pulses. This data is for 355 nm exposures with spot radii of 0.17, 0.30, and 0.50 mm. The exposure with the largest spot size has the smallest peak fluence. Over 1000 pulses were required to get any etching at all. For that sample, the peak fluence of the illumination was less than the critical fluence for pulse trains less than 1000 pulses long.

This data was fit to equation 3 above. When plotted on a log-log graph, the slope corresponds to the negative reciprocal of m, the power dependence. From the intercept at one pulse, one finds the critical dose. The fit parameters are m= 2 and $D_c = 9 \times 10^{-6}$ J^2/mm^4, which is given in the figure as a solid line.

Figure 8 is a similar plot for 266 nm exposures. Data from exposures with four different spot sizes is presented. There is more scatter in the data which may be due to the compositional inhomogeneities or other source of striae as seen in figures 2 and 3. Regardless, the data was fit by a line defined by m=2 and $D_c=6 \times 10^{-8}$ J^2/mm^4, which is plotted in the figure.

The fluence dependence is a squared term at both wavelengths, implying that the mechanism for forming the latent image is the same. The critical dose was smaller for 266 nm light which would also be expected since the absorption coefficient and thus the absorption cross section is larger at 266 nm than at 355 nm.

The result that m=2 is quite interesting, since this material can also be exposed by a low intensity lamp. Thus it is unlikely that the mechanism behind exposure is a single true two-photon process. One could speculate that it is a sequential two-step process, with a long lived intermediate state. These coupons were exposed with a laser for less than 10 seconds. Exposure in a mask aligner might take 60 minutes, depending on the intensity of the bulb. Perhaps there is an additional mechanism for exposure that can be accessed with a focused laser that would not be available to a lamp. We hope to elucidate the mechanism behind our empirical observations in future work.

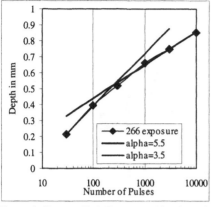

Figure 9. Foturan Etch Depth vs. Number of Pulses of 266 nm light. ω_o=0.57 mm, F_o=2.5E-4 J/mm^2.

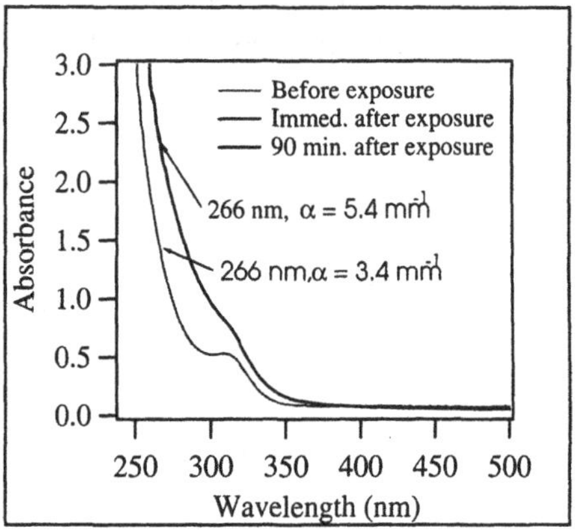

Figure 10. Absorption spectra of Foturan before and after illumination.

Figure 9 plots etch depth vs. number of pulses on a linear-log graph. If we assume that F_0 is the peak fluence incident upon the sample, and we assume that it is attenuated according to Beer's Law, the fluence at the bottom of the etched pit is $F_c=F_0\exp(-\alpha z_c)$ where α is the absorption coefficient and z_c is the etched depth. Thus the data should follow the following equation:

$$z_c = \frac{-1}{\alpha m}\left[\ln(F_o^m N) - \ln(D_c)\right] \quad (5)$$

Clearly, the data is nonlinear. The curvature implies that the absorption coefficient is increasing as a function of dose. The data can be fit with two lines. The slope of the data point with the fewest number of pulses fit a fluence dependence of m=2 and an absorption coefficient of α=3.4 mm^{-1}. The slope for longer pulse trains fits an absorption coefficient equal to α=5.5 mm^{-1}. The effect of increased absorptance wasn't seen in the previous data because the etch dimentions were measured at the incident surface. The attenuation due to absorption is only relevant in the bulk material. This suggested a new experiment to confirm the change in absorptance. Figure 10 shows the absorptance spectrum of Foturan before and after exposure to 266 nm light. The absorptance change doesn't recover in 90 minutes at room temperature. At 266 nm, the absorptance rises from 1.45 to 2.34 which corresponds to an absorption coefficient of 3.4 mm^{-1} and 5.4 mm^{-1} respectively. This is consistent with the fit of data presented in figure 9. It is not clear whether the absorptance change is associated with the photostructurable nature of this material. Silicate glasses are known to darken when exposed to UV light due to generation of defects. This phenomenon is called "solarization" and it is an open question whether or not the defects induced by the laser enhance crystallization.

One remarkable aspect of processing this material is sharp differentiation between crystalline and glassy regions. After etching, the coral layer is fairly thin. This raised the question as to whether we could create embedded structures, i.e., expose and crystallize the glass in the middle of the coupon by focusing into the center, without crystallizing the entire light path. Certainly as the focusing beam converges and diverges, the fluence increases and decreases. Where the beam exceeds the critical fluence, crystallization is possible. If z is the distance from the beam waist along the propagation axis, the fluence at the peak of the spatial profile is described as:

$$F = \frac{F_o}{\left(1 + z^2 / z_r^2\right)} \qquad (6)$$

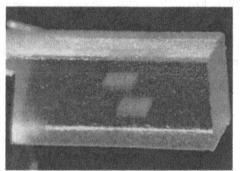

where F_0 is the peak fluence at the beam waist, z_r is the Rayleigh length [8], which is defined by the focusing geometry. The faster the focusing, the smaller the Rayleigh length. At some critical distance from the beam waist, the fluence will fall below the critical fluence and crystallization will not occur. The length, z_c over which crystallization can occur is given below:

$$Z_c = Z_r \sqrt{F_o / F_c - 1} \qquad (7)$$

Figure 11. Cross section of embedded crystalline regions in a 3mm thick coupon.

Figure 11 shows a side view of not only an embedded structure, but also a stacked, embedded structure. Exposure occurred through the top surface. The two lighter rectangles are crystalline regions. This was done in a 3mm thick sample with 355nm light focused through a 10x microscope objective. The workpiece was translated in a square pattern on an x-y translation stage. The piece was then translated 1 mm up and the pattern was repeated. Even in places where the illuminations overlapped, the total dose obviously did not exceed the critical dose, and the Foturan remained amorphous.

CONCLUSIONS

We have shown that a laser direct-write technique has some advantages over more conventional processing in creating true three dimensional structures. A systematic study was performed to test the validity of an empirical model for exposure. The results show that laser exposure depends on the square of the per-shot fluence. Lastly, this relationship was exploited to build not only embedded structures, but also stacked embedded structures.

ACKNOWLEDGEMENTS

The authors gratefully acknowledge the thoughts and efforts of Paul Adams, James Barrie, Siegfried Janson, Nick Marquez, and Joe Uht, at The Aerospace Corporation. This work was funded through the Aerospace Corporate Research Initiative program.

REFERENCES

1   S.D. Stookey, Indust. Engin. Chem. **45**(1) (1953) p.115-118

2   A. Berezhnoi, *Glass-Ceramics and Photo-sitalls*, Plenum, New York, 1970, now available at UMI, Ann Arbor, MI

3   P. Fuqua, S.W. Janson, W.W. Hansen, and H. Helvajian, in *Laser Applications in Microelectronic and Optoelectronic Manufacturing IV*, edited by J.J. Dubowski, H.H. Helvajian, E.W. Kreutz, and K. Sugioka (Soc. Photo. Opt. Inst. Engin. **3618**, Bellingham, WA 1999) p.213-220

4   W.W. Hansen, S.W. Janson, H. Helvajian, in *Laser Applications in Microelectronic and Optoelectronic Manufacturing II*, (Soc. Photo. Opt. Inst. Engin. **2991**, Bellingham, WA 1997) p.104-112

5   Foturan - A Material for Microtechnology, technical literature by Schott Glaswerke Optics Division and IMM Institut für Mikrotechnik GmBH Mainz Germany.

6   D. Hülsenberg, R. Bruntsch, K. Schmidt, F. Reinhold, *Mikromechanische Bearbeitung von fotoempfindlichem Glas*, Silikattechnik, Vol. 41 (1990), 364.

7   S.D. Stookey, Private communication. Corning also worked on UV laser exposure of photostructurable glass-ceramics, but never published the results.

8    $Z_r=\pi$ f#^{2} λ, where f# is the relative aperture. A. Siegman, *Lasers*, University Science Books, Mill Valley, CA, 1986, p.663-679

DIRECT PATTERNING OF HYDROGENATED AMORPHOUS SILICON BY NEAR FIELD SCANNING OPTICAL MICROSCOPY

Russell E. Hollingsworth*, William C. Bradford**, Mary K. Herndon**, Joseph D. Beach** and Reuben T. Collins**
*Materials Research Group, Inc. Wheat Ridge, CO
**Physics Department, Colorado School of Mines, Golden, CO.

ABSTRACT

Practical methods for directly patterning hydrogenated amorphous silicon (a-Si:H) films have been developed. Direct patterning involves selectively oxidizing the hydrogen passivated a-Si:H surface, with the oxide then serving as an etch mask for subsequent hydrogen plasma removal of the unoxidized regions. Photo induced oxidation has been extensively studied using both far field projected patterns and near field scanning optical microscopy (NSOM) for direct write patterning. Examination of the threshold dose for pattern generation for excitation wavelengths from 248 to 633nm provides indirect evidence for involvement of electron-hole recombination in optically induced oxidation. Optical exposure of a-Si:H in vacuum demonstrated that oxygen must be present in the ambient atmosphere during exposure for successful pattern generation. This suggests that oxidation of the surface may not involve removal of hydrogen, but rather breaking of Si-Si backbonds and insertion of oxygen. An additional mechanism for oxide generation was observed whereby pattern generation resulted from simple proximity of an NSOM probe within ~30nm from the sample surface. The probe dither amplitude was found to greatly affect the line width and height of patterns generated without light. Line widths of approximately 100nm, comparable to the probe diameter, were obtained.

INTRODUCTION

Scanning probe microscopies are presently under active investigation for nanolithograpic patterning of surfaces. STM and AFM have been the most extensively studied techniques. They have been used to demonstrate, for example, patterning of organic resist layers[1], selective chemical vapor deposition[1], selective chemisorption[2], selective oxidation of crystalline silicon surfaces[3], and direct scribing of lines into semiconductor surfaces.[4] Nanometer-scale devices have also been produced using these techniques. As examples, single electron transistors have been fabricated using STM induced selective oxidation of titanium.[5] AFM lithography was used to define the gate of a 100nm MOSFET.[6] Nanometer scale side-gated silicon FET's were also patterned using AFM.[7]

Recently, a-Si:H has been explored as a resist for use with many nanolithographic techniques. In general, patterning a-Si:H involves selectively oxidizing the a-Si:H surface. Selective oxidation can be achieved by the electron beam from an electron microscope, interaction with the tip of an STM or AFM, or by the light emitted from an NSOM probe. Selective oxidation is frequently attributed to the removal of surface hydrogen passivation, but the work presented here suggests that Si-Si back bond breaking with oxygen insertion is actually the mechanism in optically induced oxidation of amorphous silicon. The oxide layer formed in this way then becomes a mask for subsequent etching of the surface. An a-Si:H resist has the advantage that it can be deposited on many surfaces at low temperature. It can be deposited as a thin film, and it is compatible with semiconductor processing. Kramer et al. demonstrated 40nm metal wires using a-Si:H resists and STM.[8] Minne et al. have used parallel arrays of AFM

87

cantilevers and a-Si:H resists to pattern silicon oxide, silicon nitride, chromium and titanium metal.[9]

Studies of Near-Field Scanning Optical Microscopy (NSOM) for nanolithographic applications are fairly recent. Smolyaninov et. al. and Davy et al. used the light from the NSOM tip to pattern conventional optical photoresists.[10,11] Feature sizes of 100nm were produced. Massanell et al. patterned ferroelectric surfaces.[12] Madsen et. al. used NSOM to pattern a-Si:H.[13] In this study, a wet chemical etch was used for pattern development. In agreement with our results they observed that proximity effect lines were written by the tip in the absence of optical illumination. They did not, however, explore the dependence of this effect on write speed, dither amplitude, tip sample separation, or the dose dependence of true optical exposure as done here. Minimum feature sizes near 50nm were produced in their work.

FILM DEPOSITION

In this work, a-Si:H films were grown on crystalline silicon wafers or Corning 7059 glass to a thickness of 25-800nm. A Materials Research Group multichamber, parallel plate, capacitively coupled plasma enhanced chemical vapor deposition (PECVD) system operated at 13.56MHz was used for both deposition with pure silane and etching with pure hydrogen. The substrates were placed in a stainless steel carrier that formed the ground electrode. The same chamber was used for both deposition and etching. Film thickness was monitored *in situ* by reflecting a HeNe laser off the surface and counting interference fringes. The films were allowed to cool from the deposition temperature to room temperature in vacuum in order to minimize native oxide growth. Unless other wise specified, all processing was performed at a heater temperature of 50°C. Laser sources used for optical irradiation included a KrF excimer laser as the source of 248nm light with a nominal pulse length of 20ns and Ar ion and HeNe lasers for continuous wave (cw) irradiation from 350 to 633nm. Optical irradiation occurred with the sample in various atmospheres, including air, 5% O_2 in Ar, nitrous oxide (N_2O), and ultra high vacuum.

RESULTS

a-Si exposure

Direct patterning of a-Si:H films requires a means of modifying the surface or bulk of the film to change its resistance to etching. Two methods of modifying the a-Si:H film to generate patterns (laser crystallization and optically enhanced oxidation) have the potential to meet the throughput requirements of modern factories. Laser crystallization modifies the bulk film through localized melting, requiring high power density. Excimer lasers have sufficient power to crystallize centimeter size areas to a depth of more than 100nm, making these sources suitable for projection lithography applications. Our work in this area has been reported previously.[14,15] Other laser sources can be used for direct writing, but feature sizes will be limited by the spot size that can be realized with a scanned beam system.

Selective oxidation of the a-Si:H surface provides an alternative patterning approach that allows incident illumination power densities to be orders of magnitude smaller than the laser crystallization approach. Optical exposure at lower power densities results in the growth of a thin surface oxide, while hydrogen passivation prevents oxide growth on the unexposed areas. The oxide serves as an etch mask for subsequent pattern development. For both selective oxidation and laser crystallization the exposed region becomes more resistant to etching, making amorphous silicon a negative resist. Selective oxidation relies on hydrogen passivation of the surface to prevent native oxide growth. The surface of as-deposited hydrogenated amorphous silicon is naturally hydrogen passivated. Previous work has shown the a-Si:H surface to resist native oxide growth for periods up to one week.[16] XPS measurements of as-deposited and broad band UV exposed a-Si:H films clearly showed the growth of 5-10Å of oxide following UV exposure.[17]

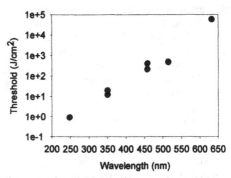

Figure 1. Threshold dose for exposing a-Si:H in air as a function of irradiation wavelength.

The optical dose required to induce oxidation of the intrinsic a-Si:H surface in air was measured as a function of illumination wavelength as shown in Fig. 1. For this study, a 100nm thick a-Si:H film was exposed to a light source through an aluminum-on- quartz shadow mask. The power densities used at each wavelength were generally kept low (<3W/cm^2) to avoid thermal heating of the sample. The threshold dose was defined as the dose which left the exposed region untouched when the unexposed region was completely removed by the etch. This threshold definition leads to an exposure that would give rise to well defined features in lithographic applications. We note, however, that partially developed patterns could be observed for doses an order of magnitude less than threshold. While we cannot rule out the possibility that high peak intensities due to pulsed operation may contribute to reduced threshold at 248nm, Fig. 1 makes it quite clear that shorter wavelengths strongly decrease the required dose.

The dependence on gas ambient during optical illumination has also been examined. Samples were irradiated with 350nm cw light under a high vacuum of 1×10^{-7} torr, and then exposed to air prior to hydrogen plasma etching. No patterns were generated even for a dose an order of magnitude greater than the threshold in air. Other samples were irradiated with pulsed 248nm light under various low pressure oxygen containing ambients. For an ambient of 5% O_2 in Ar at a total pressure of 1 torr, the threshold dose increased by roughly a factor of 10 over the value in air.[18] Pure nitrous oxide (N_2O) was also used as an ambient over a pressure range of 1-100 torr. The threshold dose in N_2O was found to be independent of pressure and comparable to the value in O_2/Ar. At this point, it is not clear whether the increased thresholds in low pressure O_2 or N_2O ambients are due to a pressure effect or the associated reduction in water concentration. However, it is clear that some oxygen containing gas must be present simultaneously with the optical irradiation in order to grow an oxide.

The sensitivity of the surface to optically enhanced oxidation has been found to depend on the deposition conditions of the a-Si:H film. Threshold dose at 248nm as a function of heater temperature during deposition is shown in Fig. 2. Due to the non-contact, radiative heating in our process system, substrate temperatures are approximately two thirds of the heater temperature at the high end and nearly equal at the low end. Optical exposure in all cases was performed with the substrate at room temperature. A factor of three increase in threshold dose was observed as the substrate temperature during deposition was increased from 30 to 170°C. (Similarly, crystallization thresholds increased by a factor of two over the same temperature range.[14]) In other process variations, the use of a 10% SiH$_4$/Ar gas mixture for deposition was found to result in films with higher threshold dose than films grown with pure SiH$_4$ at the same temperature. Qualitatively, the threshold dose appears to be determined by the band gap of the a-

Si:H film: films with larger band gap have lower threshold. For the undoped films used in this work, larger band gaps correlate with higher hydrogen contents and more disorder.

The data for low power oxidation is not consistent with dangling bonds due to hydrogen removal from the surface as the underlying oxide growth mechanism. The strongest indication is the complete lack of pattern generation following UHV irradiation since any dangling bonds generated by hydrogen removal should instantly oxidize on exposure to air. Rather, the mechanism is most likely breaking of Si-Si back bonds with simultaneous insertion of oxygen. (Note that oxygen insertion into Si-Si back bonds has been shown to be the first step in native oxide growth on hydrogen passivated

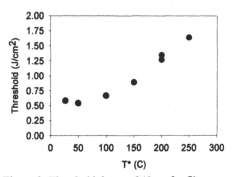

Figure 2. Threshold dose at 248nm for films grown at different heater temperatures. Exposures were performed in a 5% O_2/Ar mixture at 1 torr.

crystalline silicon.[19]) The bond breaking is probably dominated by the recombination of photo-generated electron hole pairs, although direct photo-dissociation of the most highly strained (weakest) Si-Si bonds might be possible with the highest energy photons. Dangling bond formation by free carrier recombination is one of the most widely studied aspects of the well known Staebler-Wronski degradation in hydrogenated amorphous silicon.[20]

All of the data can be qualitatively explained using the model of carrier recombination breaking of Si-Si bonds with the assumption that the dangling bonds either combine with oxygen if it is present, or the Si-Si bond reforms if oxygen is absent. This assumption provides an explanation for the lack of pattern generation following UHV irradiation. For oxide growth, only carriers that recombine at the surface can contribute. As the irradiation wavelength decreases, the absorption coefficient of the film will increase so a larger fraction of the generated carriers will be close to the surface. Hence, a lower total dose is needed at shorter wavelengths to cause the same number of carrier recombinations at the surface. Low deposition temperatures result in films with a larger fraction of strained (weak) bonds, which are then easier to break during carrier recombination.

Pattern development

Since oxide thicknesses of only a few angstroms can be expected, an extremely high selectivity is necessary in the etching process used for pattern development. Radio frequency (RF) hydrogen plasmas at 13.56MHz provide such a process with etch selectivity between oxide and a-Si:H exceeding 500 to 1. Etch selectivity was explored by comparing the etch rates of bulk a-Si:H and silicon dioxide thermally grown on crystalline silicon. Etch rates were examined as a function of RF power (10-50 watts), chamber pressure (0.5-3.0 torr), substrate temperature (25-100°C) and hydrogen flow rate (10-300 sccm). Portions of the film were covered with a stainless steel mask during the hydrogen plasma treatment to provide a reference thickness. A fixed process time was used that resulted in partial removal of the film. Thickness of the a-Si:H films was measured with a stylus profilometer, while SiO_2 thickness was measured by ellipsometry.

The dominant process parameter in determining the a-Si:H etch rate was the hydrogen flow rate. Measured etch rates as a function of flow rate for two different process pressures are shown in Fig. 3. Etch rates increased by factors of 3-5 as the flow rate increased from 20-300sccm. The RF power dependence is shown in Fig. 4 for two different heater temperatures. There was an initial increase in etch rate as the power increased, but etch rates began to drop at higher powers.

In hydrogen plasma etching of a-Si:H, the etch product will be silane. Since silane is a source gas for the growth of a-Si:H, there will be a competition between etching and

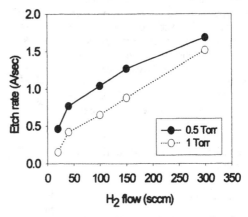

Figure 3. Amorphous silicon etch rate as a function of hydrogen flow rate for two process pressures.

redeposition during hydrogen plasma treatment. The critical silane to hydrogen concentration where deposition and etching exactly balance has been reported to vary significantly with substrate material, but to generally be in the vicinity of a few tenths of a percent.[21] Redeposition playing a larger role accounts for the decrease in etch rate seen at high plasma powers. For these experiments, the powered electrode was completely covered by silicon and therefore contributed to the silane fraction in the gas. In fact, for some conditions (high power, high pressure, low flow) silicon was etched off of the powered electrode and deposited on the grounded substrate.

The highest a-Si:H etch rate obtained in this series of experiments was 2Å/s. Silicon dioxide etch rates were nearly unmeasurable over the entire range of hydrogen plasma conditions examined. The highest oxide etch rate was approximately 2×10^{-3}Å/s, implying a selectivity on the order of 1000:1.

Because of the extremely low oxide etch rate, an indirect approach was adopted to optimize the etching conditions for the highest selectivity. A set of a-Si:H films was exposed with 350 nm light through a shadow mask with a range of exposure doses. Different hydrogen plasma etch conditions were then used on the various films. Etch conditions that have the lowest threshold dose correspond to the highest selectivity. One example is shown in Fig. 5, which shows threshold dose as a function of process pressure during etching. The error bars represent run to run variations. The large error bars are at least partly due to significant chamber history effects in the etch process. Since a single chamber was used for both deposition and etching, the quantity of silicon on the powered

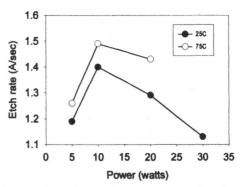

Figure 4. Amorphous silicon etch rate as a function of plasma power for two heater temperatures. The hydrogen flow rate was 300sccm.

electrode depended on the sequence of prior deposition and etching runs. Etch rates were observed to increase by factors of 2-3 when the powered electrode was clean. Chamber history effects on the etch selectivity have not been quantified.

Practical etch selectivity near 1000:1 between photo-enhanced oxide and a-Si:H has been achieved, as evidenced by the fabrication of fully developed features in 800nm thick a-Si:H films.[17]

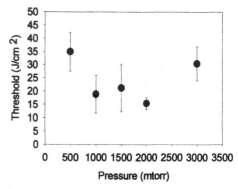

Figure 5. Threshold dose as a function of process pressure during hydrogen plasma etching.

NSOM writing

Advanced lithography has been the key technology driver for the semiconductor industry. To sustain the present rate of growth in integration level, the national technology roadmap indicates 100nm technology will ship as early as 2005. After this point there is general agreement that conventional optical lithography will no longer be adequate. As dimensions shrink, new resists, processes for making interconnects, and techniques for doping and etching must be developed. One of the goals of this research has been to test a-Si:H as a potential photoresist for sub-100nm performance. To do this we have been developing Near-Field Scanning Optical lithography as a tool for optical exposure of photoresists at characteristic dimensions in the 100nm range, with the longer term goal of moving to sub-100nm dimensions. Although scanning probe techniques will require significant development (for example, parallel writing[9]) to be capable of write speeds required for large scale VLSI *manufacturing*, they are ideal techniques for *research* focused on nanoscale patterning, process development, and device prototyping. While we have primarily used near-field lithography to study the patterning of a-Si:H, the technique has the potential to be generally applicable for research studies of nanoscale optical lithography processes. With this goal in mind, we have been exploring both the use of NSOM lithography to pattern a-Si:H and conventional optical photoresists. Preliminary results on the later will be discussed at the end of this section.

NSOM is a scanning probe technique (like STM and AFM) which allows optical excitation at spatial resolutions well below the diffraction limit. This is accomplished by scanning a tapered optical fiber with a subwavelength aperture (typically less than100nm in diameter) across a sample surface while maintaining a fixed tip-sample separation.[23] In NSOM, a shear-force interaction exists between the probe tip and the sample surface. The tip is generally dithered parallel to the surface to measure the strength of this interaction. The strength of the shear-force interaction is then fed back to the control software and used to maintain a fixed tip-sample separation, in much the same way tunneling current is used to control the tip height in an STM. The separation is much less than the wavelength of light and, hence, in the near-field. NSOM has commonly been used as a metrology tool to record nanoscale optical images as well as images of surface topography. In NSOM lithography the fiber optic probe is used as a light source to expose a photoresist, and patterns are generated by scanning the probe over the resist surface. In our measurements, an Ar-ion laser operating at 350 or 458nm, was used as the excitation source. Because this technique is not diffraction-limited, it has the potential to generate features with widths much less than those achievable by conventional far-field optical lithography.

In our studies of a-Si:H resists, we have previously discovered that proximity of the tip to the sample can lead to resist exposure in the absence of light.[17] Using proximity writing we demonstrated ~100nm patterns in a-Si:H as well as an increase in line height when optical exposure is also present as shown in Fig. 6. Although Fig. 1 indicates that write performance in a-Si:H will be improved by UV or even shorter wavelength excitation, the measurements in Fig. 6 were carried out with 458nm due to difficulties in finding suitable fibers for probes at UV wavelength. The present work has focussed on moving into the UV range, on examining the origin of proximity effect writing, and on comparing the ability to create well resolved lines in a-Si:H when features are isolated and when they are more closely spaced.

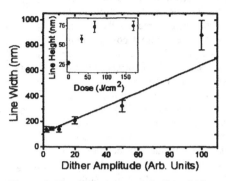

Figure 6. The dependence of the linewidth on dither amplitude in the absence of optical exposure is shown. Line width decreases to approximately 100nm as dither amplitude is decreased. The inset shows the dependence of line height on optical exposure at 458nm with fixed dither amplitude. Patterns are more fully developed as the optical exposure is increased.

Fiber optic probes are the critical element in NSOM pattern formation. NSOM tips were prepared both by chemical etching[24,25] and by taffy pulling using a Sutter Instruments Co. P2000 micropipette puller.[26] The probes were coated with approximately 100nm of aluminum leaving an open aperture at the point of the probe with a diameter of 100nm or less as measured by SEM. In our prior work we used single mode fiber designed for operation in the wavelength range of the green lines of the Ar-ion laser. Both etched and pulled fiber tips were developed. The etched fibers tend to have higher throughput because tapering of the core begins much closer to the aperture than with pulled fibers.[27] Etched fibers also have the potential for batch fabrication. The pulled fiber tips, however, have the advantage of simplicity of fabrication when only a single tip is needed. In general, single mode fibers suitable for visible wavelength operation attenuate in the UV making 350nm exposure difficult. To move into this shorter wavelength regime, two approaches were taken. In the approach used for most of the measurements presented here, 350nm light was coupled into a UV transparent multimode fiber (Thorlabs model FG-200-UEP). The UV light was recoupled from the multimode fiber into visible wavelength single mode tips in close proximity to the microscope. By minimizing the distance the light traveled through the visible fiber, attenuation, while still relatively large (80%), allowed sufficient UV transmittance to make UV exposures possible. Insertion loss into the single mode fiber was then the largest loss mechanism. The actual power reaching the probe tip was several orders of magnitude below the damage threshold of the tip. In practice, higher power levels are needed for effective UV writing. In one approach to this, direct coupling of UV fiber and extremely short visible light tips has been demonstrated by Yamamoto et al.[28] While we intend to explore this approach further, we have also been developing tips directly from the multimode, UV transparent fiber. Our standard etching procedures allow the UV fiber to be tapered to a sharp point but also leave a very scalloped and rough surface on the fiber. Other groups have made similar observations in etching of fiber tips.[29] Pulled UV tips were found to have much more uniform properties and to be suitable for UV writing. The large diameter of the multimode UV fiber, however, results in different mechanical parameters for the tips which have made it difficult to implement shear force feedback with our microscopes' present configuration. Plans are in place to modify our microscope to allow additional tests of UV writing with these tips.

Figure 7 shows the dependence of line height on optical dose for 350nm illumination. Dose is changed indirectly in these measurements. Because the UV intensity coupled into the tip was very low as discussed above, the intensity of the light coupled into the tip was held constant at its maximum value, and dose was adjusted by changing the write speed. The

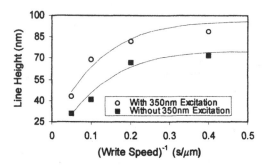

Figure 7. AFM measurements of a-Si:H line height vs. the inverse write speed both in the presence and absence of 350nm excitation. During illumination, inverse write speed is proportional to dose.

figure shows line height vs. write speed both in the presence and absence of UV excitation. In agreement with our earlier work,[15,17] proximity effect writing is observed in the absence of illumination, as well as with or without the metal coating on the probe tips. Line height is, again, found to be less than the original film thickness indicating incomplete exposure. Interestingly, the exposure for proximity effect writing was found to depend on write speed with slower speeds giving greater exposure. We also see that UV exposure results in even greater pattern height as dose is increased in agreement with observations made with 458nm excitation as shown in the inset to Fig. 6. The far-field UV dose required for full exposure in Fig. 1 is more than an order of magnitude larger than our estimates of the dose used to write the features characterized in Fig.7, which is consistent with the small amplitude difference with and without UV exposure. This again underscores the need for higher throughput UV tips.

To further explore the dependence of proximity effect exposure on write parameters, a series of lines were patterned in the absence of optical exposure while varying the tip-sample separation. As shown in Fig. 8, the line height decreases as the separation is increased. Tip sample separation can be determined as follows: the extension/contraction of the piezo tube is controlled by the feedback setpoint. Each volt that is applied results in a known length change in the z-axis piezo; contact of the tip with the sample can be detected, which provides the zero point, backing away from the zero is measured as a voltage change which is then converted into a length change. Once this calibration is done, the point above the sample at which feedback begins can be determined. The height of subsequent approaches can then be determined from the point at which feedback begins. We have previously shown that reducing dither amplitude also results in a decrease in line height. Based on the results shown in Figs.7 and 8 and our

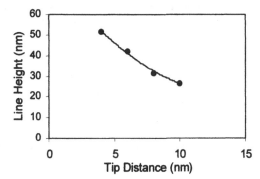

Figure 8. AFM measurements of a-Si:H line height vs. tip-sample separation during proximity induced writing.

previous work, minimizing proximity effect writing in favor of true optical exposure will require using the smallest dither amplitude possible, largest tip sample separation compatible with high spatial resolution, and fastest write speed. Higher throughput UV tips will allow us to increase write speed and are therefore a high priority. Implementing a tuning fork based feedback system should allow dither amplitude to be further reduced and will be explored in future work.[30] Optimization of processing parameters may also be useful in minimizing proximity writing. For example, from Fig. 6 it is clear that proximity features are partially removed

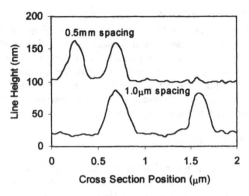

Figure 9. AFM measurements of aSi:H line set cross sections. The traces are offset for clarity.

during etching while fully exposed optical features are not. This suggests that oxidation from proximity effects is not complete. Over etching may allow the proximity lines to be completely removed while optically written structures which are more heavily protected by oxide formation remain intact.

While we have demonstrated that a-Si:H can be patterned with ~100nm feature sizes in individual lines which are well separated from one another, it is important to know whether the same spatial resolutions are possible when lines are close together. Etch rates during hydrogen plasma development may be quite bit different for regions between two closely spaced lines than for isolated features. Figure 9 shows AFM traces across a sets of lines patterned in a-Si:H with spacings of 1μm and 0.5μm. As can be seen, the 0.5μm spaced lines are well resolved with the region between the lines completely etched to the substrate. The separation of the bases of

the lines is roughly 0.15μm indicating that 100nm line spacings are likely without significant difficulties associated with incomplete plasma etching.

Finally, we have initiated a study of near-field lithography to optically pattern conventional photoresists. NSOM patterning of conventional resists has been previously demonstrated.[10] For these tests, we have initially focused on developing UV patterning capabilities. We used Shipley 1813 series positive photoresist and MicroChem Nano XFN NR2.5 negative resist which are nominally intended for mercury g-line exposure but are not specifically designed to support small features (0.5um or less). The resists were thinned and spun on a silicon wafer to a thickness of roughly 120nm followed by a standard pre-exposure soft bake. Patterns were then written using either the

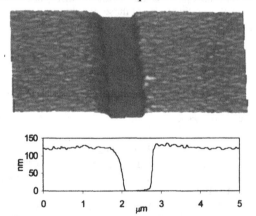

Figure 10. At the top is an AFM image of a line in positive resist. The image measures 5μm square, with film thickness of ~125nm. The plot below it shows a cross section of this line (the x axis is the cross sectional position and the y axis is height).

458nm or 350nm Ar-ion lines. Fig. 10 shows an AFM image of a typical line written in the positive resist with lateral dimensions of about 0.5um; features similar in scale (slightly narrower, but not as well defined) were attained with the negative resist. This is typical of the minimum feature size obtained with these conventional photoresists and consistent with their performance specifications. It should be noted that there is no proximity effect writing with conventional resists and both 350nm and 458nm exposure led to similar size characteristics. Estimated optical dose for Fig. 10 is consistent with the manufacturer's specifications for full exposure, and line heights indicates complete exposure of the resist.

CONCLUSIONS

a-Si:H thin films deposited by PECVD have been shown to function as a photoresist, with well defined behavior in far field exposure tests. Exposure of the a-Si:H results in oxidation in the exposed areas, which then allows selective etching and removal of the unexposed areas of a-Si:H. The dose required for exposure shows clear wavelength dependence, with the lower wavelength exposures requiring lower doses. Dose sensitivity, however, can be influenced by film deposition parameters. RF hydrogen plasma etching provides 1000:1 etch sensitivity for the development of exposed patterns, and the remaining oxide acts as a mask for further processing steps. The etching step is most sensitive to the rate of hydrogen flow in the plasma chamber.

NSOM is an effective method of direct write exposure in this resist system. Oxide can be formed from simple proximity of the NSOM probe even in the absence of light. The presence of optical excitation increases the level of exposure over and above the exposure from proximity alone. It is clear from optical versus proximity comparisons that higher levels of UV throughput in the NSOM probes are needed to facilitate further development of this capability. Closely spaced lines (500nm) were written using the NSOM, and these results indicate complete etching between the lines. It seems clear that the technique will support the construction of features in the 100nm range. In parallel with the a-Si:H exposures, the NSOM has been used to pattern conventional polymer based resist systems, both positive and negative, at or slightly below the normal operating range of those systems.

ACKNOWLEDGEMENTS

This work was supported in part by NSF grants DMI-9903718 and DMR-9704780.

REFERENCES

[1] M. A. McCord and R. F. P. Pease, J. Vac. Sci. Technol. B 4, 86 (1986); M. A. McCord and R. F. P. Pease, J. Vac. Sci. Technol. **B 6**, 293 (1988).

[2] G. C. Abeln, S. Y. Lee, J. W. Lyding, D. S. Thompson, and J. S. Moore, Appl. Phys. Lett. **70**, 2747 (1997).

[3] J. A. Dagata, J. Schneir, H. H. Harary, C. J. Evans, M. T. Posek, and J. Bennett, Appl. Phys. Lett. **56**, 2001 (1990); J. Sugimura and N. Nakagiri, Appl. Phys. Lett. **66**, 1430 (1995); T. -C. Shen, C. Wang, J. W. Lyding, and J. R. Tucker, Appl. Phys. Lett. **66**, 976 (1995).

[4] R. Magno and B. R. Bennett, Appl. Phys. Lett. **70**, 1855 (1997).

[5] Matsumoto, M. Ishii, K. Segawa, Y. Oka, B. J. Vartanian, and J. S. Harris, Appl. Phys. Lett. **68**, 34 (1996).

[6] S. C. Minne, H. T. Soh, Ph. Flueckiger, and C. F. Quate, Appl. Phys. Lett. **66**, 703 (1995).

7   P. M. Campbell, E. S. Snow, and P. J. McMarr, Appl. Phys. Lett. **66**, 1388 (1995).
8   N. Kramer, H. Birk, J. Jorritsma, and C. Schonenberger, Appl. Phys. Lett. **66**, 1325 (1995); N. Kramer, J. Jorritsma, H. Birk, and C. Schonenberger, J. Vac. Sci. Technol. **B 13**, 805 (1995).
9   S. C. Minne, Ph. Flueckiger, H. T. Soh, and C. F. Quate, J. Vac. Sci. Technol. **B 13**, 1380 (1995); S. C. Minne, S. R. Manalis, A. Atalar, and C. F. Quate, J. Vac. Sci. Technol. **B 14**, 2456 (1996).
10 Igor Smolyaninov, D. L. Mazzoni, and C. C. Davis, Appl. Phys. Lett **67**, 3859 (1995).
11 S. Davy and M. Spajer, Appl. Phys. Lett. **69**, 3306 (1996).
12 J. Massanell, N. Garcia, and A. Zlatkin, Opt. Lett. **21**, 12 (1996).
13 S. Madsen, M.. Mullenborn, K. Birkelund, and F. Grey, Appl. Phys. Lett. **69**, 544 (1996).
14 E. Betzig, P.L. Finn, and J. S. Weiner, Appl Phys. Lett. **60**, 2484(1992).
15 Russell E. Hollingsworth, Mary K. Herndon, Reuben T. Collins, J.D. Benson, J.H. Dinan, and J.N. Johnson in *Amorphous and Heterogeneous Silicon Thin Films: Fundamentals to Devices-1999*, edited by Howard m. Branz, Robert W. Collins, Hiroaki Okamoto, Subhenda Guha, and Ruud Schropp (Mater. Res. Soc. Proc. **557**, Pittsburg, PA, 1999) pp. 821-826.
16 R.E. Hollingsworth, C. DeHart, Li Wang, J.N. Johnson, J.D. Benson, and J.H. Dinan, J. Electronic Mat. **27**, 689 (1998).
17 M.K. Herndon, R.T. Collins, R.E. Hollingsworth, P.R. Larson, and M.B. Johnson, Appl. Phys. Lett. **74**, 141 (1999).
18 R.E. Hollingsworth, C. DeHart, Li Wang, J.H. Dinan, and J.N Johnson in *Amorphous and Microcrystalline Silicon Technology-1997*, edited by Sigurd Wagner, Michael Hack, Eric A. Schiff, Ruud Schropp, Isamu Shimizu (Mater. Res. Soc. Proc. **467**, Pittsburg, PA, 1997) pp. 961-966.
19 R.E. Hollingsworth, C. DeHart, Li Wang, J.N. Johnson, J.D. Benson and J.H. Dinan, J. Electron. Mat. **27**, 689 (1998).
20 Micio Niwano, Jun-ichi Kageyama, Kazunari Kurita, Koji Kinashi, Isao Takahashi, and Mobuo Miyamoto, J. Appl. Phys. **76**, 2157 (1994).
21 H. Fritzsche in *Amorphous and Microcrystalline Silicon Technology-1997*, edited by Sigurd Wagner, Michael Hack, Eric A. Schiff, Ruud Schropp, Isamu Shimizu (Mater. Res. Soc. Proc. **467**, Pittsburg, PA, 1997) pp. 19-30.
22 W. Westlake and M. Heintze, J. Appl. Phys. **77**, 879 (1995).
23 Betzig and J.K. Trautman, Science **257**, 189(1992).
24 P. Hoffmann, B. Dutoit, and R-P. Salathe, Ultramicroscopy **61**, 165(1995).
25 T. Saiki, S. Mononobe, M. Ohtsu, and J. Kusano, Appl Phys. Lett. **68**, 2612(1996).
26 G.A. Valaskovic, M. Holton, and G.H. Morrison, Appl. Opt. **34**, 1215(1995).
27 M. N. Islam, X. K. Zhao, A. A. Said, S. S. Mickel, and C. F. Vail, Appl. Phys. Lett **71**, 2886(1997).
28 Y. Yamammoto, M Kourogi, M. Ohtsu, V. Polonski, and G. H. Lee, Appl. Lett. **76**, 2173 (2000).
29 A. Sayah, C. Pilipona, P. Lambelet, M. Pfeffer, F. Marquis-Weible, Ultramicroscopy **71**, 59(1998).
30 K. Karrai and R.D. Grober, Appl. Phys. Lett. **66**,1842(1995).

LASER DIRECT-WRITE OF MATERIALS FOR MICROELECTRONICS APPLICATIONS

K. M. A. Rahman, D. N. Wells and M. T. Duignan†
Potomac Photonics, Inc.
4445 Nicole Drive
Lanham, MD 20706
†e-mail: mduignan@potomac-laser.com

Abstract

We demonstrate a laser direct-write method, a maskless process that transfers material directly from a ribbon to a substrate. This process offers the promise of fabricating passive electronic micro-components at a high speed with high spatial resolution. We are developing a workstation implementing this direct-write method, which integrates deposition, direct laser sintering, and micromachining capability on a single machine. Using this workstation we have deposited micro-patterns of conducting lines and resistors on alumina and polyimide substrates under ambient conditions that exhibit good electrical properties and substrate adhesion. From preliminary studies of laser sintering it was found that a wide range of sintering conditions may be used to arrive at silver conducting lines (~ 60 μm × 10 μm) on alumina substrate with resistivity in the range of 5 to 10 times the resistivity of bulk silver. Preliminary results also indicate direct laser sintering of cermet resistor material can yield reproducible resistance values.

Introduction

In the low intensity regime, interaction of laser energy with material results in heating, melting and evaporation. At higher intensities plasma formation can become important. Many practical applications exploit laser-material interaction phenomena. Examples include ablation, cutting, drilling and welding of a wide variety of materials [1]. Laser forward-transfer is another unique application that employs the laser interaction to deposit materials onto substrates. The absorption of pulsed laser energy leads to very rapid heating of absorbing materials to a high temperature. The high temperature conditions may be quite localized in space and time, allowing a novel method of transferring or "printing" materials onto a receiving substrate. While this is an emerging area of laser application, it has already attracted the attention of numerous investigators [2,3] because of its potential to revolutionize the microelectronics manufacturing industry [4].

A variety of techniques are presently used to produce conducting lines and passive electronic components for microcircuits. Examples include screen-printing using thick-films, thin-film metalization, and traditional PWB methods [5]. In this report we describe a matrix assisted laser direct-write (DW) method to create conducting lines, resistors, capacitors, inductors, etc., on hard (ceramic) and flexible (polyimide) substrates.

The main steps in the direct-write method are the following (see Figure 1). First, an ink is formulated with the material to be transferred, typically in powder form, as the main ingredient. The ink is then applied to a transparent backing to form a thin layer. The resulting "ribbon" is positioned closely to the receiving substrate, usually under ambient conditions. A pulsed laser beam of controlled spot size, energy and duration, irradiates the material/matrix through the transparent backing. Rapidly vaporizing matrix at the interface propels material onto the

99

substrate. Moving the beam or substrate between laser pulses permits patterning of the deposited material. The lateral dimensions of the deposited material can be as small as the laser spot size, typically of the order of tens of microns.

The main challenges in this process are: (a) high-definition transfer of the material from the ribbon to the substrate, (b) densification and acceptable properties of the deposited material, and (c) effective adhesion of the transferred material to the substrate. Post-deposition sintering may be necessary in order to achieve proper densification, because, the electrical properties of transferred features are function of their morphological details. This ability to produce direct-write electronic components on flexible substrates, presents a path to fabrication of mesoscale conformal electronic devices [4].

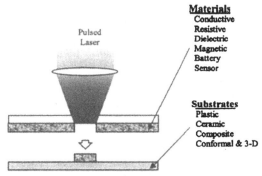

Figure 1. The basic idea of direct-write deposition process. Laser pulse energy absorbed by the material layer causes rapid vaporization and subsequent propulsion across a small gap toward the receiving substrate. If the gap is kept sufficiently small, the size and shape of the deposited material is defined by the laser spot size. A host of electronic materials can be transferred onto many different substrates.

An attractive advantage of the direct-write machine currently under development is that it integrates several functions on one platform: patterned deposition, *in-situ* sintering and/or annealing, and *in-situ* micromachining and surface treatment. The direct-write machine takes advantage of the established techniques of laser machining to produce microelectronics components significantly smaller than those achievable with standard screen-printing techniques. However, resolution of the direct-write features is also dependent on the rheological properties of the material layers on the ribbon, therefore, a better understanding of materials rheology and its tailoring for direct-write process is required.

Incorporation of precise motion control and a dynamic transfer scheme enable the direct-write technique to write patterns at a higher rate, currently about 200 mm/s (for line dimension approximately 60 μm wide and 8 μm thick). The combination of high speed and high resolution makes the direct-write technique an attractive technology for production of microelectronic components such as resistors, capacitors, inductors, conducting lines, sensors, and antennas.

Another advantage of this method is its flexibility of using many different materials for pattern generation. The basic transfer mechanisms for most material/matrix systems are similar. Deposited material properties are strongly influenced by formulation chemistry, rheology and post-deposition processing. In addition to rheology and chemistry, other essential aspects include choice of an appropriate matrix, shelf-life, and compatibility with binders and backing materials.

Yet another important requirement of system integration is to be able to sinter and/or anneal the micropatterned material on the same platform. Often the desired properties cannot be obtained from the components as deposited. For instance, silver conductor lines can be deposited on a number of substrates from a specially prepared ribbon that holds particulate silver in the form of highly viscous paste. The paste is formulated by mixing particulate silver into an organic matrix. Upon deposition, the silver particles in the pattern are not readily connected to form a current-carrying path. Thus, as transferred lines exhibit a very high resistance. However, conductivity of these lines increases dramatically (e.g., resistivity drops from some >100 times the bulk value of Ag to ~ 5 times bulk) upon high-temperature sintering. Other microcomponents such as resistors, high-K dielectric capacitors, *etc.*, would not be functional without sintering and densification. Sintering temperatures of 800-1200°C are common in traditional thick film processes. Such high temperatures are incompatible with polymer and even many kinds of glass substrates. Focused *laser* sintering permits locally high temperatures while minimizing the extent of heat-affected zone. This may permit material sintering conditions to be achieved on substrates otherwise incompatible such high temperatures. Therefore, the ability to sinter on the same platform increases productivity significantly.

In what follows, we describe some details of direct-write method applied to fabricate conducting and resistor micro-lines from formulations of particulate materials.

Experimental

In this section we first give a brief account of fabricating the "ribbon," the starting point for direct-write process. Then we describe preliminary results of silver micro-lines and resistor patterns produced by the direct-write method. Laser sintering of the direct-write components is discussed next. Finally, we report electrical properties of some components obtained by direct-write and laser sintering.

Direct-Write Workstation

The direct-write (DW) workstation under development is composed of the following main components: a laser source, beam delivery, integrated video microscope, and motion control system. All systems are interfaced to a controlling personal computer. The direct-write machine was built by modifying an in-house laser micromachining system design. It incorporates an X-Y stage motion control and disk-ribbon jog and spin control, synchronized with laser operation. A detailed discussion of various components of the workstation is beyond the scope of this article.

Direct-Write Scheme

Fig. 2 shows a schematic view of direct-write mechanism with a circular disk ribbon made from glass that supports the material layer. Upon irradiation, most of the energy of laser pulse is transmitted through glass and is absorbed by the material layer. The laser energy is carefully tailored so that the result of the interaction is to transfer the material from the ribbon to the receiving substrate. This is facilitated by utilizing an organic matrix (Fig. 3), whose function is to vaporize and act as a propellant for the material to be transferred.

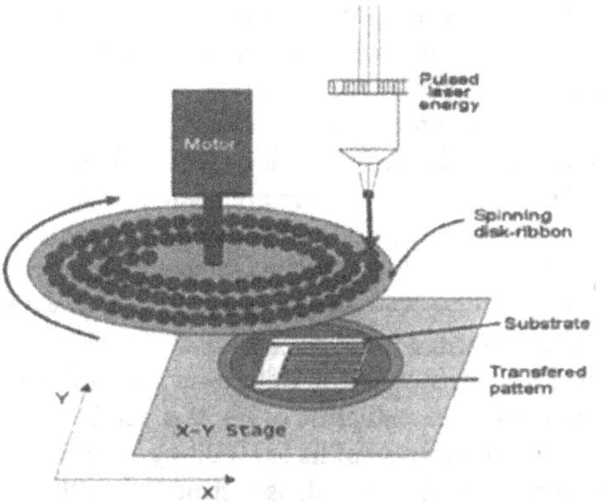

This process of "matrix-assisted forward transfer," was described by colleagues at the Naval Research Laboratory

Figure 2. Dynamic direct-write transfer scheme with a spinning disk-ribbon (not to scale).

[3], who have dubbed the process Matrix Assisted Pulsed Laser Evaporation – Direct-Write

(MAPLE-DW).

The scheme in Fig. 2 shows the main components of a dynamic direct-write process. A smart motion control system is at the heart of the speed achievable by this technique while the optics system enables one to achieve a high resolution. As can be seen from Fig. 2, a spinning disk-ribbon enhances the direct-write rate significantly; such a scheme is termed as dynamic direct-write. Another scheme of dynamic direct-write is illustrated in Fig. 4. Here, instead of a spinning disk, a reel-to-reel flexible backing is used for holding different materials on different track. This flexible ribbon system offers many advantages, including a highly compact method for storing a large amount of material, and is currently under development.

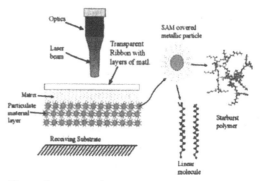

Figure 3. Matrix assisted forward transfer by absorption of laser pulse energy. Self-assembled monolayers on each particle in conjunction with a binder act as matrix as well as suspension promoter.

Ribbon

In the matrix assisted forward transfer direct-write scheme described here, a ribbon is a transparent backing on which a layer of material is applied for transfer

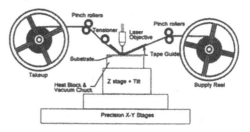

Figure 4. Wide multi-track design for very large surface area, adaptable to compact tape cassettes.

(see Fig. 3). The main physical requirements for the ribbon are: the material layer to be of optimal and uniform thickness. Material layers too thick do not ablate cleanly or reproducibly and result in poor edge definition. Material layers too thin slow the transfer process as deposition rate becomes limited by the laser repetition rate. The ribbons must be free from imperfections such as pinholes and scratches. A pinhole big enough to pass the laser beam can result in ablation of the material that has already been deposited.

With this goal in mind, we attempted a formulation with self-assembled monolayer (SAM) coated silver particles in conjunction with an organic binder as matrix (Fig. 3). That is, submicron size Ag particles were pretreated with a dilute dithiol solution in toluene; dithiols are known to form a SAM on Ag. This surface modification serves two purposes. On the one hand, it helps a better suspension of heavier Ag particles in organic solvent, a factor that is very important for formulation of ink that can be used with commercial spraying apparatus; on the other hand, because of its low boiling temperature, SAM would evaporate during direct-write deposition producing a denser structure on the receiving substrate as transferred by DW method.

Ag ink was formulated using SAM and binder with powders supplied by Superior MicroPowders, Inc. (Albuquerque, NM). The particle size distribution in the powder varies form sub micron to ~5 micron with prominent peaks at 1 and 3 micron. It was observed that SAM treatment improved the suspension of Ag particles only over a short period of time; over longer period the Ag particles were still settling, causing difficulty in spraying. However, when a small

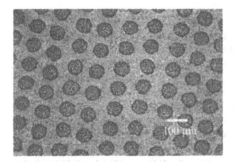

Figure 5. Holes in the material layer after direct-write deposition

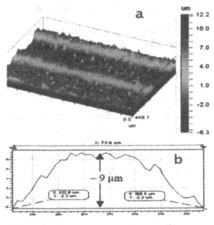

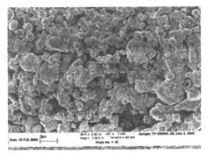

Figure 6. (a) 3-D contour plot of silver lines on alumina substrate, (b) area of cross-section ~ 340 μm^2.

amount of binder solution was added to the ink (~1% by volume of 5% ethyl cellulose solution in toluene), the ink rheology improved dramatically facilitating spraying by ordinary airbrush. A substantially homogeneous layer of silver was formed on the disk-ribbon substrate. Fig. 5 shows a photomicrograph of the ribbon after direct-write deposition. Individual holes in the material layer corresponds to transfer by each pulse. As deposited, Ag lines from this ribbon exhibited very high resistivity, which, upon sintering, improved to 5 to 10 times the resistivity of bulk Ag.

Results and Discussion

Pattern Deposition of Silver conducting lines

We deposited Ag lines and resistor patterns on alumina and polyimide substrates, respectively, using direct-write (DW) method with Potomac's Laser DW workstation and above-mentioned disk-ribbons. A 50 μm diameter focused UV beam (355 nm) was used. We used a WYKO vertical scanning white-light interferometer to accurately determine the geometry of the deposited lines (Fig. 6). This apparatus allows a narrower and finer inspection of the surface roughness of the specimen as well as an accurate measurement of the thickness profile.

Electrical properties of the lines were measured with an LCR meter at a fixed frequency of 100 Hz with an applied excitation field on the order of 1 V/cm. In general, none of the lines showed appreciable conductivity as deposited; however, upon laser sintering electrical properties improved significantly. Fig. 7 shows a typical SEM micrograph of these silver lines after baking in air at 290°C for 30 min. These samples exhibited a nominal resistivity of ~ 30 times the bulk silver.

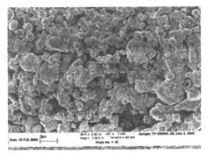

Figure 7. SEM micrograph of silver conducting line on alumina substrate after baking in air at 290°C for 30 min.

Sintering of Silver Conducting Lines

One of the major problems for IR laser sintering of silver arises from the fact that at a wavelength of 1 micron, pure silver is a poor absorber; only ~2% of the incident energy is absorbed. The energy required melt a unit volume of starting material is an upper bound to the energy required to sinter. For a 10 µm thick, 60 µm wide silver line, estimated total energy (i.e., the sum of the energy to heat to the melting point and the energy required to complete the melting process) is $\sim 2 \times 10^{-3}$ J per millimeter of line. However, since only a few percent of the delivered energy is absorbed, and loss mechanisms such as conduction and radiation cannot be ignored, significantly higher incident energy is required. As the delivered energy is highly localized in time as well, kinetics of sintering must also be considered [6]. Considering these factors, we designed experiment to conduct *in-situ* sintering of direct-write silver micro-lines with an IR laser (1.06 µm) running in the CW mode. Several parameters were varied systematically; *viz.*, laser power, spot-size, dwell-time, feed rate and number of passes. Total incident energy is then computed by summing up contributions from individual components. With an incident laser energy range of ~0.3-10 J/mm, we measured normalized resistivity to be between 5 to 10 times the resistivity of bulk silver. This range of values is suitable for many relatively low-frequency (<~100 MHz) commercial applications.

SEM cross-sections of laser sintered Ag lines (not shown) of the present study indicate that the microstructure of laser sintered lines is denser compared to that of as deposited lines (Fig. 7). Thus laser sintering can effectively dense the silver particulate material and thereby improves its electrical properties.

Direct-write Resistor pattern

Following the same procedure described above, we first made ribbons with commercial resistor inks. The ink was thinned with toluene and sprayed on disk-"ribbon." Material layer thickness was adjusted to approximately 10 µm by applying several coatings. Several line patterns were deposited on alumina and polyimide substrates. Fig. 8 shows a photomicrograph of resistor pattern deposited on a thick polyimide substrate. Fig. 9 shows a photomicrograph of laser sintered resistor on alumina substrate. As before, several adjustable parameters were varied systematically with a view to arrive at an optimum sintering condition. For a fixed set of parameters, measured resistance was reproducible, e.g., average resistance for 8 patterns was found to be 2.86 ± 0.09 kΩ.

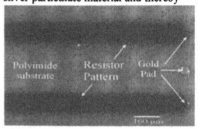

Figure 8. Photomicrograph of resistor pattern deposited on a thick polyimide substrate.

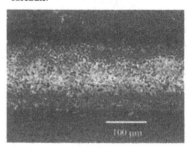

Figure 9. Photomicrograph of laser sintered resistor on alumina substrate.

Summary

In this paper we describe direct-write method and its implementation on an integrated platform for passive microelectronic applications. Using this technique we produced conducting silver lines and resistor patterns on alumina and polyimide substrates. These components were also laser sintered *in-situ* on the same workstation. Preliminary results indicate that this technique can be successfully used to deposit passive microelectronic components on both hard and flexible substrates. Preliminary data of laser sintered silver lines indicates that resistivity of these lines is bracketed within 5 to 10 times the resistivity of bulk silver. A wide range of sintering condition may be used for these silver lines to arrive at this range of resistivity.

Acknowledgement

Funding for this work was provided by the Defense Advanced Research Projects Agency. K.M.A.R. wishes to thank Dr. C. Paul Christensen for many fruitful discussions.

References

1. N. Bloembergen, in Laser Ablation: Mechanisms and Applications-II, Ed. J.C. Miller and D.B. Geohegan, AIP: New York, 1993, pp 3–10.
2. Materials Research Society Spring 2000 Meeting, Proceedings of Symposium V "Materials Development for Direct-Write Technologies," San Francisco: April 23-27, 2000.
3. A. Pique, R.C.R.A. McGill, D.B. Chrisey, J. Callahan and T.E. Mlsna, in *Advances in Laser Ablation of Materials*, Ed. R.K. Sing, D.H. Lowndes, D.B. Chrisey, E. Fogarassy and J. Narayan, MRS, Warrendale: 1998, pp 375-383.
4. W.L. Warren, "Overview of Commercial and Military Application Areas in Passive and Active Electronic Devices," to be published.
5. C.A. Harper and R.M. Sampson, *Electronic Materials and Processes Handbook, 2nd Ed.,* McGraw-Hill: New York, 1993.
6. J.A. Kittl, P.G. Sanders, M.J. Aziz, D.P. Brunco and M.O. Thompson, "Complete Experimental Test of Kinetic Models for Rapid Alloy Solidification," to be published in *Acta Materialia.*

Laser Guided Direct Writing

Michael J. Renn
Optomec, Inc., 3911 Singer NE, Albuquerque, NM 87107, mrenn@optomec.com
and
Department of Physics, Michigan Technological University, Houghton, MI 49931

ABSTRACT

Laser-induced optical forces are used to guide and deposit 100 nm - 10 μm diameter particles onto solid surfaces in a process called *laser-guided direct-writing*. Nearly any particulate material, including both biological and electronic materials, can be manipulated and deposited with micrometer accuracy. Potential applications include three-dimensional cell patterning for tissue engineering, hybrid biological and electronic device construction, and biochip array fabrication.

INTRODUCTION

Laser-induced optical forces, arising from the scattering of light by microscopic particles, are widely used for the non-contact manipulation of biological particles. Arthur Ashkin, a pioneer in optical force-based manipulation, first applied optical forces to levitate aerosol droplets and dielectric spheres [1], and later demonstrated the optical manipulation of a variety of biological materials in aqueous suspension [2-5]. Optical forces are now commonly used for noncontact manipulation of cells, subcellular components, and biomolecule-coated particles in a configuration known as optical tweezers" [6-10].

Despite the ability to control particle positioning to submicron accuracy, optical tweezers have not been applied extensively to microfabrication. The main drawback stems from the fact that optical tweezer-based surface pattering is tedious, requiring repeated cycles of particle capture in the fluid phase, transport through the fluid, deposition on a solid surface, and release. Furthermore, the small trapping volume of conventional optical traps severely limits the number of particles that can be manipulated at one time. Given these limitations it would appear that optical forces are not well-suited to provide both the micrometer-scale positioning accuracy and the high throughput deposition rates that direct-write microfabrication demands. However, by simply changing the laser beam focus, we have found that optical forces can be used to manipulate thousands of particles simultaneously and deposit them in a continuous stream onto surfaces with micrometer accuracy [11-13]. In addition, hollow-core optical fibers can assist laser guidance [11,12,14,15], and allow particles to be guided more accurately and over much longer distances than with free space beams.

Physical Basis of Laser Guidance

The model for particle-light interaction depicted in Fig. 1 was first proposed by Ashkin and is the working model for the ray optics regime where the particle is larger than the wavelength of light [1,16]. The key physical property defining the interaction between the light and the particle is the refractive index of the particle relative to that of the surrounding fluid. Larger refractive indices lead to stronger interactions.

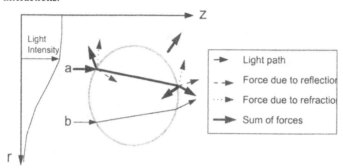

Figure 1. Optical forces on a dielectric sphere. Laser light is reflected and refracted at each interface, resulting in a redirection of the light. Since photons have momentum, their redirection by interaction with the particle results in a corresponding momentum transfer to the particle as indicated by the dashed arrows. The net result of the interactions from ray A is to push the particle along the beam axis and pull the particle radially inward. By symmetry, ray B pushes the particle axially and pushes the particle radially outward. However, ray A is stronger than ray B so it overcomes the radial force directed outward. In the absence of other forces, the particle is simultaneously pulled radially inward and pushed axially in the direction of the laser beam.

When a particle interacts with the light, it is simultaneously pulled to the center of the beam where the intensity is maximal and pushed axially in the direction the beam is travelling. Using a weakly focussed laser beam (i.e. a low numerical aperture focussing lens), Buican et al. demonstrated the guidance of living cells in an aqueous fluid for cell sorting applications [17]. If an appropriate target surface is placed in the light path, then particles can be continuously guided and deposited in a steady stream onto the target surface as shown schematically in Fig. 2.

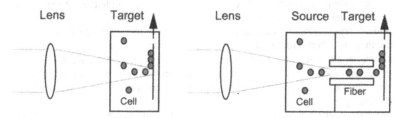

Figure 2. Laser-guided direct-writing system. At left, laser light is weakly focussed into a suspension of particles. The particles are propelled by the light through the fluid and deposited on a target surface. Translation of the target relative to the laser beam results in a line of particles being drawn. At right, light is coupled into a hollow optical fiber and particles are carried through the fiber to the target surface. The process is observed in real-time by light microscopy.

Such an approach to deposition, called *laser-guided direct-writing* has been demonstrated with a variety of organic and inorganic particles in both gas and liquid phase [13,14] and with living cells in culture medium [11,12]. In gas phase, particle fluxes as high as 10,000 Hz and placement precision below one micrometer have been achieved.

Laser-Guided Direct-Writing
As depicted in Fig. 2, light can be coupled into hollow optical fibers that allow transmission of a high intensity beam over millimeter to centimeter distances. While most fiber optics have a solid core, hollow core optical fibers (developed for infrared laser delivery) permit transmission of both light and particles. The fiber guiding geometry offers several advantages over free space guidance. First, natural convective fluid motion is often large enough to overwhelm optical forces, making free space guiding difficult. It is therefore particularly important to design chambers that suppress convective fluid motion. Hollow optical fibers alleviate this problem because the fiber interior provides a quiescent environment shielded from the external surroundings. The outside of the fiber can be exposed to air currents (or even a vacuum) and the particles within are not disturbed. Second, laser light is guided several centimeters within the hollow region of the fiber. This allows particles to be transported over longer distances than is possible with tightly focussed beams, and particle placement is accomplished by simply pointing the fiber tip toward the substrate. Third, the intensity profile inside the fiber is well-defined with the intensity being maximal at the radial center and zero at the fiber wall [18]. The intensity gradient draws particles toward the radial center of the fiber and keeps them from adhering to the fiber walls. Fourth, the fiber also allows the source and deposition regions to be isolated from each other, assuring that the direct write patterns are not contaminated by non-guided particles. Finally, fibers from several source chambers can potentially be coupled to the same chamber for co-deposition of multiple materials.

Whether accomplished with or without a fiber, laser-guided direct writing has many advantages over existing methods for surface patterning. In contrast to optical trapping, laser-guided direct-writing allows particles to be captured continuously from the surrounding fluid and directed onto the substrate. In comparison to photolithography, the process adds material to the surface (as opposed to etching material) and does not require harsh or corrosive chemicals. In contrast to robotic micromanipulator-based deposition, ink jetting, and screen printing, particles are strongly localized within the laser beam and the deposition accuracy can be below one micrometer. Most importantly, nearly any material in either liquid or aerosol suspension can be captured and deposited so long as: 1) the particle's index of refraction is greater than that of the surrounding fluid, and 2) other forces, such as convection and gravity, are weaker than the optical forces (typically in the piconewton range). Potentially many material types can be co-deposited on a single substrate, which will allow simultaneous deposition of both electronic and biological particles.

APPLICATIONS

Tissue Engineering

The long-term preservation of tissue-specific function is important if engineered tissue is to successfully compensate for organ failure. A number of studies have demonstrated the importance of three-dimensional structure on the behavior of cells in culture. For example, hepatocytes cultured as a monolayer lose many of their liver-specific functions within a few days. However, when these same cells are overlaid with a collagen gel to mimic the three-dimensional structure of the liver, they retain many of their liver-specific functions for weeks in culture [19]. Therefore, the ability to spatially organize cells into well-defined three-dimensional arrays that closely mimic native tissue architecture can facilitate the fabrication of engineered tissue. Laser-guided direct-writing potentially has this ability. In initial studies with embryonic chick spinal cord cells we found that individual cells (diameter = 9 μm) could be guided by a 450 mW near-infrared laser beam and deposited in arbitrarily defined arrays onto a glass target surface[11,12]. Importantly, cells that were exposed to the light remained viable and grew normal-appearing neurites. As depicted in Fig. 3, laser-guided direct-writing potentially allows the three-dimensional patterning of cells using multiple cell types with cell placement at arbitrarily selected positions.

Hybrid Electronic and Biological Devices

Using a laser-guided direct-writing system nearly identical to that depicted in Fig. 2, Renn et al. directly wrote clusters and lines of inorganic materials on glass surfaces [13,14]. Previous work showed that atoms could be guided in an evacuated fiber [15].

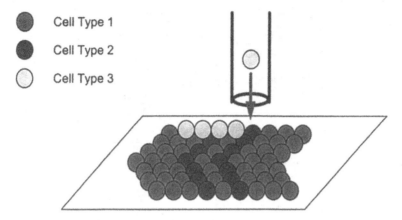

Figure 3. Schematic illustration of three-dimensional patterning of multiple cell types by laser-guided direct-writing. Potentially, multiple cell types can be placed at arbitrary positions with micrometer scale precision in an attempt to recapitulate the complex three-dimensional cellular organization of native tissues. Transport can be accomplished with or without a hollow optical fiber, depending on the degree of natural convection and the spatial separation required between the source of cells and the target.

Recent work has focused on the deposition of conducting and semi-conducting materials for electronics fabrication and rapid prototyping. As an example of the types of patterns that can be generated, Fig. 4 shows micrometer scale lines of aluminum oxide, an electrical insulator, directly written onto a glass surface. The advantage of the laser-guided direct-writing system is that it is a single system that can be used to deposit both electronic and biological materials, including living cells. The range of materials that we have successfully guided is broad and includes metals (e.g. 100 nm gold spheres), semiconductors, polymers, animal cells, bacteria, and microtubules. Also, the choice of target substrate surface material is arbitrary, as long as it is not damaged or strongly heated by the impinging laser light. The use of a single, flexible system allows rapid prototyping with a wide range of materials integrated into a single functional device

Microarray Fabrication
In addition to deposition of solid and semi-solid particles suspended in a liquid phase, liquid droplets suspended in a gas phase can also be deposited [14]. By depositing aerosols of liquid droplets containing biomolecules, such as proteins or nucleic acids, bioarrays can be generated. Individual droplets are approximately 1 μm diameter or 1 femtoliter in volume. Coalescing several droplets into a single droplet on the surface before translating the laser beam forms larger deposits.

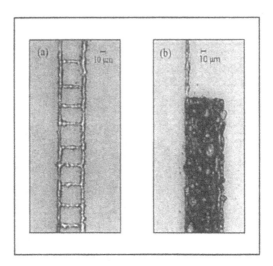

Figure 4. Optical micrographs of aluminum nitrate deposited on a glass substrate. A 500 mW, 532 nm laser was coupled into a 20 μm inner diameter, 6 mm long hollow-core fiber. The patterns were built up by the capture and deposition of aluminum nitrate droplets (1 μm diameter) while translating the substrate near the exit of the fiber. In (b) the aluminum nitrate pattern has been decomposed at higher laser power to form aluminum oxide. The smallest line width is 10 μm and in air the deposition rate can exceed 10,000 Hz.

These spot sizes are one or two orders of magnitudes smaller than those typically generated by current methods such as mechanical microspotting [20] or microjet printing [21]. For example, a 10,000 address microarray using microspotting requires about 3 cm^2, while laser-guided direct-writing would require about 1 mm^2 assuming a 10 μm spot size. An example of such a pattern generated using glycerol is shown in Fig. 5. In addition, preliminary deposition experiments have yielded droplet deposition rates in excess of 10,000 Hz, while typical deposition rates for microspotting are less than 1 Hz. The droplet size (~1 femtoliter) is orders of magnitude smaller than current dispensing techniques (~1 nanoliter), and may lead to dramatically reduced reagent consumption.

Fundamental Research

Optical trapping has been used successfully to elucidate fundamental cellular and molecular phenomena such as the discrete 8 nm steps that the molecular motor kinesin takes as it advances along microtubules [22]. The laser-guided direct-writing system is fundamentally different from optical trapping in that it provides propulsion along the beam axis instead of trapping.

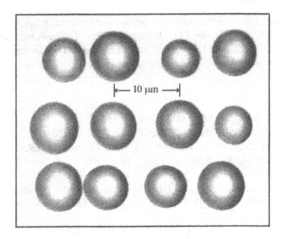

Figure 5. Array of glycerol droplets on a glass slide. The ~7 μm diameter deposits are built up from the multiple deposition of 1 μm aerosol droplets. A wide variety of biomolecules can be dissolved in glycerol (or other solvents) and then delivered to substrates as micrometer-scale droplets. The approach potentially provides a novel method for fabrication of biomolecule arrays.

However, by simultaneously coupling light into both ends of a hollow optical fiber, a trap can be set up inside the fiber [14]. The trap-in-the-fiber can be set up at a lower cost than conventional optical traps because it does not require a high numerical aperture lens. In addition, it serves as an efficient microfluidic mixer where one particle or droplet is brought in from one end of the fiber and another brought in from the opposite end of the fiber. The two particles collide in the middle and are held fixed for observation for as long as desired. One can conduct femtoliter-scale chemical reactions in droplets manipulated without any direct contact to a surface.

CONCLUSIONS

Laser-guided direct-writing is an emerging technology for high throughput deposition of micrometer and submicrometer-sized particles. It is a simple system that can be set up at low cost and will deposit nearly any material with micrometer-scale accuracy. Multiple applications are anticipated in tissue engineering, hybrid electronic/biological systems, biochip array fabrication, and basic scientific research.

ACKNOWLEDGEMENTS

The author is grateful for stimulating conversations and technical support from M. Essien, B. King, W.D. Miller, D.J. Odde, and R. Pastel. The work was supported by the State of Michigan Research Excellence Fund and the Defense Advanced Research Projects Agency under contract #N00014-99-C-0258.

REFERENCES

1. Ashkin, A. (1970) *Physical Review Letters* 24, 156-159
2. Ashkin, A. & Dziedzic, J. M. (1987) *Science* 235, 1517-1520
3. Ashkin, A., Dziedzic, J. M. & Yamane, T. (1987) *Nature* 330, 769-771
4. Ashkin, A., Schutze, K., Dziedzic, J. M., Eutenauer, U. & Schliwa, M. (1990) *Nature* 348, 346-348
5. Ashkin, A. & Dziedzic, J. M. (1971) *Applied Physics Letters* 19, 283-285
6. Berns, M. W., Wright, W. H. & Steubing, R. W. (1991) *International Review of Cytology* 129, 1-44
7. Svoboda, K. & Block, S. M. (1994) *Annual Review of Biophysics and Biomolecular Structure* 23, 247-285
8. Berns, M. W. (1998) *Scientific American* April, 62-67
9. Simmons, R. M. & Finer, J. T. (1993) *Current Biology* 3, 309-311
10. Kuo, S. C. & Sheetz, M. P. (1992) *Trends in Cell Biology* 2, 116-118
11. Odde, D. J. & Renn, M. J. (1998) *Annals of Biomedical Engineering* 26, S-141
12. Odde, D. J. & Renn, M. J. (1999) *Biotechnology and Bioengineering* submitted
13. Renn, M. J. & Pastel, R. (1998) *Journal of Vacuum Science and Technology* B16, 3859-3863
14. Renn, M. J., Pastel, R. & Lewandowski, H. (1999) *Physical Review Letters* 82, 1574-1577
15. Renn, M. J., et al. (1995) *Physical Review Letters* 75, 3253-3256
16. Ashkin, A. (1992) *Biophysical Journal* 61, 569-582
17. Buican, T. N., et al. (1987) *Applied Optics* 26, 5311-5316
18. Marcatili, E. A. J. & Schmeltzer, R. A. (1964) *The Bell System Technical Journal* July, 1783-1809
19. Dunn, J. C. Y., Yarmush, M. L., Koebe, H. G. & Tompkins, R. G. (1989) *FASEB Journal* 3, 174-177
20. Iyer, V. R., et al. (1999) *Science* 283, 83-87
21. Hayes, D. J., Wallace, D. B., Boldman, M. T. & Marusak, R. M. (1993) *Microcircuits and Electronic Packaging* 16, 173-180
22. Svoboda, K., Schmidt, C. F., Schnapp, B. J. & Block, S. M. (1993) *Nature* 365, 721-727

Photo-induced Large Area Growth of Dielectrics with Excimer Lamps

Ian W. Boyd and Jun-Ying Zhang
Electronic and Electrical Engineering, University College London,
Torrington Place, London WC1E 7JE, United Kingdom

ABSTRACT

In this paper, UV-induced large area growth of high dielectric constant (Ta_2O_5, TiO_2 and PZT) and low dielectric constant polyimide and porous silica) thin films by photo-CVD and sol-gel processing using excimer lamps, as well as the effect of low temperature UV annealing, are discussed. Ellipsometry, Fourier transform infrared spectroscopy (FTIR), X-ray photoelectron spectroscopy (XPS), UV spectrophotometry, atomic force microscope (AFM), capacitance-voltage (C-V) and current-voltage (I-V) measurements have been employed to characterize oxide films grown and indicate them to be high quality layers. Leakage current densities as low as 9.0×10^{-8} $A \cdot cm^{-2}$ and 1.95×10^{-7} $A \cdot cm^{-2}$ at 0.5 MV/cm have been obtained for the as-grown Ta_2O_5 films formed by photo-induced sol-gel processing and photo-CVD, respectively - several orders of magnitude lower than for any other as-grown films prepared by any other technique. A subsequent low temperature (400°C) UV annealing step improves these to 2.0×10^{-9} $A \cdot cm^{-2}$ and 6.4×10^{-9} $A \cdot cm^{-2}$, respectively. These values are essentially identical to those only previously formed for films annealed at temperatures between 600 and 1000°C. PZT thin films have also been deposited at low temperatures by photo-assisted decomposition of a PZT metal-organic sol-gel polymer using the 172 nm excimer lamp. Very low leakage current densities (10^{-7} A/cm^2) can be achieved, which compared with layers grown by conventional thermal processing. Photo-induced deposition of low dielectric constant organic polymers for interlayer dielectrics has highlighted a significant role of photo effects on the curing of polyamic acid films. I-V measurements showed the leakage current density of the irradiated polymer films was over an order of magnitude smaller than has been obtained in the films prepared by thermal processing. Compared with conventional furnace processing, the photo-induced curing of the polyimide provided both reduced processing time and temperature. A new technique of low temperature photo-induced sol-gel process for the growth of low dielectric constant porous silicon dioxide thin films from TEOS sol-gel solutions with a 172 nm excimer lamp has also been successfully demonstrated. The dielectric constant values as low as 1.7 can be achieved at room temperature. The applications investigated so far clearly demonstrate that low cost high power excimer lamp systems can provide an interesting alternative to conventional UV lamps and excimer lasers for industrial large-scale low temperature materials processing.

1. Introduction

As dynamic random access memories (DRAMs) are scaled down, the thickness of SiO_2 gate oxide must be correspondingly reduced and thickness control becomes a critical issue. The projected silicon dioxide (SiO_2) thickness by 2012 predicted by the semiconductor industry roadmap will reach atomic dimensions (five silicon atomic layers), or less than one nanometre [1] as indicated in Table 1 [2-3]. As can be seen, for 0.1 μm ultra-large-scale-integrated (ULSI) device technologies, the SiO_2 gate oxide thickness must be scaled below 2.0 nm and become so thin that direct tunneling effects and excessively high electric fields become serious obstacles to reliability as well as fundamental quantum mechanical difficulties. What is required is a thicker layer of a higher dielectric constant material that will have the same or similar effective capacitance when it is put into a device and enable a further decrease in device area (see table 1).

Table 1. The projected gate oxide thickness for CMOS integrated circuits for 1997-2012 [2-3]

Technology timeline	1997	1999	2001	2003	2006	2009	2012
Design rule (μm)	0.25	0.18	0.15	0.13	0.10	0.07	0.05
Wafer diameter (mm)	200	300	300	300	300	450	450
Gate dielectric (nm, ε=3.9) equivalent SiO_2 thickness	4-5	3-4	2-3	2-3	1.5-2	<1.5	<1.0
Gate dielectric (nm) Ta_2O_5 thickness (ε=19)	20-25	15-20	10-15	10-15	7.5-10	<7.5	<5

Recently, various high dielectric constant materials have been widely investigated as possible candidates to replace SiO_2 in dynamic random access memories (DRAMs) and shown in table 2. More detailed thermodynamic stability and silicon-compatibility of these dielectrics can be found elsewhere [4-6]. Amongst these dielectrics, much of focus of current high k research involves $PbZr_xTi_{1-x}O_3$ (PZT), $Ba_xSr_{1-x}TiO_3$ (BST), Ta_2O_5, and TiO_2. Tantalum pentoxide (Ta_2O_5) in particular appears to be the most promising and best candidate to replace SiO_2 because of its compatibility with ultra large-scale-integrated processing as well as chemical and thermal stability [7-10]. Various storage capacitor configuration with Ta_2O_5 dielectric films such as polysilicon/Ta_2O_5/polysilicon (SIS), metal/Ta_2O_5/metal (MIM, metal/Ta_2O_5/polysilicon (MIS or MOS)), etc. have been fabricated to study the nature of Ta_2O_5 dielectric films [11]. It was also shown that the Ta_2O_5 capacitor with the TiN/poly-Si top electrode is suitable for 256 Mbit memory devices an has applied to 256 Mbit DRAM fabrication process [9]. Recently, it has been reported the mixed Ta-Ti and Ta-Zr oxide can enhance dielectric constant up to 120 [12-14]. However, fully satisfying the demands of the microelectronics industry for present Ta_2O_5 films on Si have not yet been fulfilled, although a considerable amount of work on Ta_2O_5 dielectric films has already been done. Some obstacles need to be overcome before Ta_2O_5 can be applied to DRAMs technology, especially, the large leakage current in the as-deposited layers due to oxygen deficiency, defect and impurity contamination present. It was found that the leakage current density could be significantly reduced to acceptable levels by employing of the post-deposition annealing techniques. Therefore, approaches aimed at its reduction have received much attention. Various post-deposition annealing techniques including furnace annealing with various gases such as O_2, N_2O, N_2, H_2, NH_3, O_3 etc., rapid thermal O_2 annealing (RTA), ultraviolet generated ozone (UVO), plasma and a range of two-step process as well as several other more complicated methods [15-29], have been proposed to improve the electrical properties of the Ta_2O_5 films. Since furnace and RTA annealing are carried out at above 700°C the Ta_2O_5 films are crystallized and consequently grain boundaries lead to an increase in the leakage current [15]. In the case of UVO and plasma annealing, the leakage current can be significantly reduced compared with that obtained by other annealing techniques. Furthermore, Both annealed films remain in the amorphous phase because of low annealing temperatures (<400°C). Table 3 summarizes the properties of Ta_2O_5 films deposited and annealed by different methods. It is clearly seen that after annealing the leakage current densities significantly reduced to the range 10^{-6}-$10^{-8}A/cm^2$ from around $10^{-3}A/cm^2$ in the as-deposited films. Ta_2O_5 properties strongly depend on stoichiometry, grain size, film thickness and homogeneity, which are influenced by growth technique. A variety of chemical and physical deposition techniques have been used to deposit Ta_2O_5 films [8,16-24,30-35]. Future ULSI device architectures preclude the use of high temperature processing for the gate and so the development of low-temperature fabrication routes to these high dielectric constant materials is very important. In order to address this issue of thermal budget, a number of new low temperature growth techniques are now under study including photo-assisted methods. To date most photo-induced processing has been performed with lasers. However, their use is inherently limited by the total photon fluxes available and they are therefore not suited for large area wafer processing. Unlike lasers, lamp sources provide the potential for large area coverage. In particular, the recent development of excimer lamps [36-43], which are capable of producing high power radiation with available wavelengths in the range of 108 nm to 354 nm, has opened up new possibilities for initiating a wide range of large area low temperature photo-induced reactions.

Table 2 Alternate gate dielectrics for use in silicon MOS transistors [4-6]

Material	k	Material	k	Material	k	Material	k
$PbZr_xTi_{1-x}O_3$ (PZT)	1000	Ta_2O_5	25-28	Nd_2O_3	16-20	MgO	9.8
$Ba_xSr_{1-x}TiO_3$ (BST)	500	$LaAlO_3$	25	Y_2O_3	11-14	Li_2O	8.1-8.8
TiO_2	100	$La_2Be_2O_5$	25	Er_2O_3	12.5-13	$MgAl_2O_4$	8.3-8.6
$LaScO_3$	30	ZrO_2	22	$ZrSiO_4$	12-13	BeO	6.9-7.7
Y_2O_3-ZrO_2	29.7	La_2O_3	21	Al_2O_3	9-12	Ce_2O_3	7.0

Applications of such sources have already been demonstrated in several areas, including photo-deposition of dielectric [44-47] and metallic [48-55] thin films, photo-oxidation of silicon, germanium and silicon-germanium [56-59], surface modification and polymer etching [60-65], UV annealing [22, 24, 26, 66-67], photo degradation of a variety of pollutants [68-69], as well as large area flat panel displays [70-71] etc. Single- and multi-layered films of silicon oxide, silicon nitride, and silicon oxynitride can be deposited by photo-CVD from different mixtures of silane, oxygen, ammonia, and nitrous oxide gases [44-45, 72-75]. Fine tuning of the refractive index of the films deposited can also be controlled by varying the ratio of the precursor mixtures. Recent work on the direct photo-oxidation of silicon at low temperature (250°C) has shown the oxidation rate is more than three times greater than that for photo-induced oxidation of silicon using a typical low pressure Hg lamp at 350°C [56,58]. The fixed oxide charge number density (Q/q) has been found to be 4.5 x $10^{10} cm^{-2}$, which is comparable to some of the best values reported for thermally grown oxide on Si at a high temperature of 1030°C [76]. In this paper we briefly outline the underlying principles and properties involved in excimer lamp. A novel application towards high and low dielectric constant thin films grown on Si by photo-CVD and sol-gel processing will be reviewed while the effect of low temperature UV annealing is also presented. The physical, optical, lectrical and dielectric characteristics of Ta_2O_5 thin films are summarized and discussed. The applications investigated so far clearly demonstrate that low cost high power excimer lamp systems can provide an interesting alternative to conventional UV lamps and excimer lasers for industrial large-scale low temperature materials processing.

Table 3. Properties of Ta_2O_5 films deposited and annealed by different methods (J is the leakage current density while (a): as-deposited films and (b): annealed films)

Method and annealing conditions	Capacitor structure	k	J (A/cm^2) at 1 MV/cm	Reference
Sputtering Annealed at 850°C	Al/ Ta_2O_5/Pt Al/ Ta_2O_5/p-Si	a) 24 b) 45	a) 10^{-4} b) 10^{-7}	16
CVD at 450°C and RTA annealed at 800°C in O_2, N_2O	Al/ Ta_2O_5/n-Si TiN/Ta_2O_5/n-Si		a)10^{-3} b)10^{-8}	17
Thermal oxidation and O_2 annealed at 800°C	Al/ Ta_2O_5/p-Si	16-25	a) $>10^{-4}$ b)10^{-8}	18
CVD at 400-470°C and O_2 annealed at 700-900°C	Al/ Ta_2O_5/p-Si	19	a) $>10^{-3}$ b)10^{-8}	19-20
Plasma-CVD at 200-600°C and N_2 annealed at 700°C	Al/ Ta_2O_5/p-Si	a) 20.3 b) 19	a) $>10^{-2}$ b)10^{-6}	21
CVD at 400°C and plasma annealed at 400°C	TiN/ Ta_2O_5/n-Si		a)10^{-3} b)10^{-8}	15
Photo-CVD at 250-400°C and UV annealed at 400°C	Al/ Ta_2O_5/n-Si	a) 24 b) 20	a)10^{-3} - 10^{-7} b)10^{-8}	10,22

2. Characteristics of excimer UV sources

The principle underlying the operation of the excimer lamps relies on the radiative decomposition of excimer (excited dimer) states created by a dielectric barrier discharge (*silent* discharge) in a rare-gas gas such as Ar_2^* ($\lambda = 126$ nm), Kr_2^* ($\lambda = 146$ nm), or Xe_2^* ($\lambda = 172$ nm) or molecular rare-gas-halide complexes complexes such as ArF* ($\lambda = 193$ nm), KrCl* ($\lambda= 222$ nm), KrF* ($\lambda = 248$ nm), XeCl* ($\lambda = 308$ nm). Stevens and Hutton [78] first proposed the concept of excimers, which exist only in the excited state [77] and under normal conditions, do not possess a stable ground state, in 1960. In the last decade, the properties of excimers and kinetics of their formation have been studied extensively by Malinin et al.[79], Volkova et al. [80], Eliasson and Kogelschatz [34-40, 81-82], Neiger et al.[83], and Zhang and Boyd [41-43,85]. In the meantime it has been demonstrated that these excimer UV sources can emit high UV intensities very efficiently [27,56,71], and that large-area UV systems at high power densities, as well as different geometries and wavelengths of excimer lamps have been designed and investigated [39,50,81,86-97].

The mechanism for forming excited rare gases/rare gas halides from ions and electrons begin with dissociative attachment of the electrons to the rare gas/halogen to form positive and negative ions. We use xenon and chlorine as a specific case.

$$e^- + Xe \rightarrow Xe^* + e^- \tag{1}$$
$$e^- + Xe \rightarrow Xe^+ + 2e^- \tag{2}$$
$$e^- + Cl_2 \rightarrow Cl + Cl^- \tag{3}$$

The formation of the rare gas dimer Xe_2^*, occurs through the three-body reaction of excited Xe* with other Xe atom or buffer gas.

$$Xe^* + Xe + M \rightarrow Xe^*_2 + M \tag{4}$$

M is a third collision partner which in many cases can be an atom or molecular of the gases involved or of the buffer gases argon, helium or neon. In most XeCl* exciplexes can be created by the recombination of positive xenon ions and negative chlorine ions (5) or Harpooning reaction (6) which the excited Xe* species directly react with chlorine [98].

$$Xe^+ + Cl^- + M \rightarrow XeCl^* + M \tag{5}$$

$$Xe^* \quad + \quad Cl_2 \quad \rightarrow \quad XeCl^* + \quad Cl \tag{6}$$

These excimer molecules are not very stable and once formed decompose within a few nanoseconds giving up their excitation energy in the form of a VUV or UV photon.

$$Xe^*_2 \qquad \rightarrow \quad 2Xe \quad + \quad h\upsilon \text{ (172 nm, VUV radiation)} \tag{7}$$
$$XeCl^* \rightarrow \quad Xe \quad + \quad Cl \quad + \quad h\upsilon \text{ (308 nm, UV radiation)} \tag{8}$$

A large number of different emission spectra of excimers have been investigated from rare-gas excimers, rare-gas halide exciplexes and halogen dimers [36-43,82-83,85]. With excimer lasers only a limited number of wavelengths are available at high power levels. These include $\lambda = 193$ nm (ArF*), $\lambda = 248$ nm (KrF*), $\lambda = 308$ nm (XeCl*), $\lambda = 351$ nm (XeF* whilst much lower powers can be obtained using F^*_2 ($\lambda = 157$ nm) and KrCl* ($\lambda = 222$ nm). By contrast, more than 2 different wavelengths can be generated in dielectric barrier discharges, extending their emission bands from the VUV to the visible part of the spectrum [39,60]. Table 4 shows the main peak wavelengths and their corresponding photon energies of rare-gases and rare-gas halides, created by discharge excitation.

Table 4 Peak wavelengths and photon energies of excimer emission bands obtained from various dielectric barrier discharges

Excimer	Wavelength (nm)	Photon energy (eV)	UV range
NeF*	108	11.48	
Ar$_2$*	126	9.84	
Kr$_2$*	146	8.49	
F$_2$*	158	7.85	
ArBr*	165	7.52	VUV
Xe$_2$*	172	7.21	
ArCl*	175	7.08	
KrI*	190	6.49	
ArF*	193	6.42	
KrBr*	207	5.99	
KrCl*	222	5.58	
KrF*	248	5.01	UV-C
XeI*	253	4.91	
Cl$_2$*	259	4.79	
XeBr*	283	4.41	
Br$_2$*	289	4.29	UV-B
XeCl*	308	4.03	
I$_2$*	342	3.63	UV-A
XeF*	351	3.53	

Different geometrical configurations of excimer lamps can be designed and fabricated, such as cylindrical, with UV radiating to the outside or the inside, or planar, with UV radiating to one or both sides and windowless for VUV radiation [49-50,60]. Cylindrical lamps were used in most of our work presented here. A typical lamp consisted of two concentric quartz tubes, outer and inner metallic electrodes, an external high voltage generator, and cooling water. The measured energy conversion efficiencies (UV output/electrical input) for these lamps can be as high as 22.5% [70]. Unlike the classical discharge, in the silent discharge microdischarges generally occur. In the work described here the discharge gap was filled with either a rare-gas or a rare gas-halogen gas mixture, and the photons were emitted through a quartz and outer electrode which was transparent to the radiation generated. An alternating high voltage of typically a few kV amplitude is adequate to run the discharge. The frequency of the applied voltage can vary over a wide range from 50 Hz to several MHz. The properties of these microdischarges have been extensively studied [36,39,41-42,82]. The electron density in a microdischarge is typically of the order of 10^{14} cm^{-3} and mean energies of the electrons are a few electron volts [77]. These conditions can be influenced by gap size, gas pressure, gas mixture, and characteristics of the dielectric materials and can be optimized for the formation of the desired excimer species [41-42].

A large area system has been designed and constructed with an array of parallel excimer lamps, capable of supplying uniform radiation over a 500 cm^2 area. A theoretical model of the UV intensity distribution of such a system has been developed, which assumes the excimer lamp to be a cylindrical source taking into account the fact that it consists of many microdischarges spread over the surface of the cylinder. The UV intensity distribution for a three excimer lamp system measured by using photodiode, photomultiplier, and chemical actinometric methods [41-42,99-100]. The uniformity of he UV intensity is seen to be within ±4% over a 250 cm^2 area. The measurements are in excellent agreement with the modeling [70].

The average life of conventional lamps is governed by the decrease of luminous flux caused by the unavoidable deposition of evaporated electrode material (most usually tungsten) on the inner wall of the envelope because the electrodes are directly in contact with the discharge gas and plasma as previously mentioned. After about a 5% weight loss from the electrode, breakage of the lamp may occur. For most mercury lamps lifetime is between 500-2000 hours, depending on the precise type of lamp. For certain specialized purposes (e.g. photographic lighting) the electrode may reach a temperature as high as 3300K (melting point of tungsten: 3655K), so that the lifetime is reduced to only a few hours. On the contrary, for excimer UV lamps the electrodes are not in direct contact with the discharge gases, and thus avoid any corrosion during the discharge process, thereby providing the excimer lamp with a long lifetime. It was found that the 100% level of the original UV intensity for the 222 nm and 308 nm lamps was still output up to 4000 hours operating time [101].

It can thus be seen that the use of excimers for the generation of UV radiation offers several important advantages over other lamp systems: 1) high intensity at the defined wavelength, 2) no-self absorption, 3) a long lifetime and no contamination of the excimer gas nor electrode corrosion, 4) non-toxic materials are used (e.g. no Hg) and thus inherently there is minimal environmental problem, 5) flexibility to design in different geometries, and 6) potential for scalability to large areas. Because of these unique properties, and additionally their simplicity of construction, large emission area with high-energy VUV photons, low cost, and availability of different wavelengths, excimer sources are an attractive alternative to conventional UV lamps and lasers for large-area industrial applications. Several of their novel applications towards dielectric film formation will be reviewed in the following sections.

Photo-induced processing

Figure 1 shows a schematic diagram of large area excimer lamps system, in which comprised a set of two stainless steel chambers separated by a MgF$_2$ window transparent to the VUV radiation. The lamp chamber consisted of an array of parallel cylindrical lamp tubes, details of which are shown on the right side of Fig. 1. The lamps contained either pure xenon (Xe*$_2$, λ=172nm) or a mixture of krypton and chlorine (KrCl*, λ=222nm). Excimer UV radiation was generated in the top chamber traversed a low pressure gas phase mixture contained within the bottom chamber, and impinged upon the sample with output power, in the range of 10-200 mW/cm^2, determined using actinometric techniques [99-100].

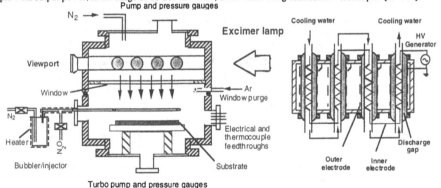

Fig. 1   Schematic diagram of photo-induced process incorporating of an array of excimer UV lamps.

N-type single crystal (100) orientation silicon (2-4 Ω·cm resistivity) wafers were used as substrates, which cleaned using a standard RCA, clean prior to use. The substrate temperature was maintained between 150 and 450°C and measured with a thermocouple attached to the heater stage. The tantalum metalorganic precursors, namely tantalum tetraethoxy dimethylaminoethoxide (Ta(OEt)$_4$(DMAE)) and tantalum ethoxide, were vaporized at temperatures between 100-130°C in a bubbler/injector and then transported into the reaction chamber by an N$_2$ carrier gas. A full description of this reactor is published elsewhere [26]. The processing chamber could be evacuated to 10^{-6} mbar by a turbomolecular pump and filled with the appropriate gas mixture for specific processing applications. UV annealing was performed on the films at different exposure times at temperatures between 350°C and 400°C for a fixed pressure of 1000 mbar in high purity oxygen (99.999%) by using 172 nm excimer lamps.

The chemical compositions in the films were determined by X-ray photoelectron spectroscopy (XPS) using a V■ ESCALAB 220i XL. The depth profiles in the samples of Ta, O, and Si were determined using Ar^+ ions (3 keV) with current of 0.8 μA at an argon pressure of 10^{-7} torr. The ratio of oxygen to tantalum in the films was calculated by pea■ deconvolution of the XPS curves. The thickness and refractive index of the films were determined using a Rudolp■ AutoEL II ellipsometer while the structure and optical properties of the layers grown on Si wafer and quartz we■ measured using a Fourier transform infrared (FTIR) spectrometer (Paragon 1000, Perkin Elmer) and U■ spectrophotometry, respectively. Surface morphology characterization was observed by using a Topometrix AF■ operating in contact mode to determine surface defects. The electrical properties of the films were measured at a frequenc■ of 1 MHz by HP4140 and HP4275 semiconductor systems (I-V, C-V) on $Al/Ta_2O_5/Si$ capacitor test structures with a evaporated Al top contact of area 8×10^{-4} cm^2 through a metal contact mask.

4. Photo-induced large area growth of dielectrics with excimer lamps
4.1. Large area growth of Ta_2O_5 by photo-induced sol-gel processing

To form a sol-gel solution, tantalum ethoxide $(Ta(OC_2H_5)_5)$ was dissolved in ethanol with a small quantity of wate■ and hydrochloric acid in ethanol. In the sol-gel process, the reaction involves two simultaneous chemical processe■ hydrolysis and polymerization. The alkoxide hydrolysis and polymerization reactions occurred over several hours, durin■ which the colloidal particles and condensing metal species linked together to become a three dimensional network, i.e., ■ slow polymerization of the organic compounds took place, leading to gelation. From this sol-gel solution, films of 10■ 300 nm thickness were prepared on Si (100) substrates by the spin-on method and then irradiated for various times a■ different temperatures, to form tantalum oxide [47].

An XPS profile of a 14 nm tantalum oxide film growth on silicon at 450°C at a fixed irradiation time of 20 min i■ shown in Fig. 2. The concentration of tantalum and oxygen is fairly constant throughout the Ta_2O_5 film. No carbon wa■ observed although it is always present in films obtained by plasma-CVD and thermal-CVD [102]. The atomic ratio o■ O/Ta was between 2.4 - 2.6 which is very close to the stoichiometric ratio of 2.5 for Ta_2O_5, and higher than the 2.2 ratic for films photo-irradiated by a low pressure mercury lamp and the 2.1 ratio reported for both heat-treated layers and photo-deposited films [102]. The average refractive index obtained at temperatures above 350°C was 2.15±0.05, which is close■ to the bulk Ta_2O_5 value of 2.2.

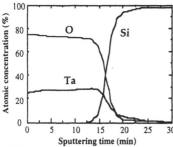

Fig. 2 XPS profile of a 14 nm tantalum oxide film formed on silicon at 450°C with a fixed irradiation time of 20 min

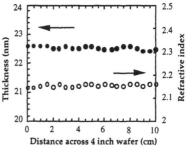

Fig. 3 Thickness and refractive index of the films formed across a 4 inch silicon wafer by irradiating at 350°C for 15 min.

Figure 3 shows the thickness and refractive index of layers formed on a 4 inch silicon wafer by irradiating spin-coated films at a temperature of 350°C for an exposure time of 15 min. As can been seen, very uniform films (about 22.6 nm) were achieved with the total variation in thickness across the 4-inch wafer being within ± 0.2 nm. The mean refractive index measured (2.15) was similar to that usually obtained for plasma-deposited Ta_2O_5 films [29]. XRD revealed the structure of films formed to be amorphous.

Table 5 Comparison of the electrical properties of the as grown Ta_2O_5 films at different temperatures

Temperature (°C)	Fixed charge density (cm^{-2})	Leakage current density at 0.5 MV/cm (A/cm^2)
150	4.0×10^{11}	1.9×10^{-5}
250	2.6×10^{11}	9.2×10^{-6}
400	1.0×10^{11}	9.0×10^{-8}

Metal oxide semiconductor (MOS) capacitors have been fabricated using 20 nm thick as grown Ta_2O_5 layers formed by this process. Table 5 shows a comparison of the electrical properties in our films formed at different temperatures. It can be seen that the fixed oxide charge density decreases with increasing temperature. The fixed oxide charge density changed from 4.0 x 10^{11} cm^{-2} at 150°C to 1.0 x 10^{11} cm^{-2} at 400°C, which is similar to those obtained in the films prepared by plasma-CVD processing [103]. The I-V characteristics of the MOS capacitors also showed that the leakage current density reduced dramatically in films grown at 400°C (see Table 5), indicating a more ideal reaction between oxygen and tantalum species at the higher temperatures. A leakage current density as low as 9.0 x 10^{-8} $A \cdot cm^{-2}$ at 0.5 MV/cm can be achieved, which is over 2 orders magnitude lower than those obtained in as grown films prepared by plasma-CVD method (see table 6) [31]. A subsequent low temperature (400°C) annealing in UV improves this to 2.0 x 10^{-9} $A \cdot cm^{-2}$ at 0.5 MV/cm (table 4). These values are comparable to those only previously obtained for films annealed at temperatures between 600° and 1000°C [21,31].

Table 6 Comparison of the leakage current densities at 0.5 MV/cm (A/cm^2) in Ta_2O_5 films obtained by different methods [31].

Plasma-CVD		Our work	
as-deposited	annealing at 700-800°C	as-deposited	annealing at 400 °C
10^{-6}	10^{-8} - 10^{-9}	9.0 x 10^{-8}	2.0x10^{-9}

.2. Thin Ta_2O_5 film grown by photo-CVD

In photo-induced CVD of Ta_2O_5 from tantalum ethoxide ($Ta(OC_2H_5)_5$) and nitrous oxide, the primary photochemistry of the N_2O involves the following reaction:

$$N_2O + h\upsilon \rightarrow O(^1D) + N_2(X^1\Sigma^+_g) \qquad (9)$$

The active oxygen species O (^1D) subsequently react with the $Ta(OC_2H_5)_5$ causing its dissociation through a series of reactions leading to Ta_2O_5 deposition on the substrate surface.

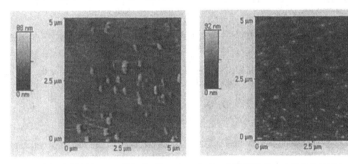

Fig. 4 AFM image of Ta_2O_5 films deposited at 200°C (left) and 350°C (right)

Figure 4 shows on AFM image of the surface features of Ta_2O_5 films deposited at 200°C (left of Fig. 4) and 350°C (right of Fig. 4).As can be seen, AFM image of the films deposited at 200°C revealed the presence of droplets varying in size from 10-500 nm which were not observed for substrate temperature over 300°C. A uniform structure down to a nanostructure scale with particle sizes of about 20 nm was observed by AFM at a temperature of 350°C.

XPS analysis showed that the atomic ratio of O/Ta, of about 2.4, is very close to the stoichiometric ratio of 2.5 for Ta_2O_5 [10]. Fig. 5 shows the evolution of the FTIR spectra in the 400-4000 cm^{-1} range for an as-deposited film, which was then annealed using a 172 nm lamp. The spectra exhibit one dominant peak centered around 650 cm^{-1} and two shoulder-like peaks near 530 cm^{-1} and in the 800-1000 cm^{-1} region agreeing with observations in previous work [24,104]. The peaks at 530 cm^{-1} and 650 cm^{-1} are assigned to the absorption of Ta-O-Ta and Ta-O stretching vibrational modes, characteristic of tantalum pentoxide [24,104]. The weak absorption band at 800-1000 cm^{-1} was attributed to the presence of suboxides [24]. It is clearly seen that UV annealing can significantly reduce this suboxide absorption and completely remove the H_2O and OH groups at 3400 cm^{-1}.

UV spectral measurements showed that the average transmittance of 90% in the visible region of the spectrum for films deposited at temperatures between 250-400°C, which is characteristic of very high quality Ta_2O_5 films [26].

The I-V characteristics of the MOS capacitors also show that after UV annealing, the leakage current density is reduced dramatically as shown in Fig. 6 where it can be seen that it decreases with increased annealing time. After 1 annealing, leakage current densities as low as 6.4×10^{-9} A/cm^2 at 0.5 MV/cm are achieved. This is two orders of magnitude lower than for as-deposited layers (1.95×10^{-7} A/cm^2) and comparable to values only previously achieved for films annealed at high temperatures (600-900°C) [11,18,21]. Several effects could cause the reduction of leakage current but here we consider three of the most likely contributors. First, the active oxygen species formed by the 172 nm light can assist in reducing or removing any suboxides present leading to improved stoichiometry as shown in the FTIR (Fig 5), where suboxides in the as-deposited films were clearly removed by UV annealing. This effect has already been reported for layers grown by pulsed laser deposition [24]. Second, it is known that the active oxygen species created can decrease the density of defects and oxygen vacancies in the as-deposited film [10]. Additionally, it has been reported and confirmed by XPS and FTIR that SiO_2 layers can be formed at the Ta_2O_5/Si interface and on the surface of Ta_2O_5, by the reaction between the active oxygen and Si during annealing, leading to improved interfacial quality [18,105]. All of these could in some measure lead to the reduction of leakage current density in our layers although it is not clear at present which of these dominates.

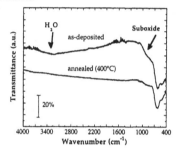

Fig. 5 FTIR spectra for Ta_2O_5 films deposited at 350°C and UV annealed at 400°C.

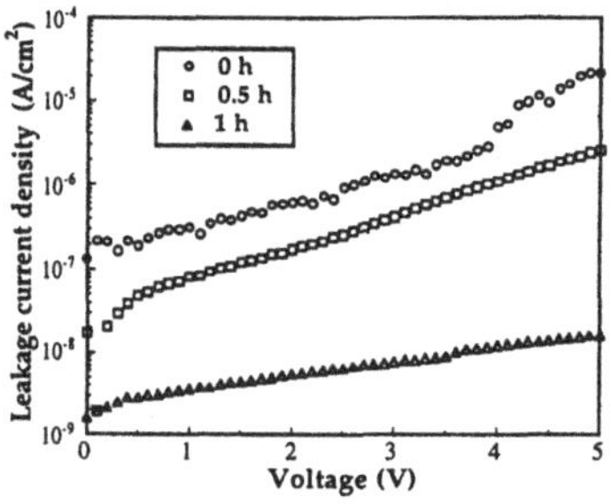

Fig. 6 Leakage current density of as-deposited and annealed films for different times

From XPS, FTIR, and electrical measurement results, we conclude that a possible mechanism for the UV annealing effect is attributed to the reactive oxygen species produced by the UV irradiation as follows:

$$O_2 + h\upsilon\,(\lambda = 172\,nm) \rightarrow \quad O\,(^3p) + O\,(^1D) \qquad\qquad (10)$$
$$O_2 + O\,(^3p) + M \rightarrow \quad O_3 + M \qquad (M\ is\ a\ third\ body) \qquad (11)$$

The ozone is decomposed by absorption of VUV light at 172 nm which produces excited state ^{1}D oxygen atoms.

$$O_3 + h\upsilon\,(\lambda = 172\,nm) \rightarrow \quad O_2 + O\,(^1D) \qquad\qquad (12)$$

The active oxygen species formed react with silicon which diffuses from the Si substrate to Ta_2O_5 surface leading to the formation SiO_2 at the surface of the Ta_2O_5 films. On the other hand, the active oxygen species can diffuse through the thin SiO_2 oxide, and react with Ta suboxides to create Ta_2O_5, and possibly also remove certain defects and oxygen vacancies present in the Ta_2O_5.

4.3. Other high dielectric constant materials (TiO$_2$ and PZT) formed by photo-induced process

Photo-induced sol-gel processing of other high dielectric constant materials such as TiO_2 and PZT using the excimer UV sources has also been demonstrated [106-107]. Single and multiple layer TiO_2 films have been successfully prepared at low temperatures by photo-induced sol-gel processing using a 172 nm excimer lamp [94]. Refractive index values ranging from 2 to 2.4 were measured for multilayers irradiated for 10 min (see fig. 7). These values compare favorably ith the value of 2.58 for the bulk material. The films formed showed good optical properties with transmittance values etween 85% and 90% in the visible range of the spectrum.

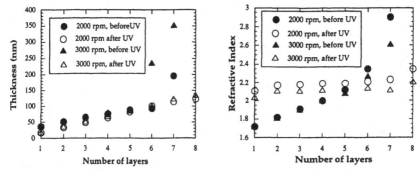

Fig. 7 *Thickness and refractive index of TiO₂ films before and after irradiation at 172 nm for different layers.*

PZT thin films have also been deposited at low temperatures by photo-assisted decomposition of a PZT metal-organic sol-gel polymer using the 172 nm excimer lamp [107]. Very low leakage current densities (10^{-7} A/cm^2) can be achieved, which compared with layers grown by conventional thermal processing. This particular photo-induced approach not only enables reduced temperatures and processing times to be used but also provides good electrical properties without the need for high temperature annealing.

4.4. Low dielectric constant materials (polyimide and porous silica)

The performance of ultra-large scale integrated (ULSI) devices becomes crucial at the metal interconnect level when the feature sizes are reduced to low sub-micron dimensions. The gain in device speed at an MOS device gate is offset by the propagation delay at the metal interconnects due to the increased RC (resistance and capacitance) time constant. This RC time delay can be reduced either with the incorporation of low permittivity dielectric materials and/or high conductivity metals. Polymeric films are one of the most promising groups of low dielectric constant materials which may eventually replace the widely used SiO₂ as an interlayer dielectric to shorten RC time delays, reduce "cross-talk" between metal lines and decrease power consumption at high signal frequencies [108]. Polyimides are particularly attractive not only because of their low dielectric constant, but also their ease of application and patterning and high thermal stability [109]. Recent work on photo-induced deposition of low dielectric constant organic polymers for interlayer dielectrics has highlighted a significant role of photo effects on the curing of polyamic acid films [110]. Compared with conventional furnace processing, the photo-induced curing of the polyimide provided both reduced processing time and temperature. In particular, I-V measurements showed that the leakage current density of the irradiated polymer was over an order of magnitude smaller than has been obtained in layers prepared by thermal processing [110].

Figure 8 presents the FTIR spectra, in the 600-2000 cm⁻¹ range for films after the initial prebake and after a 150°C cure for 20 min with the lamp off (conventional thermal curing) and under otherwise identical conditions, but with the additional UV irradiation. Figure 8a shows the characteristic bands of the carboxyl (–COOH) absorption at 1723 cm⁻¹, amide (–CONH–) groups at 1659 and 1546 cm⁻¹, and the amide stretching mode in polyamic acid at 1410 cm⁻¹. After the UV curing (Fig. 8c) all these bands completely disappear. Simultaneously, a typical doublet of a carbonyl group corresponding to an imide moiety appears at 1778 and 1726 cm⁻¹ together with the imide C-N absorption band at 1379 cm⁻¹. The small absorption at 728 cm⁻¹ has been attributed to deformation of the imide ring or the imide carbonyl groups [111]. The bands corresponding to polyimide at 1778, 1726 and 1379 cm⁻¹ for the thermally cured polyamic acid (Fig. 8b) are significantly smaller than those obtained by UV curing. Also the bands related to polyamic acid at 1659, 1546, 1410 cm⁻¹ decreased but did not completely disappear. These results indicate that the polyamic acid film is completely transformed to polyimide by the UV curing step at 150°C, whilst the thermally cured sample is only partly transformed. The degree of imidization at different temperatures for both the UV curing and purely thermal curing steps is shown in Fig. 9. The degree of imidization was calculated by comparing the 1375 cm⁻¹ imide band and the 1500 cm⁻¹ aromatic band intensities, which are known to give precise internally consistent measurements [111]. At lower curing temperatures (i.e. <150°C) the imidization characteristics of the two curing methods are markedly different. For UV curing, the films start to imidize very significantly, whilst the imidization of the films is very slow for thermal curing. The degree of imidization is 85% for UV curing at 150°C, whilst it is less than 20% for the thermal process.

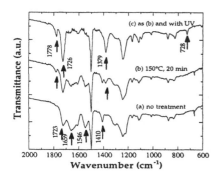

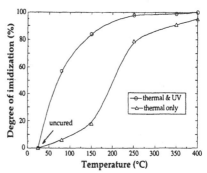

Fig. 8 FTIR spectra of polyamic acid films after a 150°C
cure for 20 min with and without UV irradiation, a) initial prebake;
b) 150°C, 20 min; c) 150°C, irradiated 20 min by 172 nm lamp.

Fig. 9 Degree of imidization of polyamic acid film
as a function of curing temperature with and without
irradiation

Very recently, low dielectric porous silica films have been successfully grown from TEOS sol-gel solutions at low
temperatures using an excimer lamp [112]. Dielectric constant values as low as 1.7 can be achieved in films prepared at
room temperature. These results indicate that this low temperature photo-induced sol-gel technique is very promising for
the preparation of low-k dielectric polymer and porous silica films or other interlayer dielectrics in future ULSI multilevel
interconnections.

5. Conclusions

Excimer UV lamps with their unique properties, providing high intensity narrow-band radiation over large-areas at a
number of different wavelengths, are an interesting alternative to conventional UV lamps and lasers for industrial large-
scale low temperature processes. The application of these sources towards low temperature deposition of high dielectric
constant (Ta_2O_5, TiO_2 and PZT) and low dielectric constant (polyimide and porous silica) materials has been successfully
demonstrated. Very uniform Ta_2O_5 films were achieved with the total variation in thickness across the 4-inch wafer being
within ± 0.2 nm. The leakage current density of the irradiated dielectric films was over several orders of magnitude smaller
than that obtained by thermal processing. UV annealing can significantly reduce the leakage current density of Ta_2O_5 thin
films incorporated in simple MOS capacitor structures. The active oxygen species, produced by 172 nm radiation, are
considered to play an important role in the reduction of the leakage current density since they can reduce the density of
defects and oxygen vacancies, remove any suboxides and impurities present in the films. This photo-induced process also
enables reduced temperatures and times to be used. Therefore, UV-induced low temperature deposition of thin films offers
a very effective method for fabrication of high quality thin films for industrial applications in electronic and optical
manufacturing.

Acknowledgement:
The authors would like to thank Dr. U. Kogelschatz (ABB, Corporate Research, Switzerland) for many stimulating
discussions. This work was partly supported by EPSRC (grant No. GR/190909).

References
1.   D.A. Muller, T. Sorsch, S. Moccio, F.H. Baumann, K. Evans-Lutterodt and G. Timp, Nature, 399 (1999) 758.
2.   Semiconductor industry Association The National Technology Roadmap for Semicond. 71-78 (Sematech Austin, 1997).
3.   M. Schulz, Nature, 399 (1999) 729.
4.   C.A. Billman, P.H. Tan, K.J. Hubbard, and D.G. Schlom, Mat. Res. Soc. Symp. Proc. 567 (1999) 409.
5.   Q.X. Jia, X.D. Wu, S.R. Foltyn, and P. Tiwari, Appl. Phys. Lett. 66 (1995) 2197.
6.   R. Singh, S. Alamgir, and R. Sharangpani, Appl. Phys. Lett. 67 (1995) 3939.
7.   S. Tanimoto, M. Matsui, K. Kamisako, K. Kuroiwa, and Y. Tarui, J. Electrochem. Soc. 139 (1992) 320.
8.   Y. Nishimura, K. Tokunaga and M. Tsuji, Thin Solid Films 226 (1993) 144.
9.   KW Kwon, CS Kang, SO Park, HK Kang, ST Ahn, IEEE Trans Electron Devices 43 (1996) 919.
10. J.-Y. Zhang, B. Lim, V. Dusastre, and I.W. Boyd, Appl. Phys. Lett. 73 (1998) 2299.
11. H. Shinriki, T. Kisu, Y. Nishioka, Y. Kawamoto, and K. Mukai, IEEE Trans. Electron Devices 37 (1990) 1939.

2. R.F. Cava, W.F. Peck Jr and J.J. Krajewski, Nature, 377 (1995) 215.
3. A. Cappellani, J.L. Keddie, N.P. Barradas and S.M. Jackson, Solid-State Electronics, 43 (1999) 1095.
4. R.J. Cava and J.J. Krajewski, J. Appl. Phys. 83 (1998) 1613.
5. S. Kamiyama, H. Suzuki, H. Watanabe, A. Sakai, H. Kimura, and J. Mizuki, J. Electrochem Soc., 141 (1994) 1246.
6. A. Pignolet, G. M. Rao, S.B. Krupanidhi, Thin Solid Films, 258 (1995) 230.
7. S.C. Sun and T.F. Chen, IEEE Electron Device letters 17 (1996) 355.
8. S.W. Park, Y.K. Baek, J.Y. Lee, C.O. Park and H.B. Im, J. Electronic. Mater. 21 (1992) 635.
9. S. Zaima, T. Furuta, Y. Koide, and Y. Yasuda, , J. Electrochem. Soc. 137 (1992) 2876.
0. S. Kamiyama, P. Lesaicherre, H. Suzuki, I. Nishiyama and A. Ishitani, J. Electrochem. Soc. 140 (1993) 1617.
1. J.L. Autran, P. Paillet, J.L. Leray, and R.A.B. Devine, Sensors and Actuators A51 (1995) 5.
2. J.-Y. Zhang, V. Dusastre, D.E. Williams and I.W. Boyd, J. Phys. D: Appl. Phys.32 (1999) L1.
3. H. Shinriki and M. Nakata, IEEE Trans. Electron. Dev. 38 (1991) 455.
4. J.-Y. Zhang, Q. Fang, and I.W. Boyd, Appl. Surf. Sci. 138-139 (1999) 320.
5. R.A.B. Devine, Appl. Phys. Lett. 68 (1996) 1924.
6. J.-Y. Zhang, B. Lim, and I.W. Boyd, Thin Solid Films 336 (1998) 340.
7. C. Isobe and M. Saitoh, Appl. Phys. Lett. 56 (1990) 907.
8. J.-Y. Zhang and I.W. Boyd, J. of Mater. Sci. Lett. 17 (1998) 1507.
9. P.A. Murawala, M. Sawai, T. Tatsuta, O. Tsuji, S. Fujita, and S. Fujita, Jpn. J. Appl. Phys. 32 (1993) 368.
0. H.O. Sankur and W. Gunning, Appl. Opt. 28 (1989) 2806.
1. I.L. Kim, J.S. Kim, O.S. Kwon, S.T. Ahn, J.S. Chun and W.J. Lee, J. Electron. Mater. 24 (1995) 1435.
2. D. Laviale, J.C. Oberlin, and R.A.B. Devine, Appl. Phys. Lett. 65 (1994) 2021.
3. M. Matsui, S. Oka, K. Yamagishi, K. Kuroiwa, and Y. Tarui, Jpn. J. Appl. Phys. 27 (1988) 506.
4. S. Oshio, M. Yamamoto, J. Kuwata, and T. Matsuoka, J. Appl. Phys. 71 (1992) 3471.
5. T. Aoyama, S. Yamazaki, and K. Imai, J. Electrochem. Soc. 145 (1998) 2961.
6. B. Eliasson and U. Kogelschatz, Appl. Phys. B 46 (1988) 299.
7. B. Eliasson and U. Kogelschatz, Proc. 40 Ann. Gas. Electron. Conf. (GEC 87), Atlanta 1987, p.174.
8. B. Gellert, B. Eliasson and U. Kogelschatz, Proc. 5 Int. Symp. on the Science & Technology of Light Sources (LS:5), York 1989, p.155 and 181.
9. U. Kogelschatz, Pure & Appl. Chem. 62 (1990) 1667.
0. U. Kogelschatz, Appl. Surf. Sci. 54 (1992) 410.
1. J.-Y. Zhang and I.W. Boyd, J. Appl. Phys. 80 (1996) 633.
2. J.-Y. Zhang and I.W. Boyd, J. Appl. Phys. 84 (1998) 1174.
3. I.W. Boyd and J.-Y. Zhang, Nucl. Instrum. Meth. Phys. Res. B121 (1997) 349.
4. P. Bergonzo and I.W. Boyd, J. Appl. Phys. 76 (7) (1994) 4372.
5. P. Bergonzo and I.W. Boyd, Appl. Phys. Lett. 63 (1993) 1757.
6. J.-Y. Zhang, L.-J. Bie, and I.W. Boyd, Jpn. J. Appl. Phys. 37 (1998) L27.
7. J.-Y. Zhang, B.-J. Bie, V. Dusastre and I.W. Boyd, Thin Solid Films 318 (1998) 252.
8. H. Esrom, J. Demny, and U. Kogelschatz, Chemtronics 4 (1989) 202.
9. H. Esrom and U. Kogelschatz, Appl. Surf. Sci. 46 (1990) 158.
0. H. Esrom and U. Kogelschatz, Appl. Surf. Sci. 54 (1992) 440.
1. J.-Y. Zhang, Qi Fang, S.L. King and Ian W. Boyd, Appl. Surf. Sci. 109/110 (1997) 487.
2. J.-Y. Zhang, H. Esrom, and I.W. Boyd, Appl. Surf. Sci. 96-98 (1996) 399.
3. J.-Y. Zhang and I.W. Boyd, J. Mat. Sci. Lett. 16 (1997) 996.
4. J.-Y. Zhang and I.W. Boyd, Appl. Phys. A 65 (1997) 379.
5. J.-Y. Zhang and I.W. Boyd, Thin Solid Films, 318 (1998) 234.
6. J.-Y. Zhang and I.W. Boyd, Electronics Letters, 32 (1996) 2097.
7. V. Cracium, B. Hutten, D.E. Williams and I.W. Boyd, Electronics Letters 34 (1998) 71
8. J.-Y. Zhang and I.W. Boyd, Appl. Phys. Lett. 71 (1997) 2964.
9. V. Craciun, J-Y. Zhang and I.W. Boyd, NATO Fundamental Aspects of Ultrathin Dielectrics on Si-based Dev. 1997, pp461.
0. H. Esrom and U. Kogelschatz, Thin solid films, 218 (1992) 231.
1. J.-Y. Zhang, Thesis, Karlsruhe University, Germany, 1993.
2. H. Esrom, J.-Y. Zhang, and U. Kogelschatz, Mat. Res. Symp. Proc. 236 (1992) 39.
3. J.-Y. Zhang, H. Esrom, U. Kogelschatz and G. Emig, Appl. Surf. Sci. 69 (1993) 299.
4. J.-Y. Zhang, H. Esrom, U. Kogelschatz, and G. Emig, J. of Adhesion Sci. and Technol. 8 (1994) 1179 .
5. J.-Y. Zhang, H. Esrom, and I.W. Boyd, Surface and Interface Analysis 24 (1996) 718.
6. V. Craciun, I.W. Boyd, D. Craciun, P. Andreazza and J. Perriere, J. Appl. Phys. 85 (1999) 8841.
7. V. Craciun, D. Craciun, P. Andreazza, J. Perriere and I.W. Boyd, Appl. Surf. Sci. 139 (1999) 587.

68. U. Kogelschatz, NATO Advanced Research Workshop on Non-thermal Plasma Techniques for Pollution Contr Cambridge University, UK, September 21-25, 1992.
69. R.S. Nohr and J.G. MacDonald, U. Kogelschatz, G. Mark, H.-P. Schuchmann and C. von Sonntag, J. Photoche Photobiol. A: Chem. 79 (1994) 141.
70. J.-Y. Zhang and I.W. Boyd, Mat. Res. Symp. Proc. 471 (1997) 53.
71. T. Urakabe, S. Harada, T. Saikatsu and M. Karino, Sci. and Tech. of light Sources (LS7) Kyoto, 1995, Eds: R. Ital nd S. Kamiya, pp159.
72. P. Bergonzo, U. Kogelschatz, and I.W. Boyd, Appl. Surf. Sci. 69 (1993) 393.
73. P. Bergonzo, U. Kogelschatz, and I.W. Boyd, SPIE, Vol 2045 (1994) 174 .
74. P. Bergonzo and I.W. Boyd, Electronics Letters, 30 (1994) 606.
75. P. Bergonzo and I.W. Boyd, Microelectronic Engineering 25 (1994) 345.
76. G. Eftekhari, J. Electrochem. Soc. 140 (1993) 787.
77. B. Gellert, U. Kogelschatz, Appl. Phys. B52 (1991) 14 .
78. B. Stevens and E. Hutton, Nature, 186 (1960) 1045 .
79. A.N. Malinin, A.K. Shuaibov and V.S. Shevera, J. Appl. Spectrosc., 32 (1980) 313 .
80. G.A. Volkova, N.N. Kirillova, E.N. Pavlovskaya and A.V. Yakovleva, J. Appl. Spectrosc. 41 (1984)1194.
81. B. Eliasson and B. Gellert, J. Appl. Phys. 68 (1990) 2026 .
82. B. Eliasson, M. Hirth and U. Kogelschatz, J. Phys D: Appl. Phys. 20 (1987) 1421.
83. M. Neiger, V. Schorpp and K. Stockwald, Proc. 41. Ann. Gaseous Electron. Conf. (GEC 88), Minneapolis p.74, 198
84. I.W. Boyd and J.-Y. Zhang, Mat. Res. Symp. Proc. 470 (1997) 343.
85. J.D. Ametepe, J. Diggs, D.M. Manos and M.J. Kelley, J. Appl. Phys. 85 (1999) 7505.
86. P. Patel, I.W. Boyd, Appl. Surf. Sci. 46 (1990) 352 .
87. F. Kessler and G.H. Bauer, Appl. Surf. Sci. 54 (1992) 430.
88. F. Kessler, H.D. Mohring, G.H. Bauer, Proc. of the 9th Conf. on Plasma Chem. 3 (1989) 1383.
89. C. Manfredotti, F. Fizotti, M. Boero, G. Piatti, Appl. Surf. Sci. 69 (1993) 127.
90. B. Bollanti, G. Clementi, P.D. Lazzaro, F. Flora, G. Giordano, T. Letardi, F. Muzzi, G. Schina and C.E. Zheng, IEE Transactions on Plasma Science 27 (1999) 211.
91. A.K. Shuaibov, L.L. Shimon and I.V. Shevera, Instr. and Experimental Tech. 41 (1998) 427.
92. P.N. Barnes and M.J. Kushner, J. Appl. Phys. 80 (1996) 5593.
93. J. Kawanaka, A. Ogata, S. Kubodera, W. Sasaki and K. Kurosawa, Appl. Phys. B-Lasers and Optics 65 (1997) 609.
94. M. Kitamura, K. Mitsuka and H. Sato, Appl. Surf. Sci. 80 (1994) 507.
95. T. Nakamura, F. Kannari and M. Obara, Appl. Phys. Lett. 57 (1990) 2057.
96. A. El-Habachi and K.H. Schoenbach, Appl. Phys. Lett. 72 (1998) 1.
97. U. Kogelschatz, B. Eliasson and W. Egli. J. Phys. IV France 7 (1997) C4-47.
98. Ch. K. Rhodes "Excimer Lasers", Vol. 30 of Topics in Applied Physics, Springer-Verlag, Berlin, 1984.
99. J.-Y. Zhang, H. Esrom, and I.W. Boyd, Appl. Surf. Sci. 109/110 (1997) 482.
100. J.-Y. Zhang, H. Esrom, and I.W. Boyd, Appl. Surf. Sci. 138-139 (1999) 315.
101. I.W. Boyd and J.-Y. Zhang,, Advanced Laser Technologies (ALT99), Potenza-Lecce, Italy, Sept 20-24, 1999.
102. T. Ohishi, S. Maekawa & A. Katoh, J. Non-Cryst. Solids 147&148 (1992) 493.
103. G.Q. Lo, D.L. Kwong, and S. Lee, Appl. Phys. Lett. 60 (1992) 3286.
104. C.H. An and K. Sugimoto, J. Electrochem Soc. 139 (1992) 1956.
105. J.-Y. Zhang, V. Dusastre, D.E. Williams and I.W. Boyd, J. Phys. D: Appl. Phys.32 (1999) L1.
106. N. Kaliwoh, J.-Y. Zhang and I.W. Boyd, Surface and Coating Technology 125 (2000) 424.
107. J.-Y. Zhang and I.W. Boyd, Jpn. J. Appl. Phys. 38 (1999) L393.
108. P. Singer, Semiconductor International, October 1994, p. 34.
109. S.P. Murarka, Solid State Technology 39 (1996) 83-90.
110. J.-Y. Zhang and I.W. Boyd, Optical Materials 9 (1998) 251.
111. C.A. Pryde, J. Polym. Sci.: Part A: Polym. Chem., 1989, 27, pp. 711-724
112. J.-Y. Zhang and I.W. Boyd, E-MRS 99 Spring Meeting (to be published Appl. Surf. Sci. 2000).

Laser direct write of conducting and insulating tracks in silicon carbide

D.K. Sengupta[1], N.R. Quick[2], and A. Kar[1]
[1] Laser-Aided Manufacturing, Materials and Micro-Processing Laboratory (LAMMMP), School of Optics, Center for Research and Education in Optics and Lasers (CREOL), University of Central Florida, Orlando, FL, 32816-2700,USA
[2] Applicote Associates, 894 Silverado Ct. Lake Mary, Florida 32746

ABSTRACT

Conventional direct write processes are multi-step requiring at least one additional process to change conductive properties. A direct conversion technique that uses lasers to irradiate silicon carbide, providing tracks which are highly conductive has been demonstrated. It was found that laser irradiation of insulating silicon carbide films could cause a drop from 10^{11} to 10^{-4} ohm-cm in a 4-point resistance test. However, in the presence of pure oxygen, laser-irradiated silicon carbide conductor and semiconductor samples exhibit insulating characteristics. Pattern formation was achieved by a computer program controlled galvo-mirror. The pads, 0.4 cm x 0.7 cm were formed by beam rastering with an overlap of 30% of the 0.025 cm beam diameter. This computer assisted processing allows the design of patterns using conventional CAD/CAE technologies and smart material behavior via selective and controlled electrical property transitions by laser irradiation.

INTRODUCTION

There has been a tremendous amount of interest in the wide bandgap semiconductors including aluminum nitride, boron nitride and silicon carbide as high power electronic and optical devices. Of particular interest is silicon carbide which has over 170 different polytypes, such as β–SiC (Zincblende structure) and 6H-SiC (6 bilayers along the hexagonal crystal direction), and exhibits electrical properties ranging from insulating to semiconducting [1]. Power semiconductor devices including diodes, thyristors and transistors are used for industrial drives and power supplies. Although silicon devices currently monopolize the industry, they exhibit high losses because of a low breakdown voltage and are limited for use below 150°C. Silicon carbide exhibits a breakdown strength approaching 4.5KV, ten times higher than that of silicon, resulting in lower losses. Silicon carbide also operates at temperatures approaching 650°C. Because of its high-saturated electron drift velocity; silicon carbide would also be favored at high or microwave frequencies [2]. These properties suggest that silicon carbide could be used in devices as a high temperature semiconductor, a blue light emitting diode, a p-doped window for solar cells, an ultra-violet radiation detector, and in high temperature and actuators [1-8]. Technology barriers to silicon carbide use include micropipe substrate defects and processing limitations (e.g., dielectric deposition, etching, oxidation, metallization, and doping [9,10,11]). In particular, use of silicon carbide for these applications has been hindered by the difficulties in metallization for source, drain, and gate contacts on devices and interconnect conductors and vias on integrating substrates. In order to fabricate silicon carbide semiconductor devices, several technologies must be developed. For example, to construct junction field-effect transistors (JFETs), other FETs or light emitting diodes, the planar technologies of doping, oxidation, and metallization are needed as well as patterning technologies. One of the limiting steps with silicon carbide based devices pertain to the metallization for the source, drain, and gate contacts. These contacts must be free of oxidation and unreactive with materials in contact, including silicon carbide for long times at 350-450°C. Metal conductors create thermal coefficient of expansion mismatch strains that can lead to debonding, or can create charge carrier traps, particularly dislocations, which decrease device efficiency. These strains are aggravated

127

within the temperature range 250-450°C, the proposed use environment for silicon carbide devices. In addition, metal conductors can oxidize and react with chemical species in hostile environments degrading their conductivity properties and further creating chemical product that can attack the device. Because of these limitations, platinum and silicide conductors are currently being developed for silicon carbide. A laser direct write technology [12,13,14] that can convert insulating silicon carbide exhibiting resistivities greater than 10^{11} ohm-cm to conductors without the need for external metallization and the effect of pure oxygen gas assist during laser irradiation of silicon carbide are demonstrated. This technology which is applicable to SiC substrates provides a solution to silicon carbide metallization problems and enables the creation of planar and three-dimensional monolithic silicon carbide devices.

LASER PROCESSING OF SILICON CARBIDE

Bulk 0.025cm thick insulating α-SiC substrates were processed using controlled boron carbide concentrations and a proprietary pressureless sintering thermal treatment cycle. The Carborundum Company processed these substrates [12,13,14]. Experiments were conducted using a modified Nd:YAG laser system with an emission wavelength of 1064 nm and pulse repetition rates from 1000-10000 Hz. The pulse duration time was 70 nsec and the TEM$_{00}$ or multimode was used. The effect of pure oxygen gas assist was studied using a continuos wave Nd:YAG laser operating at a power density of ~50W/cm^2. Pattern formation was achieved by moving the substrate with a programmable x-y table at speeds ranging from 0.1 to 2 cm/sec. For pure oxygen gas assist studies, the substrate was moved at a rate of 0.01 cm/sec. Also, 248 nm KrF excimer laser has been successfully used to obtain the conversion of electrical properties in silicon carbide. Pulse duration was 30 nsec and pulse repetition rates ranged from 1000-10000Hz. The sample to be irradiated was mounted on a stepper motor driven stage allowing the laser spot to be translated across the sample surface at a controlled rate. Evaluation of the mechanism for the electrical property conversion in these laser-processed materials requires the application of analytical procedures. Scanning electron microscopy (SEM), JOEL Super probe 733 operated at 12KV was used for the structural and chemical characterization.

Conventional direct write processes are multi-step requiring at least one addition process to change conductive properties. Laser synthesis direct-write is a one step process; the electrical property conversion is a phase transformation in the substrate resulting from selective rapid solidification, grain refinement, grain recrystallization, and chemical transitions. The fact that selective and controlled electrical property transitions can be imparted to wide bandgap ceramics, such as silicon carbide and aluminum nitride, by laser excitation is a "smart" material behavior. Laser synthesis enables the processing of prototypes from these high temperature electronic materials that can be subjected to actual field and simulated use environments. This rapid prototyping process is directly scalable to production levels.

Depth and micro-machined features can also be laser synthesized in silicon carbide because of

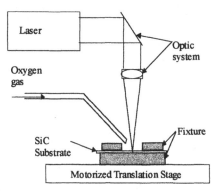

Fig. 1. The Schematic diagram of laser write system.

its transmission of infrared wavelengths. Conversion can be programmed by focusing the laser beam at plane or spot within the depth of the substrate. The substrate can be bulk, thin film and either flat or contoured; the laser can be programmed for three-dimensional surfaces. This computer assisted processing (CAP) allows the design of patterns using conventional CAD/CAE technologies. The sketch of the laser writing system is shown in Figure. 1.

RESULTS AND DISCUSSION

Silicon carbide is a line compound with little solubility for carbon or silicon. It does not melt congruently, but is characterized by a peritectic reaction around 2545°C. The peritectic melting at 2545°C should produce solid carbon and molten silicon with 27% atomic percent dissolved carbon according to the phase diagram [8]. However, because the heating is so rapid (70 nsec for Nd:YAG) Diffusion distances would be very small (< 30nm, assuming liquid like diffusivity of 10^{-4} cm^2/s) and melt of 50% carbon can be produced above 3000°C requiring little or no phase separation. Cooling is similarly rapid; therefore, phase separation is also limited during solidification. Therefore, carbon globule formation is plausible and higher laser transformed silicon carbide conductivities could be associated with size refinement of and decreased spacing between the globules. A SEM micrograph of the un-treated and the Nd:YAG laser irradiated α-silicon carbide surfaces are shown in Figure 2. Table 1 summarizes the decrease in the resistivity of α-SiC substrates as a function of laser processing parameters for the

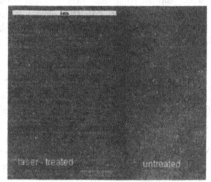

Fig. 2. SEM micrograph showing laser-treated and untreated surfaces.

sample set used in this analytical study. The Nd: YAG irradiated substrates show a slightly lower resistivity. The resistivity of an irradiated alpha-silicon carbide semiconductor substrate is thirteen orders of magnitude less than the unirradiated substrate. 4-wire probe contact resistance were measured from 15K to 300K for the conductor that was Nd:YAG laser synthesized in bulk α–SiC. The behavior of resistance as a function of temperature is indicative of metal-like conductivity [15].

Table 1. Decrease in the resistivity of α–SiC substrates due to laser-irradiation.

Sample Type	Laser System	Pulse Energy (J)	Beam Diameter (cm)	Pulse Duration (ns)	Scan Velocity (cm/s)	Un-Irradiated Resistivity (Ω-cm)	Irradiated Resistivity (Ω-cm)
Pressureless Sintered SiC	Nd:YAG 1064 nm	0.0081	0.025	70	0.3	10^{11}	5.1 x10^{-3}
Pressureless Sintered SiC	Excimer 248 nm	0.0032	0.085 (scan direction)	30	0.053	10^{11}	1.2 x 10^{-2}

The sample was further analyzed with an atomic force microscopy (AFM) as shown in Fig.3 (a) and (b) respectively. It can be seen from these images that the laser processing produce some changes and replaced larger structures with small more rounded type structures, indicative (or at least suggestive) of the globular formations. The roughness values seem to decrease due to the prevalence of these smaller globules rather than the larger (and more angular) structures seen in the preprocessed images. Between the visual inspection of the surface and the roughness value comparisons, it can be said from the AFM analysis that the laser processing caused a change in the surface character and the replacement of the larger structures with smaller, more rounded structures. It is also evident that the electrical conductivity increases with the decrease of porosity [16,]. The effect of pure oxygen gas doping during continuos wave Nd:YAG laser irradiation of silicon carbide is shown in Figures 4(a) and (b) respectively. The SEM micrograph of the oxidized surface is shown in Figure 4(a). The JOEL Super probe 733 oxygen sensitivity analysis is shown in Figure 4(b). Measurements were done from the proximity of the laser induced track and the oxygen sensitivity measurement show the oxygen level to be relatively higher closer to the laser generated tracks. The laser processing parameters could be adjusted to generate oxygen rich surfaces closer to the laser generated tracks and thereby exhibiting insulating characteristics. The present data clearly shows that the areas closer to the laser-generated tracks contains oxygen in large amounts in comparison to the areas away from the laser-induced tracks.

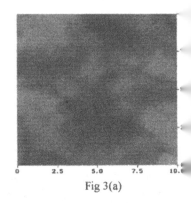

Fig 3(a)

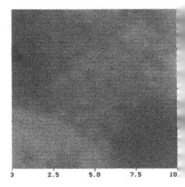

Fig. 3(b)

Fig. 3. AFM images of (a) untreated and (b) laser-treated α-SiC surfaces.

Fig. 4(a)

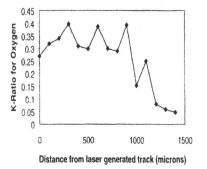

Distance from laser generated track (microns)

Fig. 4(b)

Fig. 4. Laser generated (a) SEM of the oxidized surface and (b) JOEL super probe oxygen (K-ratio) analysis of the oxidized surface.

Energy dispersive spectroscopy (EDS) of conductor synthesized by KrF excimer laser on α-SiC substrate showed the composition to be silicon-oxygen-carbon [17]. In the presence of oxygen, in more recent studies during the initial stages of SiC oxidation, a transition oxide, has been observed [18]. However, the formation of silicon oxide (like SiO) is not very likely, since that would involve dissociating Si-C bonds in the presence of oxygen and forming Si-Si bonds. Only recently has the transition oxide been attributed to silicon oxycarbides [19]. In the present study, Morse potentials were used to characterize the bonding configuration of the silicon species during oxidation (Fig. 5). Of particular interest is the energy of formation of Si-O (silicon oxide), and Si-O-C (silicon-oxycarbide) bonds. The potential energy E(r_{ij}) of two atoms i and j separated by a distance r_{ij} is given in terms of the morse potential by

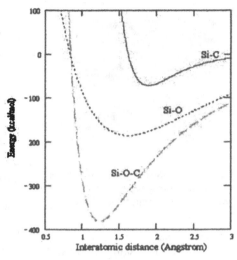

Fig. 5. Energy of formation of Si-C, Si-O and Si-O-C bonds as a function of interatomic distance.

$$E(r) = D[e^{-2\alpha(r_{ij}-r_0)} - 2e^{-\alpha(r_{ij}-r_0)}]$$

where α and D are constants with dimensions of reciprocal distance and energy, respectively, D is the dissociation energy, and r_0 is the equilibrium distance of approach of the two atoms. Suitable constant of the potential for Si-C bond (D= 71.4 Kcal/mol, α= 2.254Å^{-1}, and r_0= 1.89 Å), Si-O bond (D= 185 K cal/mol, α= 0.897 Å^{-1}, and r_0= 1.61Å) and C-O bond (D= 223K cal/mol, α= 2.48Å^{-1}, and r_0= 1.13Å) have been either taken from the literature or estimated. In the case of Si-O-C bond, the interaction energy of the system can be considered as a first approximation to be due to Si-O, C-O, and Si-C interactions. Morse calculations show preference to the formation of silicon-oxycabide in the presence of oxygen. The potential energy calculations however could not be extended satisfactorily to different silicon oxycarbides as some of the critical data for evaluation of the morse potential are not yet available. However, once the values of the constants are available, it will be a simple matter to extend the calculation including other silicon oxycarbides such as $SiOC_3$, SiO_2C_2, and SiO_3C.

Laser direct write experiments essentially consist of three parts: the laser source, the optics, and the controlling electronics. The whole system is assembled on a vibrating isolation table to ensure that the performance of the system is not degenerated by the mechanical disturbances during the writing process. The laser synthesized conductive tracks and pads, as shown in Fig. 6, were processed using a modified Nd:YAG laser system with an emission wavelength of 1064nm and pulse repetition rate of

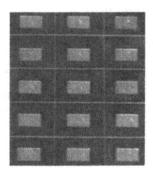

Fig. 6(a)

131

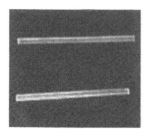

Fig. 6(b)

Fig 6. Laser direct write (a) 0.4 cm x 0.7 cm conducting pads (b) conducting 2mm wide line pattern on insulating AlN substrate.

1000Hz. The pulse duration time was 70 ns and TEM_{oo} mode was used. The pulse energy necessary to drive the conversion reaction is 3.15mJ. Pattern formation was achieved by a computer program-controlled galvo-mirror. The array (5x3) of conducting pads as shown in Fig. 6(a), each 0.4 cm x 0.7 cm were formed by beam rastering with an overlap of 30% of the 0.025cm beam diameter. The applicability of the process to other wide-bandgap semiconductors such as AlN is shown in Figure 6(b). Controlled experiments were undertaken to study the effect of annealing (or thermal stability) of the laser-irradiated α-SiC substrates. The annealing experiments were done under argon atmosphere to prevent oxidation. The changes in the resistance versus temperature are shown in Fig. 7. Low temperature (500°C) annealing of the laser-irradiated tracks leads to a small improvement in the conductivity and may be attributed to defect annealing. The decrease in the conductivity of the laser irradiated tracks upon annealing at an intermediate temperature (700°C) may also be attributed to the defect annealing plus impurity diffusion into the lattice and beyond 850°C, more impurity diffusion expected leading to alloy formation. These results additionally offer fine tuning of the electrical resistance of the laser-irradiated layers, thereby offering flexibilities in SiC integrated device design and manufacturing.

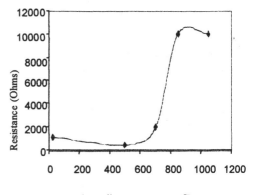

Fig. 7. Resistance vs. annealing temperature of laser Irradiated α-SiC substrate.

CONCLUSIONS

Lasers were used to directly write conducting and insulating tracks in silicon carbide. Conductor resistivity as low as 10^{-4} ohm-cm were produced from insulating substrate with an initial resistivity of 10^{11} ohm-cm. Silicon carbide substrate laser processed under pure oxygen atmosphere exhibits a large surface oxygen concentration. Resistivity measurements indicate these areas to be an insulating oxide layer. The technique of laser direct writing conductive and insulating paths on silicon carbide and other wide-bandgap semiconductors may lead to basic electronic material combinations necessary to fabricate a variety of sensors and electronic devices.

REFERENCES

[1] W.J. Choyke and G. Pensl, " Physical Properties of SiC, " MRS Bulletin, 22, 1997, pp. 25-29.

[2] E.R. Brown, " Megawatt Solid-State Electronic," 42, (Elsevier Science Ltd., Great Britain, 1998), pp. 2119-2130

[3] K. Moore, R.J. Trew, " Radio-Frequency Power Transistors Based on 6H- and 4H-SiC, " MRS Bullentin,.22, 1997, pp. 50-56.

[4] R.P. Jindal, IEEE Transactions on Electronic Devices, 46, Special Issue on Silicon Carbide Electronic Devices, 1999, pp. 441-620.

[5] V.E. Chelnokov, and A.L. Syrkin, " High temperature electronics using SiC: actual situation and unsolved problems," Materials Science and Engineering B46 (1997), pp. 248-253.

[6] J.W. Palmer et al, " High Temperature Rectifiers and MOS Devices in 6H-Silicon Carbide," US Army Research Office Contract No. DAAL03-91-0046 (April 27, 1992).

[7] D.G. Hamblen et al, " Epitaxial Growth of High Quality SiC Pulsed Laser Deposition," DOD-AF Contract No. F33615-94-C-5417, Report No. 526033 (Feburary 3, 1995).

[8] R.F. Davis and K. Das,"Silicon Carbide Semiconductor Device Fabrication and Characterization," NASA PR No. 335820, Grant No. NAG3-7825-1 (January 30, 1991).

[9] P.T.B. Shaffer, Handbook of Advanced Ceramic Materials (Advanced Refractory Technologies, Inc., Bufalo, NY 1991).

[10] W.J. Lackey et al., " Ceramic Coatings for Advanced Heat Engines-A Review and Projection," Adv. Ceram. Mat, 2, 1987, PP. 24-30.

[11] L. Matus and A. Powell, " Crystal Growth of SiC for,Electronic Applications," Ceramic Transactions, 2. Eds. J. Cawley and C. Semler 447 (American Ceramic Society, westerville, OH 1988).

[12] N.R. Quick, US Patent No. 5, 145, 441.(September 1992); US Patent No. 5,837,604(November 1998); US Patent No. 6,025,609 (February 2000).

[13] N.R. Quick, " Laser Synthesis of Conductive Phases in Silicon Carbide and Aluminium Nitride", Proceedings of the International Symposium on Novel Techniques in Synthesis and Processing of Advanced Materials. eds. J. Singh and S.M. Copley, 419 (American

[14] N.R. Quick, Proceedings of the International Conference on Lasers'94, (STS Press, McLean, Virginia, 1995), pp. 696.

[15] D.K. Sengupta, A. Kar, and N.R. Quick, " Laser generated conducting and insulating tracks in silicon carbide," Presented at the ICALEO'99 Meeting inSan Diego.

[16] K.F. Cai, J.P. Liu, C.W. Nan, and X.M. Min, " Effect of porosity on the thermal electric properties of Al-doped SiC ceramics," J. Mat. Sci. lett, 16, 1997, pp. 1876-1878.

[17] D.K. Sengupta, N.R. Quick, and A. Kar, " Laser conversion of electrical properties for silicon carbide device applications," Submitted to Journal of Laser Applications.

[18] M. Dayan, J. Vac. Sci. Technol. A. 3, 361 (1985).

[19] C.Onneby et al, " Silicon oxycarbide formation on SiC surfaces and at the SiC/SiO$_2$ interface ," J. Vac. Sci. Technol. A, Vol. 15, No. 3, May/Jun 1997.

LASER PROCESSING OF PARMOD™ FUNCTIONAL ELECTRONIC MATERIALS

Paul H. Kydd and David L. Richard*, Douglas B. Chrisey**, Kenneth H. Church***

*Parelec, Inc., 5 Crescent Ave. P.O. Box 236, Rocky Hill, NJ 08553

**Naval Research Laboratory, 4555 Overlook Ave. SW, Washington, DC 20375

*** CMS Technetronics, 5202-2 N. Richmond Hill Rd., Stillwater, OK 74075

ABSTRACT

Parmod™ is a family of materials that can be printed and thermally cured to create metallic conductors on printed wiring boards. This additive process provides a way to produce circuitry direct from CAD files without the necessity for intermediate tooling of any kind. The printed images are converted to pure metallic traces in seconds at a temperature low enough to be compatible with commonly used rigid and flexible polymer-based substrates. This simple, two-step process eliminates the hazardous wastes and employee health and safety issues associated with conventional plate-and-etch photolithographic technology.

Recently the Parmod™ technology has been extended from metals to oxides to enable printing passive electronic components such as resistors, capacitors and inductors, as well as the metallic interconnects. While thermal curing of the oxides provides useable electronic properties, particularly of resistors and capacitors, the performance of all these novel materials could be improved by laser processing. This paper discusses preliminary results on laser processing of Parmod™ conductors and components in two different systems.

INTRODUCTION

Parmod™ compositions are mixtures of fine and ultra fine powders with a Reactive Organic Medium (ROM). The mixtures can be printed by any convenient process and cured to a continuous phase at a temperature low enough to process on conventional polymer substrates. In the past year a substantial fraction of our materials R&D has been devoted to the extension of the Parmod™ chemistry from metals to oxides. The objective of this DARPA-funded [1] program was to provide materials that could be used to print passive components such as resistors, capacitors and inductors, as well as the conductors to interconnect them.

A major objective of the program was to use the Parmod™ approach to achieve continuous-phase deposits of oxides to eliminate the discontinuities at particle boundaries that degrade electrical performance. The production of a continuous phase from the discrete particles in the original ink during cure is the source of the exceptional conductivity of Parmod™ conductors relative to polymer thick film materials of equivalent curing temperature. It was hoped that the same advantages could be realized for oxide-based ferrites and ferroelectrics where the same electrical performance degradation at interfaces is important.

This objective was achieved for oxide resistor materials and for thick film, high permittivity dielectrics. The results for thick film ferrites were less satisfactory, although promising thin film materials were demonstrated. In all cases, the materials would have benefited from additional thermal processing at higher temperature to refine the crystal structure and increase mechanical strength. This must be accomplished without destroying the polymer substrate on which it is desired to deposit the thick film. The program objective is to apply these novel materials to conventional rigid and flexible electronic dielectric materials.

Mat. Res. Soc. Symp. Proc. Vol. 624 © 2000 Materials Research Society

The approach taken to combine the advantages of high temperature processing without damage to the underlying substrate is laser post processing. In this way the deposit itself can be exposed to a very high-energy fluence for a very short time, resulting in a spike in the temperature of the film without a corresponding excursion in the substrate. Two systems were used for this. The first was the MAPLE DW direct write system pioneered by the Naval Research Laboratory[2] and their industrial collaborators Potomac Photonics Inc.[3] The other was a combination of the Micropen from Ohmcraft, Inc[4] with laser post processing by CMS Technetronics.

EXPERIMENT

Materials

The materials used in these experiments were made by blending powdered ingredients with Metallo-Organic Decomposition (MOD) compounds which were purchased where possible and synthesized when not available. Preliminary performance measurements were made by screen printing the materials in test patterns and thermally curing them in air. Their electrical properties were then measured by standard methods.

Oxide Resistors

Preliminary experiments with ruthenium oxide, the standard material for semiconducting oxide resistors, were abandoned when it was found that the oxide could react spontaneously with the MOD constituents sending up a plume of sparks even in a nitrogen-inerted dry box. Better success was obtained with doped Indium Tin Oxide (ITO) compositions.

The electrical properties of the ITO-based inks can be controlled through variations in the ink composition. The resistivity can currently be varied over 3 orders of magnitude between <100 and 10,000 Ω/sq/mil. At 10,000 Ω/sq/mil we can print resistors which cover 75% of those normally used in electronic circuitry in a square format and 98+% of the requirement in a 1x10 format[5].

Higher values are possible, but have not been tested. Our objective is a resistivity range up to 1,000,000 ohms per square per mil. To date we have covered more than half of the desired range with a controllable resistive material.

Currently a temperature of >350°C is necessary for complete curing of the resistive material. It would be desirable to reduce this temperature. These films consolidate well, and have good adhesion to glass. They should adhere well to other substrates as well, but due to the high temperature currently required, other substrates have not been tested thus far.

High Permittivity Dielectrics

A substantial effort has been made to create printable high-K dielectric materials. We have found that mixed ferroelectric compositions function better than pure barium titanate. Lead titanate additions increased the dielectric constant of printed and cured barium titanate from 60 to more than 100. Because lead titanate crystallizes at about 470°C versus 750°C for barium titanate, it is easier to get it to provide a matrix with a high permittiviy. The majority of the conversion of the barium titanate is completed by 400°C, but weight loss is seen up to 600°C. The lead titanate converts completely at about 300°C. Since the dielectric properties are dependent on the crystallinity (long range order) of the material, the lower crystallization temperature of the lead titanate material means that at 350-400°C, the lead titanate material formed should have "longer range order" than the deposited barium titanate material, and thus have better properties. The only drawback is that lead titanate has a maximum dielectric constant of only about 300 compared to values in the thousands for other ferroelectrics.

The difference between pure barium titanate and lead titanate/barium titanate mixtures seen in Scanning Electron Microscope (SEM) images is in the connectivity of the barium titanate powder particles. In the barium titanate only deposits, the particles appear sharp and distinct from other particles, indicating few interconnections. The lead titanate/barium titanate deposits show a "merging" of the particles so that each particle is not completely distinct from the particle next to it. The lead titanate appears to be sticking the barium titanate particles together better than the barium titanate itself. The lead titanate is also sticking the barium titanate particles together better than polymer binders. The best polymer-based high K dielectric formulations of which we are aware have a dielectric constant of approximately 60. The preliminary MOD-based compositions we have made so far provide an advance of almost a factor of two in the permittivity available from printable dielectric materials with ample room for further improvement as this new technology matures.

Ferrites

Spinel ferrites typified by Fe_3O_4 were investigated as printable ferrite compositions because the objective was soft magnetic materials for chokes and antennas. Fe_3O_4 itself proved to be the most advantageous compound compared to mixed ferrites such as nickel-zinc-iron and manganese-zinc-iron compositions.

We have been successful in producing thin, printable ferrite deposits that were electrically insulating, mechanically strong and magnetic. These films are less than two microns thick, very well bonded to stainless steel substrates and very resistant to abrasion. The resolution of the printed images was relatively poor. No effort was made to improve it because the objective was thick films.

We were less successful in producing thicker deposits to provide substantial inductance values. The samples showed little improvement in permeability over magnetite powder bonded together with an organic binder, unlike the dielectric materials described above. The deposits were mechanically weak when cured under conditions that promoted magnetite formation, and electrically conductive when cured under conditions that resulted in metallic iron and greater strength. The magnetic moment was measured by NRL to be approximately half that of bulk magnetite, which is consistent with SEM pictures. The latter do show growth of columnar grains, which suggest that some consolidation is taking place, but more work is required to reduce the porosity and increase the bonding between magnetite particles.

RESULTS

Printed Capacitors

Capacitors were made using lead titanate/ barium titanate ink as the dielectric and screen printed silver as the top and bottom electrodes on a glass substrate. The capacitors had capacitance values in the range of 1-10 nF with a 25 micron thick layer of dielectric. The dielectric constant was in excess of 100. The dissipation factor was approximately 0.1. The dispersion in dielectric constant indicated that cure temperatures in excess of 350°C are needed. Further optimization is expected to lower this temperature.

Breakdown voltages were greater than 24 Volts for films of approximately 25 microns thickness, but in 10-micron films, the breakdown voltage was about 5 V. The main problem that remains is to improve the quality of the thin films so that a higher capacitance can be obtained with an acceptable breakdown voltage. A cross section of the capacitor structure is shown in

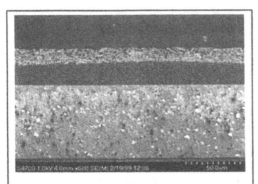

Figure 1. This Scanning Electron Micrograph (SEM)[6] shows the graininess of the screen-printed high permittivity BTO/PTO dielectric (as well as that of the Parmod™ silver electrode printed on top of it). Both could benefit from laser processing to eliminate voids and refine crystal structure.

Figure 1. Printed capacitor. From the bottom-aluminum substrate, high permittivity dielectric, printed silver top electrode, Metallographic matrix.

Preliminary Laser Processing Tests by MAPLE-DW

A total of four quartz discs were sent to NRL for laser transfer tests. Two of them were coated with silver, and two of them were coated with dilute BTO MOD compounds. These plates were used for laser transfer and annealing tests. For the experiments conducted using the silver coated quartz discs, various lines were made which were insulating after the laser transfer step, but became conductive after either a laser or a furnace annealing step. For the BTO-coated quartz discs, NRL performed transfers to silicon substrates and to magnesium oxide substrates with patterned gold interdigitated capacitors. The BTO transfers on silicon survived the tape test for adhesion, and their morphology was good as evaluated by SEM. The BTO transfer patterns over the gold interdigitated capacitors on magnesium oxide substrates showed dielectric constants around 30.

Recent Results on Silver Conductors

In later work at NRL, one cm long silver conductors were laid down on glass in successive passes at approximately 200 micron width with the following results:

Line#	Passes#	Fluence (J/cm^2)	\multicolumn{5}{c}{Avg. Thickness µm}	RLC (1 MHz)	4-point at DC
			t1	t2	t3	t4	t5		
2	5	0.75	7.4	7.9	7.3	7.2	7.0	0.61 Ω ⇒ 5.6 ρ_{Ag}	0.32 Ω ⇒ 2.9 ρ_{Ag}
3	6	1.03	8.4	7.5	8.3	7.2	7.7	0.54 Ω ⇒ 5.4 ρ_{Ag}	0.25 Ω ⇒ 2.5 ρ_{Ag}
4	4	1.03	6.5	6.6	7.3	7.4	7.0	0.70 Ω ⇒ 5.6 ρ_{Ag}	0.40 Ω ⇒ 3.2 ρ_{Ag}
5	5	0.80	6.6	6.4	6.8	5.7	7.3	0.69 Ω ⇒ 5.4 ρ_{Ag}	0.39 Ω ⇒ 3.0 ρ_{Ag}
6	4	0.80	6.6	6.5	6.3	6.4	6.1	0.89 Ω ⇒ 5.9 ρ_{Ag}	0.57 Ω ⇒ 3.8 ρ_{Ag}

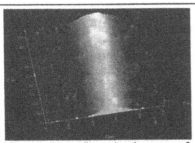

Figure 2. Three-dimensional contour of laser transferred silver conductor.

The resistivity of the conductors is approximately 5.5 times that of bulk silver at 1 MHz and three times that of bulk silver DC. The latter result is consistent with experience with printed and thermally cured Parmod™ silver. The samples were post annealed at 290°C for 15 minutes compared to the usual thermal cure at 260°C for one minute or less. The contour of the laser transferred line is shown in Figure 2. For comparison the contour of a screen-printed and thermally cured conductor is shown in Figure 3. Its resistivity was 3.0 times bulk silver at 1 MHz

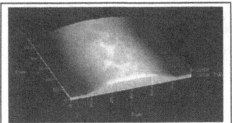

Figure 3. Three-dimensional contour of screen printed and thermally cured silver conductor

and 2.2 times bulk silver at DC.

Resistors by Laser Transfer

The same technique was applied to the ITO resistive material described above. The ITO composition was applied as a thin layer to a quartz "ribbon" and transferred across a 20-micron gap to a glass substrate. One hundred-micron wide lines 500 microns long were deposited and post annealed at 400°C. The results are shown below.

Line#	Pass#	Fluence (J/cm^2)	Thickness(μm)				RLC Meter (1 MHz)	4-point at DC
			t1	t2	t3	t4		
G1	10	1.4	21.3	13.7	20.3	13.7	320 kΩ ⇒ ρ = 1.13 Ω-m	360 kΩ ⇒ ρ = 1.27 Ω-m
G3	13	1.4	14.3	12.2	13.7	15.2	61 kΩ ⇒ ρ = 0.20 Ω-m	63 kΩ ⇒ ρ = 0.21 Ω-m

The ability to control the resistance by the number of passes of a given resistive material represents a new degree of freedom in resistor design compared to conventional printing technology in which varying thickness is possible but difficult. Clearly resistor trimming can be done simultaneously, rather than as a separate operation in which the area and aspect ratio of the resistive material are controlled.

Laser Processing of Printed Silver Conductors

Similar results have been achieved at CMS Technetronics. Figure 4 shows laser densified lines in a thin deposit of uncured silver Parmod™. Figure 5 shows a close up of the densified region showing a rather ragged structure.

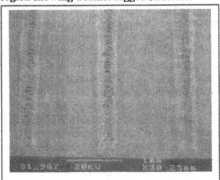

Figure 4. Laser processed lines in Parmod™ silver.

Figure 5. Close up of lines showing uneven structure.

At still higher magnification, shown in Figure 6, the individual silver flakes appear to have been melted and fused together. Figure 7 shows the appearance of unprocessed silver flakes for comparison. It should be noted that thermally cured Parmod™ itself has an appearance similar to Figure 6 [7].

Figure 6 (above). Laser processed silver flakes at the surface of the deposit.

Figure 7 (right). Unprocessed silver flakes.

When the Parmod™ is preheated to 200°C for two minutes prior to laser processing, the splattering and roughness shown in Figure 5 is much reduced and the result is a smooth deposit of good conductivity as shown in Figure 8. This is likely to be due to the decomposition and removal of most of the organic material prior to laser processing.

Figure 8. Laser processed Parmod™ silver trace preheated at 200°C for 2 minutes.

CONCLUSIONS

This work has constituted a rapid reconnaissance of the entire field to determine the technical feasibility of low temperature, printable passive components. We have been successful in developing a range of doped ITO compositions that produce good quality films with resistivities of 10-10,000 ohms/square/mil, which will satisfy 98% of the requirements for embedded passives.

Laser transfer of these materials has been demonstrated and offers the opportunity for creating resistors with a range of resistance values by depositing a variable number of layers of a single resistive material. Laser trimming to achieve a desired value can obviously be carried out simultaneously.

We have achieved our goal of a printed ferroelectric material that converts to a film with a dielectric constant of at least 100 under conditions that are compatible with polyimide substrates. This is twice the dielectric constant that is available from resin-bonded materials, confirming that the Parmod™ approach can provide materials that are not available by conventional means. We have still to demonstrate reliable 10-micron thick films to achieve higher capacitance. Materials have been provided to our colleagues to determine whether laser transfer and densification can produce such films.

We have evidence that ferrites can be consolidated to somewhat cohesive thick films by heat treatment that is compatible with polyimide substrates. Improved magnetic properties and adequate mechanical properties will require substantially better bonding of the oxides. An encouraging observation was the production of tightly adherent, nonconductive and very thin magnetic oxide films on stainless steel substrates. Such films may be useful in themselves and give encouragement that a Parmod™ approach to creating a useable thick film may yet be accomplished. Again, candidate materials have been provided for laser patterning and densification in an attempt to improve the structure of these magnetic materials without destroying the underlying substrate.

ACKNOWLEDGEMENTS

This work was Part of the MICE (Mesoscale Integrated Conformal Electronics) program sponsored by the Defense Advanced Research Projects Agency. The contract number was DAAH01-98-C-R012, administered by the U.S. Army Aviation & Missile Command, Redstone Arsenal, AL. SEM s and electrical property measurements were performed by Tim Schaefer of the Mayo Foundation and by Parelec at the Rutgers Department of Ceramics and Materials Engineering. Additional SEM s and measurements as well as the laser processing of these materials have been performed by our colleagues at NRL and CMS Technetronics.

REFERENCES

[1] Phase I SBIR contract *Materials for Printing Electronic Components*, Contract No. DAAH01-98-C-R012 administered by the US Army Aviation and Missile command, Redstone Arsenal, AL.

[2] A. Pique, D.B. Chrisey, et al *Laser Direct Writing of Circuit Elements and Sensors,* Proceedings, SPIE LASE'99, January 1999, San Jose, CA.

[3] Potomac Photonics, Inc. 4445 Nicole Dr. Lanham, MD 20706

[4] Ohmcraft, Inc., 93 Paper Mill St., Honeoye Falls, NY 14472

[5] "Evaluating the Need for Integrated Passive Substrates", Kapadia, H.; Cole, H.; Saia, R.; Durocher, K.; Advancing Microelectronics, Jan/Feb. 1999 p 12-15.

[7] "Parmod™ Metallization of Circuit Traces and Microvias" Paul H. Kydd, G. A. Jablonski, D. L. Richard, H. Gleskova and A. Singer Circuitree, August 1998 p. 104, Figure 2 a,b.

Matrix Assisted Pulsed Laser Evaporation Direct Write (MAPLE DW): A New Method to Rapidly Prototype Active and Passive Electronic Circuit Elements

J.M. Fitz-Gerald, D.B. Chrisey, A. Piqu , R.C.Y. Auyeung, R. Mohdi, H.D. Young, H.D. Wu, S. Lakeou, and, R. Chung

Naval Research Laboratory, Washington, D.C.

Abstract

We demonstrate a novel laser-based approach to perform rapid prototyping of active and passive circuit elements called MAPLE DW. This technique is similar in its implementation to laser induced forward transfer (LIFT), but different in terms of the fundamental transfer mechanism and materials used. In MAPLE DW, a focused pulsed laser beam interacts with a composite material on a laser transparent support transferring the composite material to the acceptor substrate. This process enables the formation of adherent and uniform coatings at room temperature and atmospheric pressure with minimal post-deposition modification required, i.e., $\leq 400°C$ thermal processing. The firing of the laser and the work piece (substrate) motion is computer automated and synchronized using software designs from an electromagnetic modeling program validating that this technique is fully CAD/CAM compatible. The final properties of the deposited materials depend on the deposition conditions and the materials used, but when optimized, the properties are competitive with other thick film techniques such as screen-printing. Specific electrical results for conductors are < 5X the resistivity of bulk Ag, for $BaTiO_3/TiO_2$ composite capacitors the k can be tuned between 4 and 100 and losses are < 1-4%, and for polymer thick film resistors the compositions cover 4 orders of magnitude in sheet resistivity. The surface profiles and fracture cross-section micrographs of the materials and devices deposited show that they are very uniform, densely packed and have minimum resolutions of ~10 μm. A discussion of how these results were obtained, the materials used, and methods to improve them will be given.

Introduction

There is a strong need in industry for new design and Just In Time Manufacturing (JITM) methods, materials, and tools to direct write for rapid prototyping passive circuit elements on various substrates, especially in the mesoscopic regime, i.e., electronic devices that straddle the size range between conventional microelectronics (sub-micron-range) and traditional surface mount components (10 mm-range). The need is based on the desire: to rapidly fabricate prototype circuits without iterations in photolithographic mask design, in part, in an effort to iterate the performance on circuits too difficult to accurately model, to reduce the size of PCB s and other structures (~30-50% or more) by conformally incorporating passive circuit elements into the structure, and to fabricate parts of electronic circuits by methods which occupy a smaller footprint, which are CAD/CAM compatible, and which can be operated by unskilled personnel or totally controlled from the designers computer to the working prototype. Mesoscopic direct write approaches are not intended to compete with current photolithographic circuit design and fabrication. Instead, these technologies will enable new capabilities satisfying next generation applications in the mesoscopic regime.

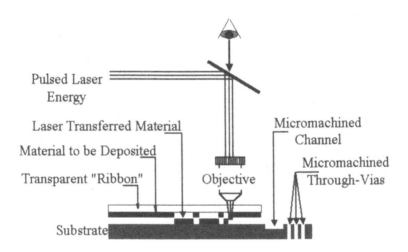

Figure 1. Schematic diagram of the MAPLE DW process. The process lends itself to both additive and subtractive processing.

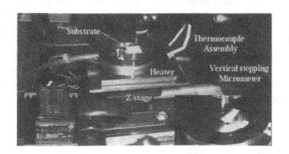

Figure 2. MAPLE DW system incorporating computer controlled X_Y_Z stages for ribbon and substrate manipulation with in-situ heating.

Many different CAD/CAM approaches exist to direct write or transfer material patterns and each technique has its own merits and shortcomings. The different approaches include plasma spray, laser particle guidance, MAPLE DW, laser CVD, micropen, ink jet, e-beam, focused ion beam, and several novel liquid or droplet microdispensing approaches. One common theme to all techniques is their dependence on high quality starting materials, typically with specially tailored chemistries and/or rheological properties (viscosities, densities and surface tension). Typical starting materials, sometimes termed pastes or inks , can include combinations of powders, nanopowders, flakes, surface coatings and properties, organic precursors, binders, vehicles, solvents, dispersants, surfactants, etc. This wide variety of materials with applications as conductors, resistors, and dielectrics are being developed especially for low temperature deposition (< 400°C).°° This will allow fabrication of passive electronic components and RF devices with the performance of conventional thick film materials, but on low temperature flexible substrates, e.g., plastics, paper, fabrics, etc.° Examples include silver, gold, palladium, and copper conductors, polymer thick film and ruthenium oxide-based resistors and metal titanate-based dielectrics. Fabricating high quality crystalline materials at these temperatures is nearly impossible. One strategy is to form a high density packed powder combined with chemical precursors that form low melting point nanoparticles *in situ* to chemically weld the powder together. The chemistries used are wide ranging, but include various thermal, photochemical and vapor, liquid, and/or gas co-reactants. The chemistries are careful to avoid carbon and hydroxide incorporation that will cause high losses at microwave frequencies or chemistries that are incompatible with other fabrication line processing steps. To further improve the electronic properties for low temperature processing, especially of the oxide ceramics, laser surface sintering is often used to enhance particle-particle bonding. In most cases, individual direct write techniques make trade-offs between particle bonding chemistries that are amenable with the transfer process and direct write properties such as resolution or speed. The resolution of direct write lines can be on the micron scale, speeds can be greater than 100 mm/sec, and the electronic material properties are comparable to conventional screen-printed materials. Optimized materials for direct write technologies result in: deposition of finer features, minimal process variation, lower prototyping and production cost, higher manufacturing yields, decreased prototyping and production time, greater manufacturing flexibility, and reduced capital investments.

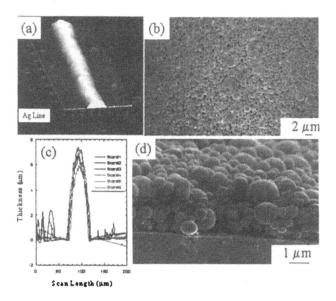

Figure 3 Ag line written on kapton by MAPLE DW, (a) shows a Tencor 3-D partial image of the 1 cm long line, (b) SEM micrograph showing the high packing density that the transferred materials exhibits, (c) Tencor 2-D line profile scans showing a 40 μm line width, (d) SEM micrograph of the Ag line further illustrating the packing density of the transferred material.

Background

Shortly after the discovery of lasers researchers representing all disciplines of science began aiming them at materials in different forms. The interaction of lasers with materials can result in a wide range of effects that depend on the properties of the laser, the material, and the ambient environment. As a function of beam energy, these effects can start from simple photothermal heating to photolytic chemical reactions and to ablation and plasma formation. We have successfully extended conventional PLD to include organic materials through a process we have termed Matrix Assisted Pulsed Laser Evaporation or MAPLE [1-4]. In this process, the excimer laser is set to a lower fluence (~0.2 J/cm²) from conventional PLD and impacts a dilute matrix target that is typically frozen to low temperatures (~77 K). The dilute matrix is made up of the organic molecules to be deposited in thin film form and a frozen solvent. Ideally, the laser is then preferentially tuned to interact with the solvent matrix, but independent of that, the laser warms a local region of the target. The laser-produced temperature rise is large compared to the melting point of the solvent, but small compared to the decomposition temperature of the organic solute. When the MAPLE process is optimized, the collective collisions of the evaporating solvent with the organic molecule act to gently desorb the organic molecule intact, i.e., with only minimal decomposition as determined by FTIR and mass spectrometry. The evaporating solvent has a near zero sticking coefficient with the substrate and is rapidly pumped away or it can be trapped for re-use.

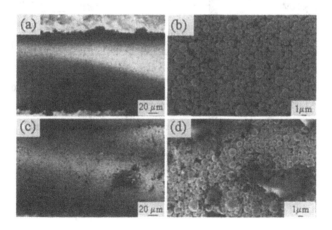

Figure 4. SEM micrographs of 1 cm Ag lines before and after scotch tape adhesion testing, written by MAPLE DW. Images a) and b) represent the Ag line prior to testing and c) and d) illustrate the lines after scotch tape removal, noting the excellent adhesion and residual polymer adhering to the surface of the Ag line.

LIFT is a simple direct write technique that employs laser radiation to vaporize and transfer a thin film (target) from an optically transparent support onto a substrate placed next to it [5-11]. Patterning is achieved by moving the laser beam (or substrate) or by pattern projection. The former is a method of direct writing patterns. There are several experimental requirements for LIFT to produce useful patterns including: the laser fluence should just exceed the threshold fluence for removing the thin film from the transparent support, the target thin film should not be too thick, i.e., less than a few 1000 , the target film should be in close contact to the substrate, and the absorption of the target film should be high. Operating outside these regime results in problems with morphology, spatial resolution, and adherence of the transferred patterns. Repetitive transfer of material can control the film thickness deposited on the substrate. LIFT is a simple technique that is used on mostly metallic targets, because the laser energy absorbed in the coated substrate atomizes the layer making LIFT inherently a pyrolytic technique. It cannot be used to deposit complex crystalline, multicomponent materials whose crystallization temperature is well above room temperature.

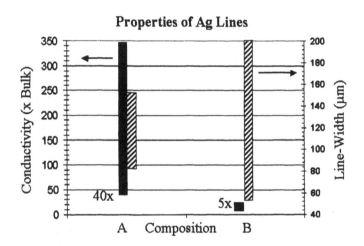

Figure 5. Schematic diagram showing the current progress writing metal lines with MAPLE DW at NRL.

The MAPLE DW technique utilizes the technical approach of LIFT and the basic mechanism of MAPLE to produce a laser driven direct write process capable of transferring materials, such as metals, ceramics, and polymers onto polymeric, metallic and ceramic substrates at room temperature and at atmospheric pressure and with a resolution on the order of 10 m [12-14]. MAPLE DW uses a highly focused laser beam that can be easily utilized for micromachining, surface annealing, drilling and trimming applications, by simply removing the ribbon from the laser path. The flexible nature of MAPLE DW allows the fabrication of multi-layered structures in combination with patterning. Thus, MAPLE DW is both an *additive* as well as *subtractive* process. MAPLE DW can also be adapted to operate with two lasers of different wavelengths, whereby the wavelength from one laser has been optimized for the transfer and micromachining operations, i.e., the UV laser, while the second laser is used for modifying the surface as well as annealing of either the substrate or any of the already deposited layers (i.e., IR or visible).

In MAPLE DW, a laser transparent substrate such as a quartz disc is coated on one side with a film a few microns thick. The film consists predominantly of a mixture or matrix of a powder of the material to be transferred and a photosensitive polymer or organic binder. The polymer assists in keeping the powders uniformly distributed and well adhered to the quartz disc. The coated disc is called the ribbon and is placed in close proximity (5 to 100 m) and parallel to the acceptor substrate. As with LIFT, the laser is focused through the transparent substrate onto the matrix coating, see Figure 1. When a laser pulse strikes the coating, it transfers the powders, nanoparticles, and precursors, to the acceptor substrate. Using MAPLE DW, the material to be transferred is not vaporized allowing complex compounds to be transferred without modifying their composition, phase, and functionality. Furthermore, there is no heating of the substrate on which the material is transferred. Both the acceptor substrate and the ribbon are mounted onto stages that can be moved by computer-controlled stepper motors. By appropriate control of the positions of both the ribbon and the substrate, complex patterns can be fabricated. By changing the type of ribbon, multicomponent structures can easily be produced.

Experimental

Fused silica quartz disks, 5.0 cm diameter x 1.5 mm thick were used as ribbon supports. Precursor Ag metal and $BaTiO_3$ (barium titanium oxide (BTO)) materials were applied to the quartz disk using conventional deposition techniques with a resultant thickness ranging from 1.5-10 μm. Ribbons are difficult to fabricate and the precursors must transfer without significant decomposition. On the other hand, by using ribbons we can effectively quantize the material transferred making MAPLE DW coatings highly reproducible. Each laser pulse deposits an identical mesoscopic brick of electronic material. In addition, the laser fires 100 thousand times a second synchronously with the computer-controlled stages thereby depositing the bricks very fast.

The size distribution of the Ag spherical particles ranged from 400 nm to 2 μm. The BTO particles ranged from 125 to 175 nm with a TiO_2 precursor. For all the transfers described in this paper, a computer-controlled stage (Z-stage) was used to control the relative travel and positioning of the substrate relative to the ribbon, as shown in Figure 2. The substrate to ribbon gap was set at 50 and 75 μm for the Ag and BTO transfers respectively. Both the substrate and ribbon were held in place using a vacuum chuck over the X-Y substrate translation stage. The third harmonic of a pulsed Nd:YAG laser (355nm, 15 ns pulse width) was used for all transfer experiments. By changing the aperture size, beam spots ranging from 20 to 80 μm were generated. The laser fluence was estimated by averaging the total energy of the incident beam over the irradiated area to be $1.6 - 2.2$ J/cm^2. Various substrates were used for the transfer experiments including silicon, glass, alumina, and polyimide. Transfer tests were performed using Ag and BTO ribbons over each of these substrates. The adhesion of the transferred was excellent on all substrate materials after proper deposition parameter optimization as determined by standard tape tests.

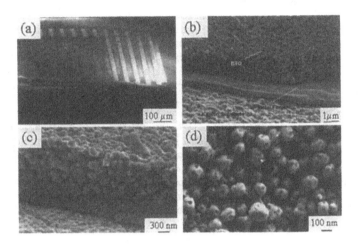

Figure 6. SEM micrographs of MAPLE DW of BTO on an interdigitated capacitor structure. Cross-section micrographs (b), (c), and (d) reveal that the material is uniform, dense and exhibits low matrix porosity.

Results and Discussion

Silver lines were fabricated as shown in Figure 1 using Ag ribbons 3 μm thick, with a size distribution ranging from 400 nm to 2 μm in diameter. Figure 3 illustrates a 40 μm wide line that is ~6 μm thick written with two passes with MAPLE DW on a kapton substrate. Figure 3(a) shows that the line morphology is uniform and the aspect ratio is within 20% as shown in (c). Figure 3(b) and (d) show that the MAPLE DW process can clearly produce lines with high density and uniform thickness. In addition to density and conductivity, the adhesion of the written devices is important as well. Figure 4 shows the results of a scotch tape testing experiment performed on Ag lines written on a glass substrate. Figure 4 (a) and (b) show SEM micrographs prior to tape application, whereas (c) and (d) show SEM micrographs after the tape has been removed. The Ag lines survived the tape test. On closer inspection, it can be observed partial polymer glue from the tape is still left on the line from the pull off, showing that the metal line has a degree of self adhesion in addition to the general concern at the glass/metal line interface.

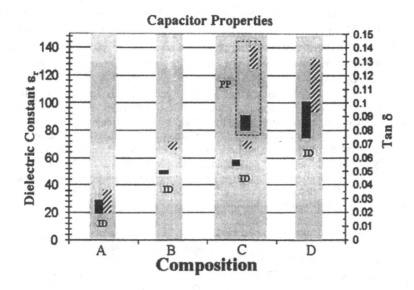

Figure 7. Properties of capacitors fabricated by MAPLE DW in both interdigitated (ID) and parallel plate (PP) geometries. A range of compositions are shown: A) commercial BTO powder 2-3 μm diameter (irregular shaped), no precursor, B) 150 nm BTO (spherical), 100 nm titania (spherical), no precursor, C) 150 nm BTO (spherical), no precursor, D) 150 nm BTO (spherical), titania precursor

Figure 5 shows the current progress at NRL in terms of conducting metal lines, without laser annealing. The two parameters listed, conductivity and line width are shown for qualification. Figure 5 compares two of the current Ag materials, compositions A, B. The differences in the materials range in amount of precursor, shape, and thickness. Currently, using MAPLE DW we have written adherent lines with 4X bulk conductivities that are uniform and reproducible on glass, alumina, kapton, and silicon substrates.

Dielectric materials were written in both line and device arrangements, such as the interdigitated capacitor structure shown in Figure 6. Interdigitated capacitor structures (1 mm x 1 mm) were fabricated by writing BTO precursor materials onto a typical interdigitated capacitor finger structure. The material composition was 70% 150 nm BTO powder (spherical), 30% titania precursor. The metal fingers were fabricated by conventional lithography techniques. Figure 6 (a) shows a low magnification SEM image of the interdigitated capacitor device structure deposited on an alumina substrate with Pd metal fingers. The density of the transferred material (post-transfer furnace anneal at 280°C) is clear from SEM micrographs in Figure 6 (b) and (c), including the Pd metal fingers. Figure 6 (d) shows a high magnification SEM micrograph of the fracture cross-sectioned region. The packing density is uniform with clear evidence of the TiO_2 precursor filling the voids between the particles. Device measurements made on these materials have shown that depending on the precursor choice, particle size, composition, and processing parameters, devices with a wide range of dielectric properties can be fabricated by MAPLE DW as shown in Figure 7.

It is clear that the material and precursor choice and amount significantly affect the final properties of the capacitor structures illustrated above. The importance of the as-deposited density is more generally critical to all electronic materials for device performance. The effect of density or microstructure on different passive components and the parallel effect on device performance are given in Table I. In all cases, the density degrades the electrical performance. For conductors and resistors this is directly proportional to the cross-sectional area, but for dielectrics the dependence is exponential and based on the relative packing of different k materials, i.e., air and BTO in our case. For ferrites this will also result in a higher coercive field, H_c.

Table I. Microstructural and devices issues for passive components.

Passive Component	Microstructure Issues	Device Issues
Metallic Conductors	Intermediate melting point, Necking, Porosity	Microwave Surface Resistance, Power Loss
Dielectrics (Ceramic Oxides)	High melting point, Difficult to Neck Particles, Oxygen Loss	Porous Material has Drastic Effect on k, Lossy
Resistors (Ceramic/Insulator)	High melting point, Most Difficult to Neck Around Insulator	Conductor/Insulator Composite
Resistors (Polymer/Insulator)	Low Processing Temperature, Necking Around Insulator	Aging, Electromigration
Ferrites	High melting point, Difficult to Neck Particles, Oxygen Loss	Porous Material has Drastic Effect on M_s, Lossy, High H_c

The temperatures that these samples and in particular the precursors are reacted is < 400°C. One way to effectively increase the processing temperature, but still maintain low temperature processing for plastic substrates, is to use *ex situ* laser surface sintering. Bulk diffusion is not going to occur for the short times that the laser sinters the sample. The low reacting temperature precursors provide chemical welding between particles (particle-to-particle adhesion). By using *in-situ* laser sintering, some of the following benefits may be realized: smaller/thinner region to sinter, less organic to remove, lower sensitivity to laser fluence, faster and better alignment, less thermal stress to the substrate, and lastly, less processing steps will result in a higher yield. We are modifying our current experimental set-up to include laser sintering. This will done in situ with an acoustic modulated CW Nd:YAG laser.

Conclusion

We have demonstrated that MAPLE DW is a rapid and versatile prototyping tool for fabricating mesoscale devices (conductors, resistors, capacitors, inductors, phosphors, sensors) in a 3-D structure on any surface. The material transfer works as predicted. It is clear that the process is probably more gentle and general than previously thought, but the MAPLE DW mechanism requires further research. There are many advantages to MAPLE-DW compared to other existing direct write technologies for 3D-structure fabrication. These advantages include the ability to do adherent depositions at room temperature and in air and the ability to rapidly change between different materials. The latter is accomplished because MAPLE-DW is a dry technique meaning there is no interval of time required for transferring one layer on top of another. We have shown that ceramic powder, nanoparticles, and precursor composites need to be densified through laser sintering. Therefore, we need low reacting temperature precursors to work with the proper laser annealing parameter both in the spatial and temporal regimes.

Acknowledgements
We gratefully acknowledge the support provided for this work from the Office of Naval Research and the DARPA MICE Program.

References

1. D.B. Chrisey and G.K. Hubler eds., Pulsed Laser Deposition of Thin Films, (New York, NY: Wily, Inc. 1994).
2. Method of producing a coating by matrix assisted pulsed laser evaporation, U.S. Pat. No. 6,025,036.
3. R.A. McGill, R. Chung, D.B. Chrisey, P.C. Dorsey, P. Matthews, A. Piqu , T.E. Mlsna, and J.L. Stepnowski, IEEE Trans. on Ultrasonics, Ferroelectrics, and Frequency Control, vol. 45, p. 1370 (1998).
4. A. Piqu , D.B. Chrisey, B.J. Spargo, M.A. Bucaro, R.W. Vachet, J.H. Callahan, R.A. McGill, and T.E. Mlsna, in Advances in Laser Ablation of Materials, MRS Proceedings, vol. 526, p. 421, (1998).
5. J. Bohandy, B.F. Kim, and F.J. Adrian, J. Appl. Phys. 60 (1986) pp. 1538-1539.
6. J. Bohandy, B.F. Kim, F.J. Adrian, and A.N. Jette, J. Appl. Phys. 63 (1988) pp. 1158-1162.
7. I. Zergioti, S. Mailis, N.A. Vainos, C. Fotakis, S. Chen, C.P. Grigoropoulos, Appl. Surf. Sci. 127-129 (1998) pp. 601-605.
8. F.J. Adrian, J. Bohandy, B.F. Kim, A.N. Jette, and P. Thompson, J. Vac. Sci. Tech. B5 (1987) pp. 1490-1494.
9. I. Zergioti, S. Mailis, N.A. Vainos, P. Papakonstantinou, C. Kalpouzos, C.P. Grigoropoulos, and C. Fotakis, Appl. Phys. A 66 (1998) pp. 579-582.

10. H. Esrom, J-Y. Zhang, U. Kogelschatz, and A. Pedraza, Appl. Surf. Sci. 86, pp. 202-207 (1995).

11. S. M. Pimenov, G.A. Shafeev, A.A. Smolin, V.I. Konov, and B.K. Bodolaga, Appl. Surf. Sci. 86, pp. 208-212 (1995).

12. A. Piqu , D.B. Chrisey, R.C.Y. Auyeung, J.M. Fitz-Gerald, H.D. Wu, R.A. McGill, S. Lakeou, P.K. Wu, V. Nguyen and M. Duignan, Appl. Phys. A 69, pp. S279-S-284 (1999).

13. D.B. Chrisey, A. Piqu , J.M. Fitz-Gerald, R.C.Y. Auyeung, R.A. McGill, H.D. Wu, and M. Duignan, Appl. Surf. Sci. 154, pp. 593-600 (2000).

14. J.M. Fitz-Gerald, A. Piqu , D.B. Chrisey, P.D. Rack, M. Zeleznik, R.C.Y. Auyeung, and S. Lakeou, Appl. Phys. Lett. 76, pp. 1386-1388 (2000).

Ion and Electron Beam
Direct Write Techniques

DIRECT FOCUSED ION BEAM WRITING OF PRINTHEADS FOR PATTERN TRANSFER UTILIZING MICROCONTACT PRINTING

David M. Longo and Robert Hull

Department of Materials Science & Engineering
University of Virginia
Charlottesville, VA 22904-4745
Longo@virginia.edu, Hull@virginia.edu

Abstract

We describe how focused ion beam (FIB) direct write technology may be combined with microcontact printing (μCP) and other pattern transfer techniques to enable nanoscale fabrication of complex patterns over both curved and planar surfaces. Nanoscale printheads are fabricated by direct sputtering or deposition (Pt or SiO_2) in the FIB. These printheads are capable of transferring 100 nm features over fields of view up to 1 mm^2, onto planar and curved surfaces. We are also investigating the concept of "programmable" printheads by fabrication of individually addressable printhead arrays, coupled with selective desorption of a relevant transfer medium (e.g. hexadecanethiol self-assembled monolayers) by heating or by application of an electric field.

Introduction

The microelectronics industry has employed optical projection lithography for several decades. The demand for ever-decreasing feature dimensions in integrated circuits will make the transition to shorter wavelength sources, i.e. extreme ultra-violet, X-Rays, electrons or ions, inevitable in the next decade. Despite formidable engineering challenges, however, these new lithographic techniques will be similar conceptually to existing optical techniques; that is they will most likely use some form of projection imaging of a mask to expose a radiation-sensitive resist. It is interesting to note, however, that outside of the microelectronics industry, the vast majority of "pattern transfer" in technology is achieved using direct contact of a master pattern with the target surface to transfer an "inking" medium. Such printing techniques, have of course, been in existence for over half a millennium, and have evolved into a sophisticated technology. Minimum feature sizes transferred by conventional inking processes, however, are in the tens of microns range. In this paper we describe part of a major project at the University of Virginia to develop an ultra-high resolution analogue of conventional printing technology, applicable to both planar and curved surfaces.

The focused ion beam (FIB) is now an indispensable tool in both nanoscale characterization and fabrication of electronic materials and devices.[1-5] A main advantage of using such an instrument is the short timescale in which nanoscale characterization and fabrication can be completed. In our research, the FIB is utilized to fabricate "printhead master patterns" by direct write technology (sputtering).[6-7] The FIB allows for rapid prototyping (< 1 day total turnaround) of complex patterns with spatial definition of order 100 nm and fields of view up to 1 mm^2. This "direct writing" is accomplished by using the ion beam (30 kV Ga^+) to sputter predefined patterns onto a single crystal Si substrate. These patterns then serve as a master template from which duplicate molds are cast, Figure 1. The process of duplicating and printing these patterns is called microcontact printing (μCP) and was developed by G. Whitesides' research group at Harvard University.[8]

µCP is a non-photolithographic technique for printing micron-scale and submicron-scale features and devices.[9-11] It utilizes self-assembled monolayers (SAMs) as an "ink," and polydimethylsiloxane (PDMS) as an elastomeric "stamp." Patterns are cast as molds into PDMS from "masters" (such as traditional photolithography masks, e-beam masks, and now FIB patterned printheads). The PDMS replicas of the original patterns are then coated with SAMs and dried. The PDMS is brought into conformal contact with an Ag-coated single crystal Si substrate, and the SAMs transfer from the PDMS to the Ag film. When the Ag film is etched, it will etch only in regions not patterned by the SAMs, which acts as an etch barrier. Thus, the patterns from the original master are reproduced on the target substrate. (Materials other than Ag have been "printed," such as Al, Au, Cu, Si, etc.).[12-17]

The main advantages of µCP as a soft-lithographic technique are low cost, ease of reproducibility, and application to both planar and curved surfaces. Conventional lithographic techniques are limited generally to planar surfaces because of small depth of focus; using µCP we have been able to print deep sub-micron features on both planar and curved substrates (radius of curvature 3-5 cm), Figure 2. Most results herein are shown for planar surfaces; note that similar results have been realized for experiments on curved surfaces. Whitesides, et al., have also used µCP for curved surfaces.[18-19]

We have developed a new avenue in both µCP and FIB fields by fabricating masters for µCP in the FIB. A main advantage is the rapid prototyping of printhead master patterns in the FIB, enabling a much quicker turnaround time from concept and design to printing. In addition, use of the FIB allows for printhead fabrication on both planar and curved surfaces, which cannot be done easily by conventional lithography. This paper discusses our recent developments in FIB fabrication of masters and applications of µCP.

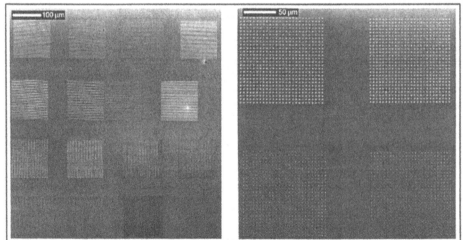

Figure 1: (a) FIB micrograph (left) showing 15 patterns out of a 37-pattern printhead. The individual patterns were milled at 1000 pA and 2700 pA beam currents in an FEI 200 FIB. Line patterns are in the top half of the figure, and dot patterns in the bottom half. (b) FIB micrograph (right) showing a magnified view of the lower left dot patterns in (a). Each pattern is 26 x 26 dots.

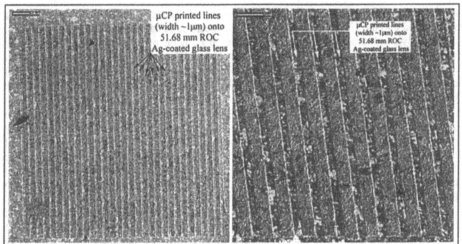

Figure 2: (a) FIB micrograph (left) showing μCP surface of a Ag film (150nm thick) on a 5cm radius of curvature glass lens. (b) FIB micrograph (right) showing a magnified view of the image in (a). Arrows indicate μCP lines.

Experiment

Printheads are fabricated in the FIB on single crystal Si (100) wafers. The FIB employed is an FEI Series 200, with a Ga$^+$ liquid metal ion source, minimum spot size ~10 nm, ion energy 30 keV and ion current density ~10 A/cm^2. The field of view is up to 1 mm^2 at 30 keV, and calibrated specimen drift rates are as low as 1.5 nm/min. The depth of focus is of order 200 μm. Feature sizes less than 0.1 μm may be fabricated by both milling and deposition.

FIB patterns are fabricated over fractions of a square mm and are added, or stitched, together to produce larger overall patterned areas (currently up to ~2 mm^2 area). Figure 1 (~0.75

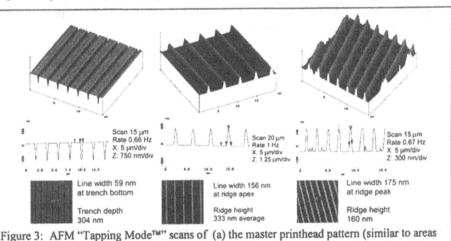

Figure 3: AFM "Tapping Mode™" scans of (a) the master printhead pattern (similar to areas in Figure 1) (left), (b) PDMS mold (elastomeric stamp) (middle), and (c) μCP surface (right)..

159

mm^2) is an example of part of a FIB printhead master containing a series of patterns. Patterns were milled into the Si with 2700 pA and 1000 pA beam currents (120 nm and 80 nm spot sizes, respectively), with times ranging from 20 minutes to 1.5 hours per pattern. Lateral feature sizes varied from micron scale to deep sub-micron scale. Vertical feature depths ranged from approximately 200 nm to greater than 1 µm. Figure 3 shows atomic force microscope (AFM) data (Digital Instruments Nanoscope III) from the FIB-patterned surface as an example of the pattern trench dimensions, along with data from the PDMS mold, and final µCP surface.

Figure 4 outlines the µCP procedure. Features milled in the FIB exhibit a trapezoidal geometry, with the base of the feature smaller than the top, due to the nominally Gaussian profile of the ion beam. This is an ideal situation for casting PDMS stamps, as the PDMS replicas will incorporate this geometry from the trench patterns in the FIB master. This serves to both reduce feature dimension on the PDMS stamps, and improve stability of the PDMS features when combined with careful control of aspect ratios. This is to prevent very thin or high aspect ratio PDMS features from crushing during µCP.

Following casting of the PDMS mold, it is coated with a 2 millimolar SAMs solution of hexadecanethiol in ethanol and dried gently under a stream of N$_2$ gas. Conformal contact between the inked PDMS and the Ag film surface is allowed for approximately 5 seconds by lightly placing the stamp onto the Ag film with tweezers. The Ag films used in this experiment were 150 nm thick uniform depositions by e-beam evaporation (therefore they are assumed to have bulk Ag resistivity). Then the SAMs-patterned Ag film is etched for ~ 25 seconds during which the printed pattern is developed. (The Ag etch solution consists of 0.001 molar potassium ferrocyanide(II)trihydrate, 0.01 molar potassium ferricyanide, and 0.1 molar sodium thiosulfate pentahydrate in deionized H$_2$O. Follow MSDS precautions for handling these harmful chemicals.) This pattern corresponds directly to the regions protected from etching by the SAMs. The sample is then removed from the etch solution, rinsed in both de-ionized water and ethanol, and dried. The patterned areas are those protected by the SAMs, evident in the Ag features left behind on the Si substrate. Figure 3c shows AFM measurements of µCP features with line widths down to 175 nm.

In addition, we are exploring the fabrication of multiple level printheads, which will enable alignment between successive levels of printing. In these multiple level printheads, independently addressable arrays will be activated separately. Each array consists of an independently wired set of resistively heated elements of FIB-deposited Pt, such that passing a current through any wired array will induce heating of the elements only within that array. Then

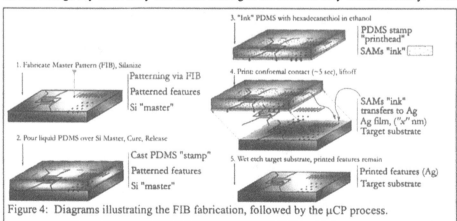

Figure 4: Diagrams illustrating the FIB fabrication, followed by the µCP process.

a given array may be selected for printing, for example through a thermal transfer mechanism, or by selective desorption of SAMs from all other arrays. By aligning the multiple level printhead with the substrate, one can print the required features, re-align, and then print another set of features—all from the same printhead. This will greatly facilitate alignment between successive printing levels. A working prototype multiple level printhead has been fabricated with two independently wired arrays of nine elements each, Figure 5a. Initial tests have verified the electrical continuity of each separate wire array. Utilizing the procedure outlined in Figure 4, the patterns for interconnections between heating elements were fabricated in the FIB and printed via μCP. After printing, these interconnections are Ag. After μCP produces these Ag wires, the specimen was placed back in the FIB for further fabrication. (Resistive) Pt elements were deposited in the FIB along the pre-defined gaps in the μCP Ag wires. In addition, "bridges" of alternating Pt/SiO$_2$/Pt were FIB-deposited to complete the wiring of each level. The Pt was used to connect each wire level, and the SiO$_2$ was deposited between to electrically isolate between levels, Figure 5b. FIB-deposited Pt exhibits much higher resistivity than pure Pt, due to the fact that the FIB-deposited material contains a percentage of the organic residue from the original precursor gas. Exploiting this property allows us to use this material as a resistive heating element. Based on preliminary measurements, the FIB-deposited elements have a resistivity approximately 30 times higher than the μCP Ag lines. This concentrates resistive heating within the Pt heating elements, rather than within the Ag interconnect lines. In addition, we are exploring the feasibility of co-depositing Pt and SiO$_2$ as a means of further increasing resistivity of the FIB-deposited printhead elements.

Conclusion

μCP has been performed successfully on both planar and curved substrates utilizing patterns of lines and dots fabricated in the FIB. Printed line widths range from micron scale down to 175 nm. FIB fabrication of printhead masters is advantageous because it enables rapid

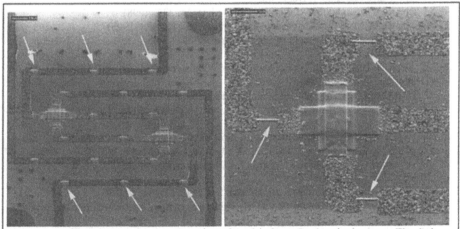

Figure 5: (a) FIB micrograph, central portion of the 2 independently wired arrays. The dark gray regions are μCP Ag wires; the lighter-gray regions are FIB-deposited Pt heating elements (arrowed), with two Pt/SiO$_2$/Pt bridges. (b) FIB micrograph centered on a wire bridge and three Pt elements. The horizontal bar is Pt for one wire level, with SiO$_2$ on top, followed by a top layer of Pt for the second wire level.

prototyping of specimens for printing via µCP, with the ability to pattern and print deep sub-micron features. Initial concept and design of patterns can be taken from fabrication through printing in less than 24 hours. In addition, the large depth of focus of the FIB allows for printhead patterning on both planar and curved substrates. Multiple level printhead prototypes are being fabricated currently, with promise for future application shown in initial tests. These printheads utilize a combined approach of FIB fabrication, µCP, and post-µCP FIB deposition, and will facilitate inter-level alignment between successive printings.

Acknowledgements

This work was supported under the DARPA Molecular Level Printing Program (Grant # N66001-98-1-8917). The authors gratefully acknowledge their collaborators: G.M. Whitesides and T. Deng at Harvard University, and T. Chraska and Y. Liu in the Dept. of Materials Science at the University of Virginia.

References

1. J. Orloff, Rev. Sci. Instrum., **64**, 1105 (1993).
2. D.N. Dunn and R. Hull, Appl. Phys. Lett., **75**, 3414 (1999).
3. S. Matsui and Y. Ochiai, Nanotechnology, **7**, 247 (1996).
4. K. Gamo, Nucl. Instr. and Meth. in Phys. Res. B, **121**, 464 (1997).
5. A.A. Iliadis, S.N. Andronescu, W. Yang, R.D. Vispute, A. Stanishevsky, J.H. Orloff, R.P. Sharma, T. Venkatesan, M.C. Wood, and K.A. Jones, J. Electronic Mat., **28**, 136 (1999).
6. D.M. Longo and R. Hull, in *Proceedings of the 1999 International Semiconductor Device Research Symposium*, (University of Virginia, Charlottesville, VA, 1999), pp. 33-36.
7. R. Hull and D.M. Longo, in *Proceedings of the Tenth International Workshop on the Physics of Semiconductor Devices*, edited by Vikram Kumar and S.K. Agarwal (Solid State Physics Laboratory, Delhi, India, 2000), pp. 974-981.
8. J. Tien, Y. Xia, and G.M. Whitesides, Thin Films, **24**, 227 (1998).
9. Y. Xia and G.M. Whitesides, Annu. Rev. Mater. Sci., **28**, 153 (1998).
10. H.A. Biebuyck, N.B. Larsen, E. Delamarche and B. Michel, IBM J. Res. Develop., **41**, 159 (Jan-Mar 1997).
11. T. Deng, L.B. Goetting, J. Hu, and G.M. Whitesides, Sensors and Actuators, **75**, 60 (1999).
12. L.B. Goetting, T. Deng, and G.M. Whitesides, Langmuir, **15**(4), 1182 (1999).
13. Y. Xia, N. Venkateswaran, D. Qin, J. Tien, and G.M. Whitesides, Langmuir, **14**(2), 363 (1998).
14. Y. Xia, E. Kim, and G.M. Whitesides, J. Electrochem. Soc., **143**(3), 1070 (1996).
15. Y. Xia, E. Kim, M. Mrksich, and G.M. Whitesides, Chem. Mater., **8**(3), 601 (1996).
16. D. Wang, S.G. Thomas, K.L. Wang, Y. Xia, and G.M. Whitesides, Appl. Phys. Lett., **70**(12), 1593 (1996).
17. Y. Xia and G.M. Whitesides, J. Am. Chem. Soc., **117**(11), 3274 (1995).
18. J.A. Rogers, R.J. Jackman, and G.M. Whitesides, Adv. Mater. (Comm.), **9**(6), 475 (1997).
19. Y. Xia, D. Qin, and G.M. Whitesides, Adv. Mater., **8**(12), 1015 (1996).

ION BEAM INDUCED CHEMICAL VAPOR DEPOSITION OF DIELECTRIC MATERIALS

H.D. WANZENBOECK, A. LUGSTEIN, H. LANGFISCHER, E. BERTAGNOLLI
Institute for Solid State Electronics; Vienna University of Technology; A-1040 Vienna, Austria

M. GRITSCH, H. HUTTER
Institute for Analytical Chemistry; Vienna University of Technology; A-1060 Vienna, Austria

email: Heinz.Wanzenboeck@tuwien.ac.at

ABSTRACT

Direct writing by locally induced chemical vapor deposition has been applied to direct-write tailor-made microstructures of siliconoxide for modification and repair of microelectronic circuits. Focused ion beam (FIB) tools are used for locally confined deposition of dielectric material in the deep sub-μm range. State-of-the-art procedures typically provide insufficient dielectrics with high leakage currents and low breakdown voltage. The detailed investigation of the deposition mechanisms in this study proposes an approach to significantly improve dielectric material properties. Siloxane and oxygen as volatile precursors introduced in a vacuum chamber are used to deposit siliconoxide at ambient temperatures on various substrates such as Si, GaAs, or metals. The deposition process was initiated by a focused Ga^+-beam. As elementary electronic test vehicles for a systematic electrical investigation ion beam induced depositions in of capacitor architectures are applied. The chemical composition of the layers is investigated by secondary ion mass spectroscopy (SIMS) and reveals effects of atomic mixing at the interfaces. The variation of process parameters such as ion energy and ion dose, scan time and delay time lead to a better understanding of the mechanisms. The composition of the precursor gas mixture is of significant influence on insulating properties. The results demonstrate that optimized FIB-induced deposition of dielectrics offers a new window for in-situ post-processing of integrated circuits.

INTRODUCTION

The advancement and refinement of Direct-Write Technologies has triggered a steadily increasing interest of semiconductor fabrication and microelectronic circuit design in these techniques [1,2]. Improvement [3] of the reliability of the process and the properties of materials deposited by direct-writing have qualified direct-write technologies suitable for practical production line applications. Since direct-write technologies allow complex modifications on VLSI devices and ASICs [4,5] without using a lithographic transfer step the semiconductor fabrication has a fast and versatile tool in its hands.

A majority of direct-write methods such as laser writing [6] or inkjet deposition provide a reliable technique for fabricating mesoscopic structures [7]. The continuous decrease of feature size in microelectronics generates a growing demand for methods that allow to directly write structures in the deep sub-μm range. Direct-writing of nanostructures is applied for fabrication of novel device prototypes or for designing advanced integrated circuits using a mix and match approach [7]. The availability of direct-write technologies for the sub-μm range, however, is limited. Electron beam induced deposition suffers from slow writing speed. AFM direct-writing appears to be a promising approach [8], but still lacks the necessary refinement and reliability of a fabrication tool. On the other hand focused ion beam (FIB) technology has already been well established as in-situ modification tool and is widely accepted in industry [8]. Apart from sputtering and chemically enhanced etching for patterning [9] the FIB direct write deposition of metals [10] has been strongly promoted during the last decade.

The benefits of FIB are the higher writing speed in comparison to e-beam direct -writing and the capability of deposition on non-flat surfaces directly from the vapor phase. This is presently not feasible with an AFM. The FIB system is also highly suitable for mix-and-match techniques as this technique integrates 2 major functions in the same tool: (i) imaging of the sample for alignment of structures to be deposited and (ii) the local deposition of material in multiple length scale ranging from nm up to the mm range.

In principle the FIB is capable of depositing a wide variety of materials on bias of established CVD processes. The deposition of dielectric material [13,14,15] has been demonstrated by the formation of siliconoxide utilizing a FIB tool. However, the deposition of dielectrics is rarely applied in technology mainly due to the unsatisfactory quality of FIB-deposited insulators. Such dielectric films fabricated by FIB consist of Si and O with significant concentrations of Ga and C [16]. In this study the deposition process was investigated and a chemical characterization of the grown dielectrics was performed. With a better understanding of the entire process the deposition parameters can be adapted for fabrication of better dielectric material.

With a reliable dielectric material this FIB direct- writing approach may significantly reduce time and costs of prototype development of advanced integrated circuits, MEMS, photonic lattices and novel electronic devices. The 3-dim integration of passive devices (capacitors, resistors, inductors, interconnect dielectrics) into complex 3-D architecture becomes feasible by combining FIB deposition of metals and insulators. The achieved advance of focused ion beam technology is extending the potential of direct-write technologies from mesoscopic methods towards a reliable micro- and nanostructure direct-write fabrication.

EXPERIMENTAL

Fabrication of siliconoxide insulators was performed by chemical vapor deposition of siloxane and oxygen utilizing a Ga^+ ion beam. The deposition chamber had a base pressure of 10E-7 Torr. Introduction of precursor gases increased the total pressure to the 10E-6 Torr regime. Metal structures were fabricated using Tungstenhexacarbonyl. For deposition of dielectric material Tetramethylcycloterasiloxane and oxygene were coadsorbed on the substrate surface. The surface reaction leading to deposition of solid material was locally induced by an impinging Ga^+ beam thus confining the grown layers on a defined area with arbitrary geometry and a maximum diameter up to 1 mm. The ion beam could be focused down to a diameter of 5nm but smallest features deposited were in the 100nm range due to overspray effects by sputtering a redeposition of previously deposited material.

The Ga ions were generated by a liquid metal ion source accelerated with 50 kV. An electrostatic lense system for focusing and a deflection system for scanning operation provided a focused beam with a ion current adjustable between 4 pA to 2 nA. The pixel spacing during scanning could be adjusted in the nm to the μm range. For deposition the ion beam was directed perpendicular to the sample surface. Cross sections were sputtered with a perpendicular beam using a ion current above 200 pA. The prepared cross sections were imaged with an 45° tilted sample surface.

Substrates used were semiconductor materials such as (100) Silicon and GaAs. In order to approximate conditions on microelectronic devices also insulator surfaces obtained by thermal oxidation of the Si substrate and metal coated semiconductors were used. For electrical testing of materials, multilayer samples with a metal-insulator-metal (MIM) setups were produced.

The chemical composition of deposited materials was determined by secondary ion mass spectroscopy (SIMS) using a CAMECA IMS 3f sector field system. The SIMS-tool was operated in depth profiling mode to determine changes of chemical composition with varying layer thickness. A 15 nA Cs+ ion beam was focused on a 60 μm diameter spot of the 150x150 sample area. The recorded signal as counts/min account for the concentration in the deposited material.

Simulations of ions impinging on substrate material were performed using "TRIM". This Monte-Carlo simulation software developed by Ziegler and Biersack [17,18] makes detailed calculations of the energy transferred to every target atom collision. The presumption for this software of an amorphous material simply characterized by the elemental composition and the density can be assumed to be widely fulfilled for the non-crystalline materials typically deposited by FIB processes. In order to approximate the experimental results, calculations were performed with 50 kV Ga ions at a 90° angle to the surface.

RESULTS

For fabrication of microelectronic devices typically dielectric structures are demanded for insulation purpose. A functional stacked capacitor as an exemplary device has been used to demonstrate the feasibility of applications of FIB direct-writing fo dielectrics. This basic microelectronic device was built in a metal-insulator-metal (MIM) setup using a mix-and-match approach. The bottom electrode was obtained by standard optical lithography using conventional sputter-deposition of a 150 nm Al layer (60x60 μm²) on an insulating surface. This Al pad also acted as contact for posterior electronic measurements. The dielectric spacer of siliconoxide (25x25 μm²) and the top electrode of tungsten (16x16 μm²) were deposited by focused ion beam induced deposition of chemical precursors. The FIB tool allowed alignment of these stacked layers with other surface structures.

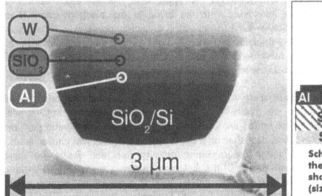
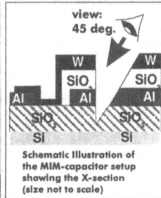

Schematic Illustration of the MIM-capacitor setup showing the X-section (size not to scale)

Fig. 1 Cross-section through a MIM-capacitor with a 670 nm thick siliconoxide layer deposited by FIB. The cross section was prepared by sputtering with FIB and shows the sample with a 45° tilted surface.

A typical MIM structure is illustrated in Fig. 1 showing a homogeneous thickness of the layer over the entire area. The dielectric spacer layers were deposited with a thickness in the range from 70 nm to 1.2 μm. The material appears to be dense without any cavities and the interface to metal layers is clearly visible. From a mechanical point of view the stacked setup can be processed without significant problems.

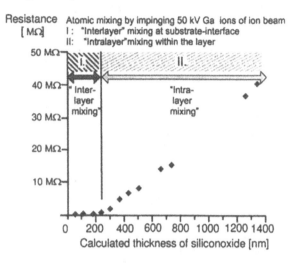

Fig. 2 Correlation of thickness of the deposited dielectric layer and obtained resistance

Electrical measurements of the insulating properties of the dielectric layers were made with structures fabricated by FIB-deposition of the siliconoxide as insulator and tungsten as top electrode. It must be considered that deposition occurred under 50 kV ions impinging on the surface. For thin dielectric layers below 200 nm thickness a total resistance in the kΩ range was measured rendering those devices as insufficient for application. As explained lateron a heterogeneous intermixing of the siliconoxide with the metal at the electrode-interface is proposed. Exceeding a critical thickness the FIB-deposited dielectric provides a satisfactory insulation. For dielectric layers with a thickness between 250 nm and 1.2 μm a resistance in the range of 5 to 50 Mega-Ohm was measured. It is assumed that intermixing with the metal substrate has deceased to an insignificant level and only homo-intermixing within the insulating layer itself occurs.

In order to clarify the unsatisfactory resistivity of thin siliconoxide layers chemical analysis of the deposited material was made by secondary ion mass spectroscopy. It was assumed that the poor electrical properties are corresponding with the composition of the material. A thick siliconoxide layer was deposited on a GaAs substrate. GaAs was chosen as this conductive material allows to discharge the incoming primary ions and shows no mass interference with siliconoxide. A depth profile of the dielectric layer was made [Fig 3] without reaching the interface to GaAs. As for SIMS spectra typical the first cycles cannot be interpreted. The number of sputter cycles corresponds with the depth of the investigated surface. Within this initial altered layer a sputter equilibrium depending on the atomic mass and the elemental composition of the matrix is adjusting and due to the dynamic change of composition values cannot be counted for chemical evaluation.

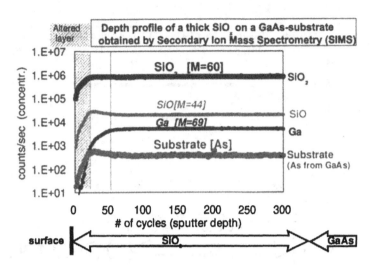

Fig. 3   Depth profile showing the chemical composition of a FIB
deposited siliconoxide layer on a GaAs substrate

The spectrum shows a very homogeneous composition of the sample over the entire analyzed thickness. However, the chemical composition respects a mean value of a 60 µm diameter area and does not directly reflect the real chemical composition as the signal depends on the sputter yield of the element or molecule ion in the actual matrix.

The investigated film of FIB-deposited siliconoxide showed strong signals of [SiO2] and [SiO] ions. Furthermore, ion signals of [Si] and [O] not displayed in the shown spectra for the clarity of the illustration were detected in the mass spectra. Also impurities of carbon were found. The signal of Arsene in Fig. 3 originating from the underlying substrate is negligible as displayed in this logarithmic scaled illustration. Hence, the Ga signal is predominantly attributed to implantations by the Ga ion beam itself. An intersting detail is the different behavior of the Ga signal to all other signals such as [SiO2] in respect to the altered layer. The a adjustment of an equilibrium concentration for Ga was observed in a greater depth than with [SiO2], [SiO] or [As] originating from the substrate. This gives rise to the assumption that the original surface layer may not contain as much Ga as deeper layers. Concluding, the mass spectra revealed a homogeneous composition of the siliconoxide with significant impurites of Ga, C and As.

In accordance to the fabricated MIM-capacitor structures a thin SiO2 layer deposited on a aluminum layer obtained by sputter deposition on a silicon substrate. A depth profile of this sample [Fig.4] includes not only the siliconoxide film but also the underlying Al layer. The mass spectra of the dielectric layers shows a constant composition of the siliconoxide as in Fig. 3 and the comparable delayed adjustment of an equilibrium situation for Ga. The main difference, however, is the signal of Aluminum in the siliconoxide film. The spectrum indicates a detectable Al concentration already at the initial siliconoxide surface and an exponential increase of the Al content with decreasing distance to the dielectric-metal interface. On the other hand signals of siliconoxide were detected in what is proposed to be the aluminum layer. This result is interpreted as the outcome of atomic mixing of the dielectric layer and the metal substrate.

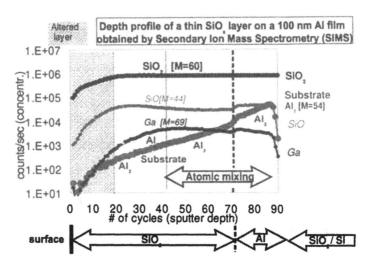

Fig. 4   Depth profile showing the chemical composition of a thin film
of FIB deposited siliconoxide layer on a Al-covered substrate

The chemical analysis suggests significant impurities of metal atoms in thin FIB deposited dielectric layers. With increasing thickness of the siliconoxide film the Al-content becomes negligible. This observation correlates with the low resistance experimentally shown for thin FIB-deposited insulating films

This interpretation was supported by calculations of transport and distribution of ions in matter using "TRIM" [17,18] Monte-Carlo-Simulation Software. The results for 50 kV Ga+ ions approaching the surface of SiO_2 at a 90° angle are illustrated in Fig.5. The simulation suggests that siliconoxide is amorphized down to a depth of 100 nm beneath the surface due to recoils of silicon and oxygen atoms effected by the high energy of the impinging ions. With 50 kV acceleration voltage little Ga is found in the first layers up to 20 nm thickness off the surface. This result proposes that ion energy initially is so high that Ga penetrates the first 20 nm without significant amounts being stopped. It is only in greater depths between 20 and 80 nm that Ga is stopped. The suggested Ga concentration profile near the surface may be an explanation for the delayed steady state documented in the SIMS depth profiles. The result is in correlation with low resistivity of the thin layers and the observed mixing of the Al substrate having a similar atomic mass than Si and O.

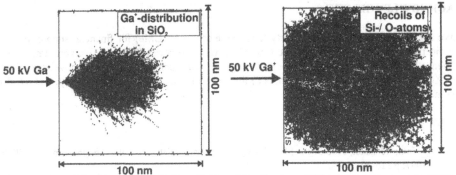

Fig. 5. Path and final location of Ga ions (left) and distribution of Si and O recoil atoms in SiO2 hit by a 50 kV Ga beam.

The ion beam effect was also simulated with a target material of a higher atomic mass and a higher density than SiO2 and was calculated to be restricted to a layer thickness only 40% of that obtained with SiO2. Hence, with a metalization layer consisting of a metal with a high atomic mass such as W the atomic mixing can be restricted to 40 nm penetration depth. This simulation suggests, that by using heavy metals the thicknesss of the dielectric layers may ber reduced by 60% reduced thickness. It is assumed that W absorbs the impact energy of Ga ions much faster than SiO2. With this clearer interface between metal and dielectric also the insulating properties of the FIB-deposited siliconoxide may become better.

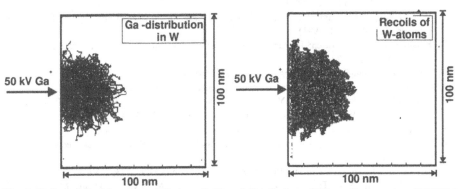

Fig. 6 Path and final location of Ga ions (left) and distribution of W recoil atoms in solid W hit by a 50 kV Ga beam.

CONCLUSIONS

The results of this investigation demonstrate that good material properties are closely connected with good electrical properties. The insulating properties of FIB-deposited siliconoxide were shown to be closely connected with chemical impurities and mixing with substrate material. A solution using heavy elements for the substrate has been suggested. Further improvements of materials and the FIB technique will allow to deposit fully functional microelectronic devices on a deep sub-μm range.

ACKNOWLEDGEMENTS

We acknowledge the Austrian Society for Microelectronics (GMe) for financial support of this project. We thank E. Gornik and the Center for Microstructures (Mikrostrukturzentrum) of the Vienna University of Technology for providing this clean room facility. We acknowledge James F. Ziegler (IBM-Research, Yorktown, NY) for providing the "SRIM" software for public use for simulating ion/target interactions.

REFERENCES

[1] K Jain; TJ Dunn, Proc of the SPIE (The-International-Society-for-Optical-Engineering) vol.3046; 1997; p.316-20.

[2] T. Drummond; D. Ginley; US patent document 5,132,248/A/

[3] TM Bloomstein; ST Palmacci; RH Mathews; N Nassuphis; DJ Ehrlich; Laser Processing: Surface Treatment and Film Deposition. Proceedings of the Advanced Study Institute (Kluwer Academic Publishers), Dordrecht, Netherlands; 1996; 895-906.

[4] C Schomburg; B Hofflinger; R Springer; R Wijnaendts van Resandt; Microelectronic-Engineering. 35 (1-4), 509 (1997)

[5] RF Miracky; Laser- and Particle-Beam Chemical Processes on Surfaces Symposium. Mater. Res. Soc; (1989),547

[6] A Wassick; Proceedings-of-the-SPIE (The-International-Society-for-Optical-Engineering) 1598,141 (1991)

[7] J Paufler; H Kuck; R Seltmann; W Doleschal; A Gehner; G Zimmer; Solid-State-Technology. 40(6), 175 (1997); p.175-6, 178, 180, 182.

[8] WW Flack; DH Dameron; Proceedings-of-the-SPIE (The-International-Society-for-Optical-Engineering) 1465, 164 (1991)

[9] K Birkelund; EV Thomsen; JP Rasmussen; O Hansen; PT Tang; P Moller; F Grey; J. Vac. Sci. Technol. B15(6),2912 (1997)

[10] J. Glanville; Solid-State-Technology 32(5), 270 (1989)

[11] JMF Zachariasse; JF Walker; Microelectronic-Engineering. 35(1-4), 63 (1997)

[12] AA Iliadis; SN Andronescu; W Yang; RD Vispute; A Stanishevsky; JH Orloff; RP Sharma; T Venkatesan; MC Wood; KA Jones; J. Electronic-Mat. 28(3), 136 (1999)

[13] H Komano; Y Ogawa; T Takigawa; Jap. J. Appl. Phys., 28(11), 2372; (1989)

[14] S Lipp; L Frey; C Lehrer; B Frank; E Demm; S Pauther; H Ryssel; J. Vac. Sci. Technol. B14(6), 3920 (1996)

[15] BH King; D Dimos; P Yang; SL Morissette; J. Electroceramics. 3(2), 173, 1999

[16] A.N. Campbell, D. M. Tanner, J. M. Soden, D. K. Stewart, A. Doyle, E. Adams, M. Gibson, M. Abramo; Proc. 23rd Internat. Symp. For Testing and Failure Analysis, 1997, p. 223

[17] J. P. Biersack JP and L. Haggmark; Nucl. Instr. and Meth., 174, 257 (1980)

[18] J. F. Ziegler; "The Stopping and Range of Ions in Matter", 2-6, Pergamon Press, 1977-1985

FOCUSED ELECTRON BEAM INDUCED DEPOSITION OF GOLD AND RHODIUM

P. HOFFMANN*, I. UTKE*, F. CICOIRA*
*Institute of Applied Optics, Swiss Federal Institute of Technology Lausanne,
CH-1015 Lausanne-EPFL, Switzerland, Patrik.Hoffmann@epfl.ch

B. DWIR**, K. LEIFER**, E. KAPON**
**Institute of Micro- and Optoelectronics, Swiss Federal Institute of Technology Lausanne,
CH-1015 Lausanne-EPFL, Switzerland

P. DOPPELT***
***ESPCI-CNRS, 10 rue Vauquelin, 75231 Paris Cedex 05, France

ABSTRACT

Electron beam induced deposition with two noble metal precursors (Rhodium and Gold) having the same halogeno and trifluorophosphine ligands is presented. The deposit geometry of lines and freestanding bridges is discussed with respect to electron energy, beam shape, and backscattered electron distribution. Electron beam heating effects are estimated to be negligible in our deposition conditions. Using PF_3AuCl, lines of percolating gold grains were deposited with electrical resistivities as low as $22\mu\Omega cm$ at room temperature (Au: $2.2\mu\Omega cm$). Auger electron analysis shows about 60at% Rh in deposits obtained with $[RhCl(PF_3)_2]_2$, however the resistivity of $1\Omega cm$ is high compared to $4.5\mu\Omega cm$ of pure Rh.

INTRODUCTION

Focused electron-beam induced deposition (EBID) offers high flexibility compared to classical lithographic microelectronics processing. Deposition of metals and insulators of high aspect ratio columnar structures, air-bridges and other 3D features is possible. Limiting for most applications is the low deposit purity resulting in low electrical conductivity and low average grain density of deposited metal structures. Increasing the substrate temperature during EBID with Me_2Au-(tfa) as precursor increased the metal/carbon ratio to 22at% [1]. The high carbon content in the deposits results from both the oil vapors in the electron microscope pumping system and to a higher extend from the carbon rich precursor ligands. For improvement the addition of efficient carbon etching partner gases in combination with noble metal precursors was proposed: adding oxygen into the deposition chamber together with Me_2Au-(tfa) as gold precursor resulted in 50at% Au [2]. The carbon free precursor WF_6 tested for EBID resulted in almost pure W [3].

In this article we present EBID of Rh- and Au-rich structures from carbon and oxygen free trifluorophosphine chloro complexes without substrate heating nor adding reactive gases.

EXPERIMENTAL

Deposition

Our EBID system is based on a scanning electron microscope (SEM) Cambridge S100 with tungsten filament, operating in steps between 3kV and 25kV. For deposition the electron beam is controlled by a Nabity NPGS lithography software running on a PC. Dwell times and point-to-point distances can be varied between 1µs to 1s and 5nm to 5µm, respectively. Horizontal lines were deposited as single slow scan with 10nm point-to-point distance and a dwell time of 1s. Further reduction of the scan speed results in three-dimensional tip deposition with different inclination angles (vertical if scan speed is zero). Alternatively, multiple scans using the SEM line scan mode (scan speed 0.2mm/s, scan frequency 50Hz) were applied to write lines.

The specimen chamber background pressure is $2 \cdot 10^{-6}$ mbar obtained with an oil free pumping system. For EBID, solid precursors are filled into a stainless steel syringe reservoir inside a dry nitrogen glove box and then transferred to the SEM. The precursor vapor is directed to the sample surface by pointing the nozzle (0.8mm diameter) to the substrate. The electron beam is conveniently focused by means of a dot deposited in the SEM spot mode at highest magnification on the substrate surface. The working pressure in presence of a precursor can increase up to $2 \cdot 10^{-5}$ mbar. As substrates mostly SiO_2 - coated Si wafers with Au electrodes for resistivity measurement were used.

Characterization

Transmission electron microscope images were taken with a Philips CM 300 microscope. Auger Electron Spectra were obtained in a PHI·660 system (electron probe: 5nA at 3kV) equipped with an argon ion gun for sputtering. Scanning electron microscope images were taken with a JEOL 6400.

Monte Carlo Simulation

Monte Carlo simulations were carried out with a commercial software from L. Reimer based on models described in his book [4]. The program allows the calculation of secondary electron (SE) yields and backscattered electron (BSE) yields, Bethe ranges, and trajectories in layered structures. 5000 electrons were simulated in bulk silicon covered with a 100nm SiO_2 layer (density: $2.2 gcm^{-3}$).

RESULTS AND DISCUSSION

EBID from PF₃AuCl

Deposition with the inorganic carbon free volatile gold precursor PF_3AuCl resulted in high metal content deposits. Lines with electrical resistivities down to 22µΩcm could be deposited [5]. The influence of irradiation conditions on electrical properties is reported elsewhere [6]. The precise vapor pressure determination of this thermally fragile precursor is presently under investigation. In order to gain better understanding of the decomposition process and deposit geometry we varied several parameters. Changing the electron energy is the most striking difference observed. Line deposition in the single slow scan mode, as shown in Fig. 1 results in conducting deposits. Additional to the line deposit, we also observed gold grains deposited relatively far away from the incident beam with distances up to several microns, see Fig. 1a. Secondary electrons generated at the surface by backscattered electrons and/or thermal effects must be considered as nucleation processes of these gold grains with respect to the easy decomposition of the precursor and its thermal fragility. Monte Carlo simulations where carried out to quantify this phenomena. Electron trajectories are shown in fig. 2 and electron ranges and yields are summarized in table 1.

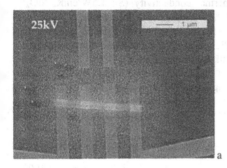 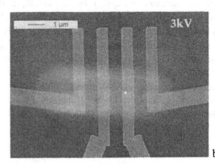

Fig. 1: SEM images of EBID gold lines on Si wafers covered with 100nm SiO$_2$ and pre-deposited lift off gold electrodes. Both lines were written as single slow scan with 10nm/s. (a) At 25keV and 425pA beam current a 250nm wide line surrounded by gold grains is obtained. (b) With 3keV and a comparable beam current of 500pA, the line width is 1.5μm due to an increased electron beam spot size at low energy, however, the gold grains spread less far around the line deposit.

For both beam energies about 15% of impinging electrons are backscattered to the surface. There they generate secondary electrons, which decompose adsorbed precursor molecules. The Bethe electron range, being the average distance traveled by an electron within the specimen, is 7.5μm at 25keV and corresponds well with the vertical "cloud" diameter in Fig.1a. The line is 250nm wide and corresponds approximately to the beam diameter of 200nm measured by the knife edge method. At 3keV and 500pA, in our electron microscope the beam has a diameter of about 1.4μm explaining the shape of the deposit in Fig. 1b where a homogeneous 1.5μm wide line is observed. The region of surrounding gold grains from the beam edge is about 400nm in width, which is more than the calculated 97nm. This might be qualitatively explained by a charging effect since the total yield of BSE and SE is 105% according to table 1, i.e. more electrons leave the sample than arrive. This will be compensated by a positive charge distribution on the sample, which widens the beam profile. In fact a positive sample current was measured during this deposition, which means that most impinging electrons do not penetrate through the insulating SiO$_2$ cover layer into the charge dissipating Si substrate.

	Bethe range R	BSE yield η	SE yield δ	Beam diameter (knife edge)
25keV	7.5μm	0.15	0.14	200nm
3keV	195nm	0.14	0.9	1.4μm

Table 1: Results of Monte Carlo simulations with 5000 vertically impinging electrons on a bulk silicon substrate covered with 100nm SiO$_2$. The measured electron beam diameter of our electron microscope at 500pA (3keV) and 425pA (25keV) is also indicated.

The local temperature increase ΔT in the beam center on a plane homogenous substrate surface can be estimated by [4]:

$$\Delta T = 3 \cdot f \cdot U \cdot i / 2\pi \cdot k \cdot R ,$$

with U acceleration voltage, i beam current, k thermal conductivity (Si: 1.5W/cmK, SiO_2: 0.014W/cmK at room temperature), R the Bethe electron range, and f the fraction of electrons which does not leave the sample as SE or BSE. The implanted power $f \cdot U \cdot i$ is assumed to dissipate homogenously from a hemisphere of radius R/2. At 25keV R/2=3.7μm, hence the 100nm thick SiO_2 cover can be neglected and a Si-bulk substrate assumed with f=0.71 from table 1. The calculated temperature increase is 0.006K. At 3keV R/2=97nm which is comparable to the SiO_2 cover thickness. Assuming a bulk SiO_2 sample with f=0.1 gives ΔT=0.55K. Since the underlying Si substrate with its 100 times better heat dissipation is neglected, this estimation represents a maximum value. Thus we can conclude that thermal effects are negligible in our deposition experiments.

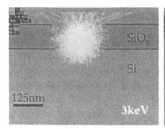 

Fig. 2: Trajectories of 1000 electrons impinging on bulk silicon covered with 100nm SiO_2 to illustrate the lateral expansion of backscattered electrons. Note the different scale.

Lines deposited as single slow scan or as multiple scan (50Hz sweep) show the same spread of gold grains when the line dose was kept constant, i.e. the same number of electrons was used to write both lines, see fig. 3.

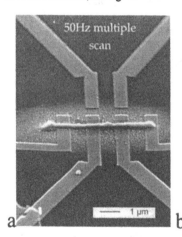 

Fig. 3: Tilted (75°) SEM images of EBID gold lines deposited at 25keV and 425pA on gold electrodes. In (a) the line is written by multiple scans at 50Hz for 8.3min. The central bump is due to a dot deposited for prior focusing and the bumps at the line ends are due to an increased beam residence time. In (b) the single scan with 10nm/s was applied.

The effective dose D for precursor decomposition in the "cloud" area is due to SE2 (SE of type 2 generated by BSE) and can be estimated by $D_{cloud} = \eta \cdot i \cdot t / \pi R^2$. As a first approximation a homogenous spatial distribution of BSE with yield η and a ratio SE2/BSE=1 is assumed, i.e. each BSE generates 1 SE2. Entering η=0.15, i=425pA, t=8.3min, and R=7.5μm at 25kV gives

$D_{cloud}=18mC/cm^2$. This value can be regarded as a rough estimate for the dose necessary to induce noticeable precursor decomposition and formation of separated gold grains. In contrast, the line deposit is due to SE of type 1 generated by impinging probe electrons. Hence, the effective dose for the line can be estimated by $D_{line} = \delta \cdot i \cdot t/d \cdot l$, where $\delta=0.14$ the SE yield, $d=200nm$ the beam diameter or line width, and $l=5\mu m$ the line length. With i and t as above, the effective line dose $D_{line}=3C/cm^2$. This dose is about 170 times larger than the "cloud" dose and leads to the formation of a three dimensional deposit composed of percolating gold grains.

This is confirmed by TEM studies shown in fig. 4. The line deposit is absorbing, however, beside the line separate gold grains with diameters ranging from 20nm to 5nm can be clearly distinguished. It should be mentioned that cloud formation can be suppressed by optimizing the deposition conditions.

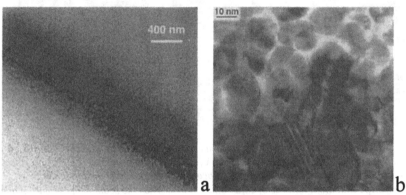

Fig. 4: Bright field transmission electron microscope images of gold deposits on TEM grids. In (a) the dark diagonal is the EBID deposited gold line absorbing the 300keV electrons. In part (b) a closer look to the edge of the line reveals percolating gold crystals.

EBID from $(PF_3)_4Rh_2Cl_2$

Important for EBID precursors is on one hand a good chemical stability and on the other hand a sufficient volatility. In contrast to PF_3AuCl with its relatively low stability, $(PF_3)_2Rh_2Cl_2$ represents both a suitable vapor pressure of $7.4 \cdot 10^{-2}mbar$ at room temperature and long time stability (days) under vacuum conditions [7]. EBID with $(PF_3)_4Rh_2Cl_2$ was carried out at constant electron energy of 25keV varying mainly the electron beam current and the dwell time of the beam. The electrical conductivity of rhodium containing lines deposited in the same way as with the gold precursor could not be measured due to the appearance of local breaks in the deposit. This is due to a volume change after deposition, the origin of which is not known at the moment, however, when depositing freestanding bridges as shown in Fig. 5, the deposit is much less constricted to the rigid substrate and can adapt freely to any volume change without breaking. For three-dimensional depositions, the lateral scan speed must be comparable to the vertical growth rate, strongly depending on the precursor vapor pressure and the precursor molecule density distribution on the substrate. Different deposit inclination angles can be obtained according to the adjustment of these parameters as can be seen in fig. 5a: on the right side of the image a freestanding bridge was deposited with 530pA but when doubling the lateral scan speed a compact triangular deposit was obtained (left image side). In fig. 5b, three arcs were deposited under identical conditions (point-to-point distance of 4nm and 0.5s dwell time) with a smaller beam current of 250pA. Additionally, a tip was deposited. The difference in height and shape reveals drifts on the sub-micron scale in our current EBID system. The diameter of the arc legs ∅=330nm in fig. 5c, deposited with the lowest current of 65pA but otherwise same parameters, is smaller than the diameter produced at higher currents ∅=670nm for 530pA and 250pA. The arc bending observed in fig. 5c could be attributed to drifts during deposition.

  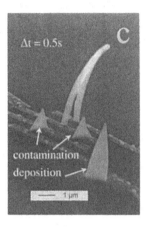

Fig. 5: SEM images (tilt 75°) of Rh-rich arc deposits obtained from EBID with (PF₃)₄Rh₂Cl₂.
The electron energy is 25keV, beam currents 530pA (a), 250pA (b), and 65pA (c). The insert in
(a) demonstrates the arc deposition: The electron beam scans cube1 from the left to the right
edge in 4nm steps with dwell times Δt of either 0.25s or 0.5s. Then the beam jumps to cube 2 and
scans from the right edge to the left. Cube 3 is deposited starting in the middle of cube 1 and
moving the beam to the right and so on.

The vertical growth rate seems to be independent of the electron current in the examined range. Resistance measurements with a two-point probe give as lowest value $4 \cdot 10^5 \Omega$ resulting in a resistivity of approximately $1 \Omega cm$ at room temperature compared to $4.51 \mu \Omega cm$ for metallic Rhodium. Cloud deposition was not observed for these 3D structures since the impinging electrons are mostly backscattered by the growing deposit and exit through its periphery. This leads to a radial deposit growth with deposit diameters larger than the beam diameter. However, the tip apex diameter is less affected by the radial distribution of BSE and apex diameters down to 16nm of carbon tips were obtained in our microscope [5].

Line deposits with $(PF_3)_4Rh_2Cl_2$ at high beam currents of about 5nA at 25kV also show a noticeable cloud deposit within the exit distance of BSE. The separated Rh grains embedded in a phosphorous rich matrix can be resolved in TEM and have a diameter of about 8nm. The dose for cloud formation with $(PF_3)_4Rh_2Cl_2$ is about 10x higher than for PF_3AuCl.

The chemical composition of the Rh containing deposits is characterized by Auger Electron Spectroscopy (AES). A characteristic AES spectrum is shown in Fig. 6. The averaged values of several spectra result in the following atomic percentages: 60% Rh, 16% P, 9% N, 7%Cl, 7% O, 1% F. The nitrogen and oxygen uptake of the deposits is probably taking place during exposure to air.

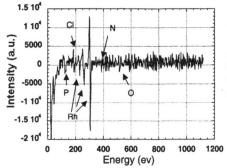

Fig. 6: Auger Spectrum of a Rh- rich line with 1μm width and a maximum thickness of about 500nm. The deposition was carried out on naturally oxidized (111) silicon with 25 kV at 6nA beam current. Dwell time 1s, point to point distance 10nm; precursor $(PF_3)_4Rh_2Cl_2$.

CONCLUSIONS

The chloro- trifluorophosphine complexes of Rh and Au are excellent candidates for deposition of metal rich lines and three dimensional submicrometer structures. EBID with the chemically less stable gold compound results in percolating nanocrystalline grains of almost pure gold and deposits of low electrical resistivity, whereas with the Rh precursor deposit metal contents of 60at% are achieved but the deposit remains electrically insulating. The dose for noticeable precursor decomposition of the gold compound is about 18mC/cm², which leads to deposition inside the exit area of BSE. Minimization or suppression of this effect is obtained by optimizing the beam energy and dose. Further investigations will be carried out on chemical and physical EBID parameters to decrease lateral deposit dimensions, to increase the structure aspect ratio, and to improve the electrical properties.

ACKNOWLEDGEMENTS

For Auger Electron Spectroscopy we thank H.-J. Mathieu and N. Xanthopoulos from DMX/EPFL.

REFERENCES

1 C. Schoessler, J. Urban, and H. W. P. Koops, J. Vac. Sci. Technol. B, **15**, pp. 1535-1538 (1997)
2 A. Folch, J. Tejada, C. H. Peters, M. S. Wrighton, Appl. Phys. Lett., **66**, pp. 2080-2082 (1995)
3 S. Matsui and T. Ichihashi, Appl. Phys. Lett., **53**, pp. 842-844 (1988)
4 L. Reimer in Scanning Electron Microscopy, Springer Series in Optical Sciences 45, Springer Verlag, Berlin, 1985
5 I. Utke, B. Dwir, K. Leifer, F. Cicoira, P. Doppelt, P. Hoffmann, E. Kapon, J. Microelectronic Engineering **53**, pp. 261-264 (2000)
6 I. Utke, B. Dwir, P. Doppelt, P. Hoffmann, E. Kapon, accepted in J. Vac. Sci. Technol. B Nov/Dec 2000, proceedings EIPBN 44
7 T. Ohta, F. Cicoira, P. Doppelt, L. Beitone, and P. Hoffmann, submitted

Other Techniques for
Direct Write Processing

Thermal Spray Techniques for Fabrication of Meso-Electronics and Sensors

S. Sampath, H. Herman, A. Patel and R. Gambino
Center for Thermal Spray Research
State University of New York
Stony Brook, NY 11794-2275

R. Greenlaw
Integrated Coating Solutions
Huntington Beach, CA

E. Tormey
Sarnoff Corporation
Princeton, NJ.

Introduction

THERMAL SPRAY is a directed spray process, in which material, generally in molten form, is accelerated to high velocities, impinging upon a substrate, where a dense and strongly adhered deposit is rapidly built. In the case of ceramic deposits, it is necessary to bring the particles to well above the melting point, which is achieved by either a combustion flame or a thermal plasma arc. The deposit microstructure and, thus, properties, aside from being dependent on the spray material, rely on the processing parameters, which are numerous and complex. In recent years, through concerted, integrated efforts of the Center for Thermal Spray Research at the State University of New York at Stony Brook and others, significant fundamental understanding of the process has been achieved, allowing for an enhanced control of the process.

Thermal spray coatings (i.e., thick films > 5 micrometers) are crucial for the economical, safe, and efficient operation of many engineering components. Numerous industries, in recognition of thermal spray's versatility and inherent economics, have introduced the technology into the manufacturing environment. The technology has emerged as an innovative and unique means for processing and synthesizing of high performance materials. The main advantages of the process are: (1) versatility with respect to feed materials (metals, ceramics and polymers in the form of wire, rod or powder); (2) capacity to form barrier and functional coatings on a wide range of substrates; (3) ability to create free-standing structures for net-shape manufacturing of high performance ceramics, composites and functionally-graded materials; and (4) rapid solidification synthesis of specialized materials. Opportunities exist for many novel applications in advanced materials synthesis and deposition. The technology is rapidly becoming the process-of-choice for the synthesis of advanced functional surfaces, such as *electrical conductors, magnetic components, dielectrics, ferrites, bio-active materials*, and *solid-oxide fuel-cells*. Thermal spray offers advantages for manufacture of deposits on large area substrates and for the creation of complex conformal functional devices and systems. A more complete overview is given in a recently published MRS Bulletin Issue [1].

The *virtues and unique advantages* of thermal spray with respect to direct write and related processes are:
- High throughput manufacturing as well as high speed direct writing capability
- *In situ* application of metals, ceramics, polymers or any combinations of these materials;
  - without thermal treatment or curing
  - incorporation of mixed or graded layers
- Useful materials properties in the as-deposited state
- Cost effective, efficient and ability to process in virtually any environment
- Limited thermal input during processing, allowing for deposition on a variety of substrates
- Adaptable to flexible manufacturing concepts
- Robotics-capable for difficult-to-access and severe environments (portable)
- Readily available for customizing special sensor systems (i.e., prototyping)
- Green technology *vis-à-vis* plating, lithography, etc.
- Can apply on wide range of substrates and conformal shapes
- High aspect ratio conductors and capability for vias production
- Rapidly translatable development to manufacturing (using existing infrastructure).

Mat. Res. Soc. Symp. Proc. Vol. 624 © 2000 Materials Research Society

Though there has been limited published, there has for a number of years been some deliberation on using thermal spray for the production of electronics components, sensors, ceramic superconductors, waveguide components, insulated metal substrates, various magnetic deposits, photochemical coatings, etc. There has been limited success in the past to achieve high quality functional multilayers by thermal spray. This can be attributed to several deficiencies, key among these being: Lack of fundamental understanding of the process and the ensuing process-materials-property relationships; Absence of diagnostic tools to evaluate and to optimize the highly dynamic processes; Insufficient process control; Limited personnel expertise in advanced materials processing.

The capabilities of thermal spray technology, even as recently as five years ago, were deficient for meeting the stated direct write needs. A number of important changes have occurred. Cost-driven application developments in the automotive industry (electrical applications), availability of sophisticated, affordable diagnostic tools, enhanced process control and reliability, and, finally, improved fundamental understanding through integrated, interdisciplinary research, such as at Stony Brook's NSF Center for Thermal Spray Research. Our understanding of the process and the ability to control material's microstructures now offer unique opportunities with which to synthesize functional surfaces of a variety of complex systems. The current technology remains limited in its capacity to satisfy the needs of direct write. The present limitation, however, is a classic case of a technology on the verge of a needs-driven upheaval. Thermal spray, therefore, represents a potential breakthrough or disruptive technology.

Thermal spray techniques offer new opportunities for hybrid microelectronics, sensors, superconductors, insulated metal substrates and other applications, including:

- Embedded sensors in coatings and structural parts for condition-based maintenance engineering, processing and system monitoring. [2]
- Variety of thick film materials and sensors including magnetoresistive Sensors, thermistors, thermocouples etc.
- High Frequency Inductive Components — Inductors, transformers and antennas using low loss ferrites and sprayed windings, interconnects and dielectrics.

The currently available thermal spray techniques are briefly described below. More detailed descriptions are available in the literature [3-6].

High Velocity Oxy-Fuel Spraying (HVOF), a novel variation on combustion spraying, has had a dramatic influence on the field of thermal spray. This technique is based on special torch designs, in which a compressed flame undergoes free-expansion upon exiting the torch nozzle, thereby experiencing dramatic gas acceleration, to perhaps over Mach 4. By properly injecting the feedstock powder from the rear of the torch, and concentrically with the flame, the particles are also subjected to velocities so high that they will achieve supersonic values. Therefore, upon impact onto the substrate, the particles spread out very thinly, and bond well to the substrate and to all other splats in its vicinity, yielding a well adhered, dense coating, comparable, if not superior to plasma-sprayed coatings. It should be noted, however, that the powder particles are limited in the temperature they can achieve due to the relatively low temperature combustion flame. It is, therefore, not currently generally possible to process high temperature ceramics using this technique (e.g., zirconia-based systems). However, of particular importance here is HVOF's ability to create dense deposits of alumina, spinel, etc.

Plasma spraying is relatively straightforward in concept but rather complex in function [4,6]. The plasma spray torch operates on direct current, which sustains a stable non-transferred electric arc between a cathode and an annular water-cooled copper anode. A plasma gas (generally, argon or other inert gas, complemented by a few percent of an enthalpy enhancing gas, such as hydrogen) is introduced at the back of the gun interior, the gas swirling in a vortex and out of the front exit of the anode nozzle. The electric arc from the cathode to the anode completes the circuit, generally on the outer face of the latter, forming an exiting plasma flame, which axially rotates due to the vortex momentum of the plasma gas. The temperature of the plasma just outside of the nozzle exit is effectively in excess of 15,000K for a typical dc torch operating at 40 kW. The plasma temperature drops off rapidly from the exit of the anode, and, therefore, the powder to be processed is introduced at this hottest part of the flame. The powder particles, approximately 40 micrometers in diameter, are accelerated and melted in the flame on their high speed (100-300 m/sec) path to the substrate, where they impact and undergo rapid solidification (10^6 K/sec). Depending on the torch design and the powder particle size distribution, plasma spray can form deposits of greater than 5 micrometers of a wide range of materials, including nickel and ferrous alloys, refractory ceramics, such as alumina and zirconia-based ceramics. Stony Brook's program emphasizes modified torches with de Laval type nozzles, implementing specially designed collimators, with a goal towards achieving a narrowly focussed plasma-particle stream. An additional development has demonstrated the feasibility of fine-feature spray (down to 100 micrometers). This involves a novel non-masking automated collimation system, which will permit addressing break-through line resolutions.

Cold Spray Deposition and related solid state kinetic energy processes are a new family of spray devices (not strictly "thermal spray") [7]. These systems, through special convergent-divergent nozzles use continuous gas pressure to accelerate ductile materials to supersonic velocities to impact onto virtually any substrate (e.g., plastic, glass, ceramic, metal) where an unusually high adhesive bond is achieved. This unique process can gain a fully dense deposit, for example, of copper and silver, leading to a high level of electrical conductivity. We have found that cold spray deposits achieve copper conductivity of better than 80% bulk. It is clear that direct write can be readily facilitated by cold spray methods.

Materials and Microstructural Characteristics:

Thermal spray offers the ability to deposit virtually any material that can be softened or melted. This encompasses metals, ceramics, polymers and combinations thereof to produce composites and functionally graded materials (FGMs). A wide range of materials are used commercially principally for protective coating applications. Limited applications for thermal spray have been found in the area of functional materials. Typically, materials are injected in powder form into the spray plume. The particle sizes range from 5 to 100 micrometers depending on the process and materials considerations. In order to obtain a high quality deposit it is critical to carefully control the feedstock characteristics. Numerous issues relative to feedstock need to be considered, especially as related to direct write technologies. These include particle size and distribution, the ability to feed fine particles, morphology, flow characteristics and chemistry [8].

A typical thermal spray microstructure is composed of an array of cohesively bonded micrometer-sized splats, which are the result of individual particle impingement and subsequent accumulation to produce the deposit. The splats form through impact and spreading followed by rapid solidification. The splat-based layered microstructure leads to an anisotropic microstructure, this having clear implications on properties. The properties of the deposit depend on a number of factors, principally affected by particle impact conditions (e.g., velocity, temperature), Substrate conditions (e.g., roughness, temperature, chemistry) and ambient environment. Control of these variables is key to achieving requisite microstructural quality [9,10].

Figure 1. HVOF Spinel Cross-section
200x, 100 microns scalebar

The advent of high velocity deposition processes in recent years has considerably expanded the quality and utility of thermal sprayed deposits. **Figure 1** shows the microstructure of a spinel dielectric deposited using HVOF oxy-fuel spray processes. These deposits are dense and smooth and retain the phase structures of the powder, and have electrical properties, which are acceptable for component applications. Other examples of microstructures of electronic materials deposited are shown in **Figure 2**. These include silver conductors, magnetoresistive metallic alloys (permalloy Ni-Fe) 2(a) and BaTiO₃ 2(b). The properties of these deposits are discussed below.

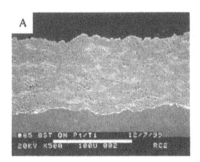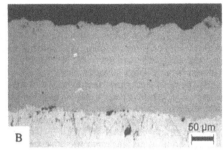

Figure 2. Cross-sections of HVOF sprayed BaTiO$_3$ and Permalloy

Table 1 summarizes the properties of various electronic materials fabricated using thermal spray methods to date. Also included in the Table are typical results that can be observed in bulk materials of equivalent compositions.

Table 1: Summary of Materials Property Results

Component	Materials/Processes	Properties-in as-sprayed state	Typical Bulk
Base Dielectric Dielectric Constant (K) Loss Tangent (tan δ) Surface Roughness (μm) Breakdown voltage (V/mil)	Spinel or Alumina HVOF and Plasma Spray	8-9 0.005 (10 kHz); ≤0.007(1.5 GHz) 1.6 350	~ 8 < 0.005
Conductor Traces Resistivity (μΩ-cm) Surface Roughness (μm) Linewidth (microns) Thickness (μm)	Copper and Silver Plasma and Cold Spray	~4.5 (Ag) to ~6.2 (Cu) ~ 1.1 –1.2 200 microns 25-30	Bulk Copper ~ 1.8 Bulk Silver ~ 1.6 Thin film conductor s ~ 4-5
Resistors Sheet Resistance (Ω/sq.) TCR (ppm/ °C) Surface Roughness (μm)	NiCr/Al$_2$O$_3$;NiAl/ Al$_2$O$_3$ HVOF and Plasma	HVOF NiCr/ Al$_2$O$_3$ 17 Ω/sq to 54 KΩ/sq. 175-300 at 10 Hz ~ 1.8	NA
Capacitors Dielectric Constant (K) Loss Tangent (tan δ) TCC (ppm/ °C) Surface Roughness (μm)	BaTiO$_3$ and BST (68/32) HVOF and Plasma Spray	120-175 ~ 0.01 < 500 ppm/°c ~ 2-3	500 to 10,000 0.005 to 0.01
Ferrites Sat. magnetization (Gauss) Coercivity (Oersteds) Resistivity (Ω-cm)	Mn$_{0.27}$Zn$_{0.26}$Fe$_{2.47}$O$_4$ Mn$_{0.35}$Zn$_{0.17}$Fe$_{2.48}$O$_4$ HVOF and Plasma Spray	3500-4000 50-70 70	Bulk Fired Crystals 4000-5000 2 200

Several important observations can be noted from Table 1:
1. Conductors:
   The typical resistivity values of plasma and solid state deposited materials are approximately 2-3X (i.e., times bulk) for Ag and 3-4X for copper. These were produced under ambient conditions and clearly there is considerable room for improvement. However, these conductor lines are useful for electronic components at their current levels.

2. Dielectrics:

Magnesium aluminate spinel was chosen as the base dielectric material for our studies. The choice of spinel over alumina was based on the formation of a metastable gamma-phase in plasma sprayed alumina under rapid solidification conditions. Gamma-alumina is hygroscopic and can play a deleterious role in the dielectric properties of the material. The spinel material is observed to have a dielectric constant of ~ 8 in the as-deposited state with tan □ of < 0.005. This low loss is maintained over a range of frequency from 100 mHz to 5 GHz. The dielectric roughness in the as-sprayed state is between 1 and 2 micrometers.

3: Capacitors:

BaTiO₃-based capacitor materials were examined under a variety of operating conditions. These materials yielded the most significant challenge in the sprayed deposits in terms of achieving requisite properties. Several important findings emerged. The deep eutectic nature of the BaTiO3 compositional phase field tends to kinetically suppress the formation of the crystalline perovskite phase, promoting the formation of an amorphous phase. This amorphous phase is typically found at the interface between the substrate and deposit and creates a series capacitive network, thereby reducing the through-thickness dielectric constant. **Figure 3** shows the X-ray diffraction patterns of several HVOF sprayed BaTiO3 samples with varying degree amorphicity. The dielectric constants illustrate the strong dependence on amorphous phase content.

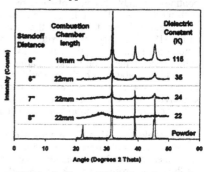

Figure 3. X-ray diffraction of several HVOF sprayed BaTiO₃ deposits

D: Ferrites:

Ferrites are important for the development of magnetic layers for high frequency circuits. They represent complex chemistries, making it difficult to thermal spray these materials. Mn-Zn ferrite compositions with different Mn/Zn ratios were selected for high frequency permeability. Two compositions of ZnO/Fe₂O₃/MnO were examined; one with a slightly higher zinc content, in the event that some of the zinc is lost from the thermal spraying. The average particle diameter for both compositions 25 micrometers. Optimization of the deposition parameters for this material was carried out using a conventional plasma system as well as two different HVOF systems; the Sulzer-Metco DJ2700 and Praxair HV2000.

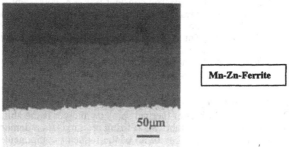

Figure 4. Optical Micrograph of the HVOF Ferrite deposits.

Table 2. Process-Properties for Thermal Sprayed MnZn Ferrite

Ferrite	Process	Thickness μm	σ emu/g	Sat. Mag. Gauss	Hc Oe	ρ Ω-cm
High Zn	Plasma. (air)	60	51.3	3223	66	2
Low Zn	Plasma. (air)	60	64.0	4021	72	2
High Zn	HVOF (DJ)	60	64.9	4080	51	78
High Zn	HV2000	250	55.8	3506	177	1

High Zn Composition: $Mn_{0.27}Zn_{0.26}Fe_{2.47}O_4$
Low Zn Composition: $Mn_{0.35}Zn_{0.17}Fe_{2.48}O_4$

The results of a process study are shown in the **Table 2**. As-deposited microstructure of HVOF sprayed ferrite is shown in **Fig. 4**. The resistivity is low for the APS and HV2000 coatings, probably because of some reduction of trivalent iron to divalent. The ferrite deposited with the HVOF (DJ) torch has the highest resistivity, perhaps because the propane fuel used in the DJ gun is less reducing than the hydrogen fuel for the HV2000. The coercivity is also lowest in the HVOF (DJ) sprayed ferrite. The coercivity is increased by porosity and by poor coupling among the grains, so optimizing spray parameters can be expected to further reduce coercivity. Initial measurements on thicker coatings with the DJ torch indicate a resistivity of about 250 Ω-cm, which is about as high as would be expected for this ferrite. The best ferrite properties we have observed so far show that one can prepare cores for high frequency transformers, but the coercivity must be reduced for filters, antennas and inductors. This can be accomplished with improved powder size, morphology and enhanced density deposition.

Spinel Dielectric High Frequency Measurements
Ring resonators were produced using conventional thin-film metallization (Ti/Pt/Au) on 175 micrometers and 625 micrometers thick HVOF-spinel on Ti substrates Preliminary measurements on the 175 micrometers thick sample indicated that the fundamental frequency is ~ 4 GHz and the Q (quality factor- energy stored/energy dissipated per cycle) of the resonator is reasonably high. From these data, a relative dielectric constant of ~ 8 was calculated at 4 GHz. The associated loss tangent was 0.005. A ring resonator measurements on another sample (with a non-optimal parameters) gave a relative dielectric constant of ~ 9, calculated at 1.5 GHz, with a loss tangent of 0.007.

Measurements were also made on the ring resonators produced on 500 micrometers thick spinel base dielectric deposited by thermal spray on a Ti substrate. Well-defined resonances were observed From the quality factor of the first resonance curve, a loss tangent of 0.0095 was calculated at 2.1 GHz. The dielectric constant was ~ 8, consistent with previous measurements. The loss tangent value includes the effect of the spinel surface roughness; R_A = 1.85 micrometers. Note that for a 175 micrometers thick specimen having a surface roughness of R_A = 1.05 micrometers, the loss tangent is 0.005. The effect of surface roughness has been modeled to extract the material loss tangent. Removing the effect of the surface roughness, the material loss tangent is 0.0065. Based on these measurements, it is concluded that thermal sprayed spinel is a good base-dielectric material at microwave frequencies.

Mutlilayer Circuits by Thermal Spray:
One of the capabilities of thermal spray techniques is the ability to fabricate multi-layered systems rapidly, with no post-fabrication firing required. As a demonstration of this, an inductor was produced using thermal spray of all of the inductor components: dielectric, ferrite core and metal conductors. A Mn-Zn ferrite-based inductor with the 10 turn coil was fabricated using plasma spray to form the insulator (spinel) and ferrite (Figure 5). Complete connectivity in the conductor was achieved. This enabled preliminary evaluation of the inductor. The high frequency performance of the inductor was evaluated from 40 Hz to 110 MHz. Using a series equivalent

circuit model, a plot of the series inductance Ls and series resistance Rs shows a well defined resonance at 35.2 MHz. The inductance below the resonance is about 165 nH and above the resonance the inductance was 50 nH. The series resistance is a measure of all the losses in the circuit, mainly core losses at high frequency. At 40 Hz Rs is less than one ohm, consistent with the dc resistance of the coil. The core loss has a hysteresis term, linear in frequency, and an eddy current term with a frequency-squared dependence. The series resistance increases to about 15 ohms at 110 MHz. By fitting the frequency dependence of Rs to a sum of f and f^2 terms we find that at 110 MHz most of the loss is from eddy currents, as expected for a ferrite with a resistivity of one ohm-cm. The resistance of the ferrite can be increased to 100 ohm-cm, which will significantly decrease the eddy current loss.

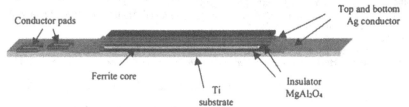

Figure 5. Schematic diagram of high frequency Mn-Zn ferrite based inductor.

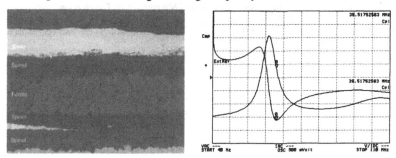

Figure 6. A cross-section of Mn-Zn ferrite-based inductor and its behavior.

The Mn-Zn ferrite composition was selected because of its' high permeability at high frequencies. The relatively low hysteresis loss at high frequency is consistent with high initial permeability. The coercivity of the ferrite core is 50 Oe, so in high power applications approaching saturation the hysteresis loss may become significant. Therefore, for some applications it will still be necessary to find process conditions that produce lower coercivity ferrite cores. The inductor has a Qmax of 143 at 33mhz.

Future activities include, increasing the number of coils with finer line widths, which will allow larger current capability as well as a fabrication of the flux gate magnetometer.

Summary and Concluding Remarks:

Using both traditional and hybrid thermal spray methods it has been possible to produce multilayer deposits of ceramics and metals having properties required by the electronics industry. Exemplifying this early success are the values of resistivity obtained for cold spray copper verses bulk values: that is, about 3x. In the case of the ceramic insulator-dielectric, spinel yields excellent insulative properties, as does alumina. The same trend can be attributed to ferrites, which display reasonably good values for low temperature deposition and can clearly be improved with further process optimization. Components ranging from insulated substrates, capacitors, conductors, resistors, to inductors can be fabricated using a suite of established and novel thermal spray processes. Versatility relative to materials, both for feedstock and substrates, processing convenience, together with cost-efficiency, makes the family of thermal spray devices particularly suitable for the production of electronic components and circuits.

Acknowledgements:
This project was supported by DARPA/ONR award N000140010654 under the auspices of Dr.W.Warren and Dr.C.Wood. Use of shared facilities supported partially by the MRSEC program of NSF under award DMR 9632570 is duly acknowledged.

References
1. MRS Bulletin, Guest Editors: S. Sampath and R. McCune, No.7, **25**, 2000, pp. 12-53.
2. M. Fasching, F. Prinz and L. Weiss, J.Thermal Spray Tech., 4(2) (1995) 133.
3. American Welding Society, Thermal Spraying: Practice, Theory, and Application, 1985; Library of Congress No: 84-62707, ISBN 0-87171-246-6
4. H. Herman, Plasma Sprayed Coatings, Scientific American, **256** (1988) 113.
5. H. Herman, MRS Bulletin, No.4, **13**, 1988, 60.
6. D.A. Gerdeman and N.L Hecht, *Arc Plasma Technology in Materials Science*, Springer, New York, 1972
7. A.P. Alkhimov, A.N. Papyrin, V.F. Kozarev, N.I. Nesterovich and M.M. Shushpanov, U.S. Patent No. 5,302,414 (April 12, 1994)
8. H. Herman, Powders for Thermal Spray Technology, KONA- Powder Science and Tech., No.9 (1991) pp.187-199.
9. S. Sampath and H. Herman, 'Rapid Solidification and Microstructure Development during Plasma Spray Deposition', J.Thermal Spray Tech., 5(4) (1996) 445.
10. H. Herman and S. Sampath, in Metallurgical and Protective Coatings, ed. K.Stern (Chapman and Hall, NY, 1996) p.261

Electrical Characteristics of Thermal Sprayed Silicon Coatings

S.Y. Tan, R.J. Gambino, R. Goswami, S. Sampath and H. Herman
Department of Materials Science and Engineering, SUNY at Stony Brook, 11794-2275

ABSTRACT

Polycrystalline silicon deposits were formed on a monocrystalline silicon substrate by thermal spraying. The resulting structure exhibits a device characteristic. Pressure-induced transformations of silicon, namely, Si-III (BC-8) and Si-IX are identified by X-ray diffraction in a Si-I matrix on deposits formed by vacuum plasma spray. The presence of the Si-III and Si-IX indicates that the pressure-quenched silicon deposit is highly conductive, as determined by four-point van der Pauw resistivity measurement. Hall mobility measurements, combined with photoconductivity results, indicate that the highly conductive silicon deposit displays the same range of mobility as a polycrystalline deposit containing only Si-I. The silicon deposit, with or without metastable phases, displays the same photoconductivity properties. The silicon deposit on a monocrystalline silicon substrate exhibits rectifying I-V characteristics, possibly caused by band bending of trapping states associated with impurities segregating at the polycrystalline deposit/monocrystalline substrate interface.

INTRODUCTION

There has been intense theoretical and experimental work concerning the high-pressure and metastable phases of silicon [1-4]. High-pressure compression transforms diamond-cubic silicon to a tetragonal form known as β-Sn (Si-II) [1,2]. Si-II then transforms to the BC-8 phase (Si-III) on slow depressurization or to Si-IX on fast decompression [3]. Si-IV (hexagonal diamond) is formed from Si-III when the temperature is greater than 450°K [4]. The BC-8 phase (Si-III) is known to be a semi-metallic at room temperature [5].

As observed in previous studies [6,7], thermal spraying has been used to form metastable phases of silicon deposit by shock synthesis. Silicon powder is injected into a high-energy plasma flame where the silicon particles are melted, the molten droplets being accelerated towards a rotating substrate. Upon impact, the kinetic energy of the molten silicon generates a shock wave, which propagates through the underlying silicon layer, causing a phase transformation to a high-pressure form of silicon [8]. The molten droplets then solidify on the substrate, forming a polycrystalline deposit. Impurities tend to segregate at the grain boundaries, resulting in the formation of trapping states which generate barrier potential energy [7,9-10].

The goal of this study was to examine the electrical and photoconductive characteristics of thermal sprayed polycrystalline silicon deposits by depositing on monocrystalline silicon substrates.

EXPERIMENTAL

The substrates employed are monocrystalline polished <100> p-type and <111> p-type silicon wafers with a wafer thickness of about 530 μm and 340 μm, respectively. In addition, a <100> n-type silicon wafer with thickness of about 340 μm is used as a reference for X-ray diffraction and resistivity analysis. The resistivity of the p-type wafer is approximately 10 Ω-cm and 1 Ω-cm for the n-type wafer. The silicon deposit is produced by vacuum plasma spray (VPS) in a reduced pressure/inert environment chamber [11,12]. Silicon powder is injected into a high-energy flame, where the molten silicon droplets impact the samples with a velocity of

189

approximately 400-600 m/sec, producing a shock wave with a Hugoniot pressure of about 12 GPa [6].

For photoconductivity studies, the wafers were cut into $2\times1\,cm^2$ rectangles. About half of the surface on the substrate was covered with a metal mask so that the other half surface receives a 25-30 μm sprayed silicon coating. Ohmic contacts were formed by abrading molten indium on the samples. The sample geometry is given schematically in figure 1. A 20 mW He-Ne laser (Newport U-1345) and a 20 mW GaAlAs laser diode (SDL 1401) were the light sources. The He-Ne laser for photoconductivity measurements has a beam diameter of 0.8 mm with a wavelength (λ) 633 nm and photon energy ($h\nu$) of 1.9 eV. The GaAlAs laser has a beam focused to manufacturer specification of about 0.8 mm diameter with a wavelength (λ) 840 nm and photon energy ($h\nu$) of 1.48 eV. The laser beam was scanned away from the contacts to avoid injection of carriers (see figure 1). A bias current of 10 mA was applied to the sample. The photocurrent was detected as a DC voltage change across a 1Ω standard resistor by a standard lock-in amplifier (EG&G 5210). The light was chopped at a frequency of 165 Hz.

Hall mobility up to ± 0.2 Tesla and resistivity measurements was carried out by the four-point van der Pauw method [13] at room temperature. Electrical contacts were made with silver epoxy.

To test the *I-V* characteristics, the sample was prepared by directing the thermal spray onto the sample protected by a metal mask with an opening in the center of $1\times1\,cm^2$. The sample is ¼-section of a 7.6 cm diameter polished silicon wafer. Figure 2 depicts the photograph of a top view of the device.

X-ray diffraction measurements have been carried out on a diffractometer (Philips 172a System) using Cu K_α radiation.

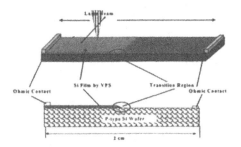

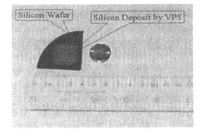

Figure 1. Schematic cross-section showing the sample geometry and the photoconductivity measurements. A layer of silicon film is deposited by VPS on half of the substrate. The specimen was illuminated with light of 1.9 eV (λ=633 nm) and 1.48 eV (λ=840 nm) scanning across the surface.

Figure 2. The photograph of a 1-cm square of silicon deposit on a monocrystalline silicon substrate by VPS process.

EXPERIMENTAL RESULTS

X-ray Diffraction (XRD) Patterns

Figure 3a, 3b, and 3c show the XRD diagrams of the deposit on <100> p-type, <100> n-type and <111> p-type monocrystalline silicon substrates, respectively. Figure 3a and 3b exhibit a broad peak in the 2θ range, 31.7-38.6, which originates from metastable phases of Si-III (BC-8) and Si-IX, in addition to Si-I [6]. However, figure 3c shows that the silicon deposits on <111> p-type contain no significant metastable phases. The sharp peak associated with the hump in <100> p-type and <100> n-type, as shown in figure 3a and 3b, is from Si-III (BC-8) [6].

Electrical Data

For the case of the silicon deposits on a monocrystalline p-type silicon substrate, since the Hall measurements and resistivity determine average values, it is prudent to consider the mathematics in terms of parallel conducting slabs, within each of which current flows uniformly. The equation for a two parallel layer model is given by [14]:

$$\sigma = \frac{(\sigma_1 T_1 + \sigma_2 T_2)}{(T_1 + T_2)} \tag{1}$$

where σ is the conductivity of the composite material, σ_1 is the conductivity of the deposited layer, σ_2 is the conductivity of the substrate layer, T_1 is the thickness of the deposited layer, and T_2 is the thickness of the substrate layer.

The carrier type of the silicon deposits is p-type as measured by the Hall effect and also by the sign of the thermal EMF. The apparent resistivity at room temperature is reported in Table I. Comparing the <100> and <111> substrate samples, the normalized resistivity ratio of the silicon deposits is about 1:4.6. The Hall mobility of silicon deposits on p-type <100> and <111> substrates at room temperature is not significantly different. Photoconductivity measurements as reported on Table II demonstrate the same range of photocurrent for deposits on the p-type <100> and <111> substrate.

Table I. Resistivity summary of the VPS coatings.

Silicon Substrate	Substrate Resistivity (Ω-cm)	Deposit Thickness (μm)	Deposit Resistivity (Ω-cm)	Normalized Resistivity Ratio
P (100)	10	30	0.6±1%	1
P (111)	10	30	2.8±4%	4.6
N (100)	1	30	0.45±1%	0.66

Table II. Experimental results of Photoconductivity measurements.

Silicon Substrate	Substrate Resistivity (Ω cm)	Deposit Resistivity (Ω-cm)	Deposit Conductivity (1/Ω cm)	Highest Photocurrent (μA) hv=1.9 eV	Highest Photocurrent (μA) hv=1.48 eV
P (100)	10	0.6	1.6	0.9±20%	3.1±5%
P (111)	10	2.8	0.35	0.9±20%	3.4±5%

Figure 4 shows *I-V* characteristics of the pressure-quenched deposits on monocrystalline substrate of <100> p-type and <111> p-type. The forward voltage is dominated by contact resistance due to a small epoxy contact which is about 500 Ω. The actual rectifier forward voltage is estimated to be about than 0.6 volts. The reverse biased current is somewhat leaky causing by unintentional metallic impurities. Note that the monocrystalline substrate is not chemically etched before deposition. In addition, the area of the deposit is fairly large, almost 2x2 cm², with the perimeter not very well defined (see figure 2). Neverthless, we do detect rectifier-like *I-V* characteristics.

The photocurrent generated at a junction is given by a heterogeneous model [15]:

$$I_p = q \, \Phi(X) \left[1 - \frac{\exp(-\alpha W)}{1 + \alpha \sqrt{D \, \tau}} \right] \qquad (2)$$

$$\Phi(X) = \Phi_0 \, \exp(-\alpha X) \qquad (3)$$

where $\Phi_0 = P_{opt} \, (1-R) \, / \, (h\nu)$, Φ_0 is the incident photon flux at the surface of the sample, $\Phi(X)$ is the photon flux at X penetrating depth within the sample, P_{opt} is the incident optical power, R is the reflectivity of the incident photon energy, α is the absorption coefficient, W is the depletion width, D is the diffusion coefficient, and τ is the lifetime of minority-carriers.

Reported here is the photocurrent generated by two photon flux sources, namely 1.48 eV and 1.9 eV on two p-type samples with different pressure-quenched deposit, while the resistivity of the monocrystalline substrates is of the same level (≈ 10 Ω-cm). A scan is made from the top of the deposit through the transition region to the bare substrate (see figure 1). The highest photocurrent occurs at or near the transition region, as reported in Table II. The photocurrent is in the same range for all samples in spite of different levels of conductivity.

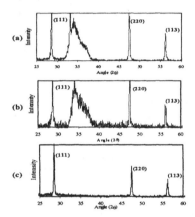

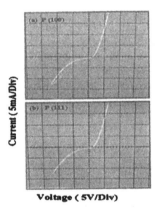

Figure 3. *X-Ray diffraction patterns of VPS silicon deposits. (a) deposits on <100> p-type silicon; (b) deposits on <100> n-type silicon. (c) deposits on <111> p-type silicon.*

Figure 4. *Room temperature I-V characteristics of the VPS silicon deposit/silicon substrate. (a) deposits on <100> p-type silicon; (b) deposits on <111> p-type silicon*

DISCUSSION

The relative XRD intensities indicate that the VPS deposits exhibit anisotropic shock-induced transformations. It was reported that a pressure of 18 GPa is required for a complete fraction volume of Si-I to Si-II transformation [5]. Since the VPS process produces a shock wave pressure of about 12 GPa [6] the phase transformation of the deposit is therefore partial. It is interesting to note that the deposit on <100> p-type and <100> n-type monocrystalline silicon substrates contain polymorphs Si-III and Si-IX in a Si-I matrix, but no metastable phases are detected for deposits on <111> p-type substrate. Perhaps the silicon phase transformation is anisotropic and requires higher pressures to transform Si-I to Si-II on <111> silicon [6,16].

There are models predicting the electrical behavior of grain boundaries in polycrystalline silicon [9,14]. We intend to discuss our electrical data qualitatively only. The silicon powder used in our VPS processing contains metallic and doping impurities [17], and the resulting silicon deposit is p-type by the sign of thermal EMF. Thus, we have a p-type deposit on <100> p-type and <111> p-type substrate of 10 Ω-cm. The two parallel conducting layers model was used for the four-point van der Pauw method for resistivity calculations, assuming that the current flows uniformly within each layer. We also use a <100> n-type substrate of 1 Ω-cm to form a p-n junction for comparison. However, all samples are produced by an identical VPS process. The resulting calculation demonstrates that the resistivity of the deposit on <100> p-type and <100> n-type is in the same range (see Table I), but the deposit on <111> p-type shows a resistivity that is four times as high. As for the Hall mobility, overall the samples for both <100> p-type and <111> p-type are in the same range of less than 50 cm^2/V-s as previously works [7]. The photocurrent of the samples is also in the same range (see Table II). This implies that the density of the trapping states or defects on sprayed coatings is about at the same level.

It is not surprising that the apparent resistivity of the deposit on a <111> p-type substrate is four times higher than that on <100> p-type, while the mobility remains in the same range. The apparent resistivity of the deposit on <111> p-type is about 2.8 Ω-cm and 0.6 Ω-cm on <100> p-type. These, in turn, convert to a doping concentration of ≈5x10^{18} atoms/cm^3 [9]. The critical dopant concentration N* of approximately $10^{16} - 10^{17}$ atoms/cm^3, at which all the grain-boundary traps are filled with carriers from substitutional dopant atoms, is given by [9]:

$$N^* = \frac{N_T}{L} \qquad (4)$$

where N_T ($\approx 10^{12}$ cm^{-2}) is trap density of grain boundary area [18], and L is the grain size

For the Si-I grain size ranges 1-0.1 μm as measured by TEM, the intergrain barrier potential, ϕ_B, is estimated to be less than 0.02 eV using [9]:

$$\phi_B = \frac{q\, N_T^2}{8\varepsilon_s N} \qquad (5)$$

where N is doping concentration, and ε_s is the semiconductor permittivity.

Thus, the mobility in the deposits on both <111> p-type and <100> p-type should be in the same range of less than 50 cm^2/V-s. At such a high doping level, the grain boundaries are saturated with segregated impurities and the excessive impurities pass into the grains, nucleating at defects such as crystal imperfections and contributing to the conductivity [9]. Although of the same doping level, the pressure-quenched silicon deposit on the <100> orientation substrate is slightly more compacted and more stressed, as demonstrated by the presence of Si-III and Si-IX, which behave as highly disordered microcrystalline conductors with highly localized nucleation [5]. The apparent resistivity of the deposit on the <100> p-tpe and <100> n-type is clearly lower than that on the <111> p-type, which shows no metastable phases. Therefore, we can postulate that with the same doping level, the presence of Si-III and Si-IX produce lower apparent resistivity.

Another issue is the pressure-quenched silicon deposit/monocrystalline substrate interface, where a barrier potential may exist. This barrier potential is caused by impurities segregating to the interface [9,19,20]. Since the monocrystalline substrate is much lower in doping concentration, the barrier potential, ϕ_B, at the interface is high enough to exhibit rectifying-like I-V characteristics [9-10,21-22].

CONCLUSIONS

Pressure-quenched silicon VPS deposits on monocrystalline <100> silicon substrates form polycrystalline layers which display polymorphs, such as Si-III and Si-IX in a Si-I matrix. The pressure-quenched deposit contains low resistivity Si-III and Si-IX which behave as a highly disordered conductive microcrystalline deposit, resulting in lower apparent resistivity. Photoconductivity measurements indicate that a light-detecting device may be fabricated form the polycrystalline silicon deposit. Also, the resulting structure exhibits that the pressure-quenched deposit on monocrystalline substrate forms a rectifier characteristic by thermal spray techniques.

ACKNOWLEDGMENTS

This work was supported by the MRSEC program of the National Science Foundation under Award No. 9632570 through the Center for Thermal Spray Research.

REFERENCES

1. S.J. Duclos, Y.K. Vohra and A.L. Ruoff, Phys. Rev. Lett. 58, 775 (1987).
2. M.I. McMahon and R.J. Nelmes, Phys. Rev. B 47, 8337 (1993).
3. Y.X. Zhao, F. Buehler, J.R. Sites and I.L. Spain, Solid State Comm. 59, 679 (1986).
4. R.H. Wentorf and J.S. Kasper, Science 139, 338 (1963).
5. J.M. Besson, E.H. Mokhtari, J. Gonzalez and G. Weill, Phys. Rev. Lett. 59, 473 (1987).
6. R. Goswami, J.B. Parise, S. Sampath and H. Herman, J. Mater. Res. 14, 3489 (1999).
7. S.Y. Tan, R.J. Gambino R. Goswami, S. Sampath and H. Herman. (to be published).
8. J.M. Houben, in *Proc. 2nd National Thermal Spray Conf.*, Long Beach, CA (ASM, Metals Park, OH, 1984), p. 1.
9. J.Y.W. Seto, J. Appl. Phys. 46, 5247 (1975).
10. M.S. Tyagi, in *Metal-Semiconductor Schottky Barrier Junctions and Their Applications,* edited by B.L. Sharma (Plenum Press, New York, 1984), p. 8.
11. H. Herman and S. Sampath, *Metallurgical and Ceramic Protective Coatings*, (Chapman and Hall, London, 1996), p. 261.
12. K. Mailhot, F. Gitzhofer, and M.I. Boulos, in Proc. 15th International Thermal Spray Conf. , Nice, France (1998) 1419.
13. F.M. Smits, Bell Syst. Tech. J. 41, 387 (1962).
14. P.Blood and J.W. Orton, *The Electrical Characterization of Semiconductors: Majority Carriers and Electron States*, (Academic Press, London, 1992), p. 109.
15. S.M. Sze, *Physics of Semiconductor Devices*, (Wiley, New York, 1981), p. 756.
16. T.Y. Tan, H. Foll and S.M. Hu, Philosophical Magazine A 44, 127 (1981).
17. R. Suryanarayanan and G. Zribi, Journal de Physique. Colloque C-1, 375 (1982).
18. T. Kamins, *Polycrystalline Silicon for Integrated Circuits*, 2nd ed. (Kluwer Academic Press, Bosten, 1998), p. 208.
19. J.W. Orton, B.J. Goldsmith, J.A. Chapman and M.J. Powell, J. Appl. Phys. 53, 1602 (1982).
20. J.W. Orton, and M.J. Powell, Rep. Prog. Phys. 43, 1263 (1980).
21. D. Stievenard and D. Deresmes, Appl. Phys. Lett. 67, 1570 (1995).
22. R.H. Bube, L.E. Benatar, M.N. Grimbergen and D. Redfield, J. Appl. Phys. 72, 5766 (1992).

DIRECT-WRITE TECHNIQUES FOR FABRICATING UNIQUE ANTENNAS

ROBERT M. TAYLOR,[1] KENNETH H. CHURCH,[1] JAMES CULVER,[2] AND STEVE EASON[2]

[1]CMS Technetronics, Inc., 5202-2 North Richmond Hill Road, Stillwater, OK, U.S.A. 74075.
[2]Raytheon Systems Corp., P.O. Box 12248, St. Petersburg, FL, U.S.A. 33733-2248.

ABSTRACT

The current fabrication methods used to produce many antennas are limited by variances in the precision and skill levels of individual laborers. These variances slow production and often create inconsistent results. As radio-frequency transmitter and receiver design moves towards higher operating frequencies, the physical dimensions of the supporting antennas decrease. Smaller sizes add new complexities to the fabrication of these antennas. Several designs that may be considered high-performance antennas are difficult to reproduce; many times, they cannot be fabricated at all due to the sophisticated patterning and precision necessary for successful function. Direct-write technologies provide the tools necessary to fabricate unique patterns in two and three dimensions. A demonstration of a directly written antenna, constructed from a silver-based thick-film paste pen-deposited onto cylindrical alumina substrates, is presented for review.

INTRODUCTION

Market trends in wireless communications and Global Positioning System (GPS) devices are requiring increasingly smaller, more affordable, and higher-performance antennas. To design small antennas to operate at lower frequencies requires mostly an empirical approach. Established theoretical analysis and simulation tools can provide the basis for a new antenna design. However, for these tools to be effective, several simplifying assumptions must be made about the design. The antenna must still be fabricated and tested to demonstrate its true performance. For the design process to be efficient, rapid-prototyping methods are required. The ability to efficiently write an antenna directly onto a device substrate can significantly reduce design time during the iterative process described above. The direct-write method described herein advances the field of rapid prototyping, specifically, as applied to such antenna designs as the half-wavelength cylindrical coupled slotline (CCS) antenna of Culver, King, and Weller [1], or the fed-slot and folded-slot CCS antennas of Scardelletti [2].

EXPERIMENT

The antennas presented herein were written using the OhmCraft MicroPen™ dispensing system. They are constructed from a silver-based thick-film paste, deposited onto cylindrical alumina substrates (1.27 cm diameter), then oven-fired after deposition. The thickness of the silver conductor film was held at 37 μm.

RESULTS

The direct-write deposition of half-wavelength slot antennas was demonstrated on cylindrical alumina substrates as shown in Figure 1 (*next page*). The radiation pattern is similar to that of a simple dipole, with the exception of a partial null that exists along one side of the

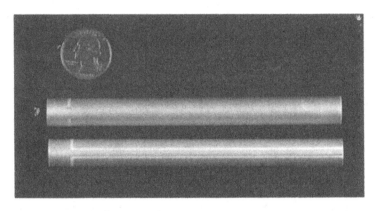

Figure 1: Cylindrical Slot Antenna

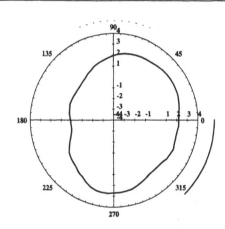

Figure 2: Measured *H*-Plane Gain Pattern for the Cylindrical Slot Antenna

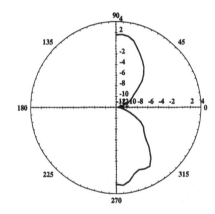

Figure 3: Measured *E*-plane Gain Pattern for the Cylindrical Slot Antenna

cylinder. The measured *E*- and *H*-plane gain patterns are shown in Figure 2 (*above*) and Figure 3 (*above*), respectively. The return loss measurement for this antenna, shown in Figure 4 (*next page*), indicates that the 3-dB bandwidth is approximately 11%.

The use of a cylindrical substrate eliminates the problems associated with reflections at the dielectric boundary. The slot radiates into free space, not into the substrate. Another advantage of the cylindrical substrate is that the parasitic even-mode on the coplanar waveguide feedline is naturally suppressed.

The success of the initial antenna fabrication led to the generation of a more complicated fractal design. A fractalized slot would increase the antenna bandwidth and would provide a more substantial test of the direct-write capabilities. The fractalized slot antenna shown as

Figure 5 (*next page*) was written on the same substrate as the antenna described in Figure 1. Figure 6 (*next page*) shows a magnified view of the slot that highlights the fine detail achieved.

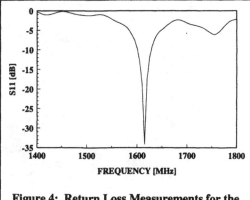

Figure 4: Return Loss Measurements for the Cylindrical Slot Antenna

CONCLUSIONS

Why use direct-write methods? Antenna designers are excited about the ease of fabrication and the freedom of flexibility provided by the direct-write techniques currently under development. The ability to write an antenna directly onto a substrate of choice, very quickly, is a significant improvement over the techniques currently used in the industry. The time saved in writing a single antenna element will allow the fabricator to write all of the design iterations available in a shorter time. The only limitation to achieving a preferred antenna response will be the designer's imagination.

The flexibility now available through direct-write technologies can be expressed in terms of an expanded choice of materials and the use of three-dimensional substrate geometry. The cylindrical antennas presented above are examples of three-dimensional antenna fabrications that are difficult and time-consuming to achieve with conventional fabrication techniques. Geometries that are more complex are possible and can be expected to influence the manner in which antennas are designed in the future. The low-processing-temperature metals and dielectrics that are becoming available allow an expanded choice of substrates for deposition. The ability to use substrates with a higher relative dielectric constant allows the physical size of an antenna to be reduced while maintaining the same electrical footprint. Smaller antennas are imperative as the physical size of the supporting electronics continues to shrink. Lastly, the ability of the direct-write process to produce layered structures will have an instant impact on the antenna designer. Patch antennas are currently assembled as alternating layers of metal and dielectric. Direct-write methods allow the elimination of the assembly step as the alternating layers are written in place.

FUTURE WORK

The authors are presently continuing work on their direct-write rapid-prototyping methods. Their main efforts are focused on the incorporation of lasers for the *in situ* firing of the deposited pastes and on automation.

ACKNOWLEDGMENTS

The authors would like to thank Lowell Matthews of CMS for manuscript assistance, and DARPA and the U.S. Navy for funding support. They would also like to thank William Warren of DARPA specifically for his support of research on direct-write technologies.

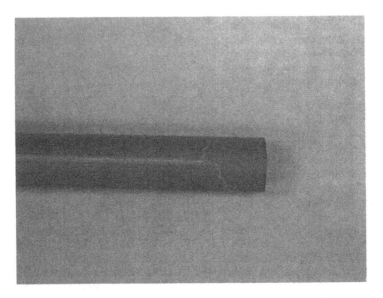

Figure 5: Cylindrical Fractal Slot Antenna

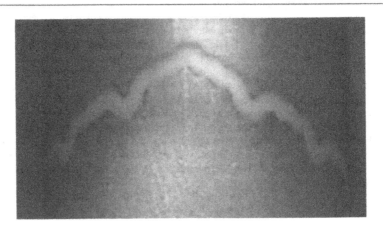

Figure 6: Close-Up of the Fractal Slot

REFERENCES

1. J. Culver, B. King, and T. Weller, "A 1.6 GHz Slot Antenna on a Cylindrical Alumina Substrate," 2000 International Symposium on Antennas and Propagation, in press.

2. M. Scardelletti, "Power Dividers and Printed Antennas Utilizing Coplanar Transmission Lines," M.S. Thesis, University of South Florida, Tampa, FL, April, 1999.

SYNTHESIS AND RADICAL POLYMERISATION OF PYROCARBONATE FUNCTIONALIZED MONOMERS: APPLICATION TO POSITIVE-TONE PHOTORESISTS

S.O.Vansteenkiste[*], Y. Martelé[*], E.H. Schacht[*], M. Van Damme[**], H. van Aert[**], J. Vermeersch[**].
*Polymer Materials Research Group, Department of Organic Chemistry, Faculty of Science, University of Gent, Krijgslaan 281 S4bis, B-9000 Gent, Belgium, stefan.vansteenkiste@rug.ac.be
**Agfa-Gevaert N.V., Septestraat 27, B-2640 Mortsel, Belgium.

ABSTRACT

The synthesis of two new *tert*-butyl carbonic anhydride monomers, 4-vinylbenzoic and methacrylic *tert*-butyl carbonic anhydride was achieved successfully. Radical polymerization at 45°C yielded *tert*-butyl pyrocarbonate protected materials. Thermographic analysis showed that both polymers decompose cleanly at 135°C. The lithographic performance of both materials was evaluated in presence of 2,4,6-tris-(trichloromethyl)-s-triazine as photoacid generating species. It was demonstrated that the large polarity change resulted in a chemically amplified positive photoresist system.

INTRODUCTION

The chemical amplification concept proposed by Ito and coworkers [1,2] has been an important contribution to the design of advanced photolithographic imaging systems. In chemically amplified processes, a cascade of acid-catalyzed deprotection steps is light-induced by conversion of photoacid generating species (PAGs). The most widely used acid-labile protecting group is the *tert*-butyloxycarbonyl (*tert*-BOC) group which is readily removed by H^+-catalyzed thermolysis. This process is accompied by a large polarity shift from a nonpolar to a polar state, inducing the necessarily solubility changes required for practical imaging applications. Poly[4-(*tert*-butyloxycarbonyl)oxystyrene] (PBOCSt) has been studied extensively as the model for *tert*-BOC-resist systems [3-6]. Acidolysis of the latter material results in the formation of polar phenol moieties, easily developed in alkaline medium, affording a positive working resist sytem. *tert*-Butyl esters such as *tert*-butyl methacrylate or 4-*tert*-butylcarboxystyrene also have been evaluated successfully as photoresist materials [7]. They offer an even larger shift in polarity, but only at higher temperature as compared to the *tert*-butyl carbonate analogues. Moreover, the latter polymers are often characterized by a high glass transition temperature Tg, requiring elevated processing temperatures.

Therefore, in this study a synthetic route was devised yielding the *tert*-butyl carbonic anhydrides of 4-carboxystyrene and methacrylic acid. This new class of thermo- and acid labile *tert*-butyl protected monomers and corresponding polymers retains a high thermosensitivity while still resulting in a carbonic acid after deprotection, providing a significant polarity change. In this paper we report the results of the polymer synthesis as well as an evaluation of the performance of poly(4-vinylbenzoic *tert*-butylcarbonic anhydride) and poly(methacrylic *tert*-butylcarbonic anhydride) as novel positive working UV-photoresist materials. Their thermal properties as well as their lithographic application as compared to *tert*-butyl ester and carbonates are discussed.

EXPERIMENT

Measurements

The average molecular weights $\overline{M}_w$ and $\overline{M}_n$ of the copolymers were determined by gel permeation chromotography (GPC) on a PL-Gel Mixed D-5μ 7.8x600 mm column, calibrated without correction by polystyrene standards, using chloroform as eluent and a MELZ LCD-212 refractive index detector. ^1H NMR and ^{13}C NMR measurements were carried out on a Brücker AM 360 MHz instrument with tetramethylsilane as an internal standard. Infra-red measurements were made on a Perkin Elmer 1600 Series FT-IR spectrometer. Thermographic analysis was performed on a Hi-Res TGA-2950 apparatus of T.A.-instruments at a heating rate of 3°C/min. under nitrogen atmosphere. Differential Scanning Calorimetric (DSC) data were obtained using a Modulated DSC-2920 apparatus of T.A.-instruments at a heating rate of 10°C/min. applying a nitrogen flow.

Materials

Dimethyl sulfoxide (Acros Chimica, Geel, Belgium) was dried over calcium hydride and distilled *in vacuo* prior to use. Toluene and tetrahydrofuran (THF) (Acros Chimica, Geel, Belgium) were refluxed over sodium metal for 12 h and then distilled. 4-bromomethylbenzoic acid was supplied by Aldrich, Bornem, Belgium and used without further purification. 2,4,6-Tris-(trichloromethyl)-s-triazine was obtained from P.C.A.S., Longjumeau, France. All other chemicals were purchased from Acros Chimica, Geel, Belgium and were used as commercial grade. Bis(4-*tert*-butylcyclohexyl)peroxydicarbonate (Perkadox 16) was kindly supplied by Akzo Chemicals International B.V., Amersfoort, The Netherlands. 4-vinylbenzoic acid was prepared as described in literature [8]. 4-vinylbenzoyl chloride was synthesized following the method of Ishizone et al [9].

Synthesis

Synthesis of 4-vinylbenzoic tert-butylcarbonic anhydride (**1**). A solution of dry *tert*-butanol (25 mL, 0.265 mol) in 80 mL THF is treated under nitrogen at 0°C with 80 mL nBuLi (2.5M solution in hexane, 0.200 mol). After diluting with one equivalent volume of dimethyl formamide (DMF), the reaction mixture is saturated with CO_2. To this mixture a solution of 4-vinylbenzoyl chloride (25g, 0.150 mol) in 25 mL DMF is added dropwise (5 min). The reaction is completed within 10 min. at room temperature. Subsequently, the reaction mixture is poured into a mixture of ice and pentane/diethyl ether (1/1) and extracted twice with saturated $NaHCO_3$. Finally, the organic layer is dried over anhydrous magnesium sulfate, and concentrated under reduced pressure. TLC-analysis (eluent hexane/acetone, 90/10) showed a single spot (R_f = 0.37). Yield: 35.3g, 95%.
FT-IR (neat): ν = 2984 (CH), 1799 and 1740 (C=O, anhydride), 1625 (C=C), 1607, 1566, 1508 and 1476 (C=C, phenyl ring), 1403 and 1372 cm^{-1} (CH$_3$, *tert*-butyl). ^1H NMR (360 MHz, CDCl$_3$, TMS): δ = 8.02 and 7.50 (2xd, 4H, phenyl ring; J_{A-B}=9.0Hz), 6.78 (dd, 1H, -C*H*=CH$_2$, J=18.0 and 11.2Hz), 5.90 and 5.43 (2xd, 2H, -CH=C*H*$_2$), 1.59 (s, 9H, -C*H*$_3$).

Synthesis of methacrylic tert-butylcarbonic anhydride (**2**). A solution of dry *tert*-butanol (30 mL, 0.318 mol) in 200 mL THF is treated under nitrogen at 0°C with 100 mL nBuLi (2.5M solution in hexane, 0.250 mol). After diluting with one equivalent volume of DMF, the reaction mixture is saturated with CO_2. To this mixture a solution of methacryloyl chloride (33.5g, 0.32

mol) in 25 mL DMF is added dropwise (5 min.). The reaction is completed within 10 min. at room temperature. Subsequently, the reaction mixture is poured into a mixture of ice and pentane/diethyl ether (1/1, 500 mL) and extracted with a saturated aqueous solution of NaHCO$_3$. Finally, the organic layer is dried over MgSO$_4$, and concentrated under reduced pressure. TLC-analysis (eluent hexane/acetone, 90/10) showed a single spot (R$_f$ = 0.40). Yield: 46.0g, 99%.

FT-IR (neat): ν = 2982 (CH), 1800 and 1739 (C=O, anhydride), 1636 (C=C), 1396 and 1372 cm^{-1} (CH$_3$, tert-butyl). ^{1}H NMR (360 MHz, CDCl$_3$, TMS): δ = 6.23 and 5.78 (2xd, 2H, -C=CH$_2$), 1.96 (s, -C-CH$_3$, methacryl), 1.55 (s, 9H, -CH$_3$, tert-butyl).

Free radical polymerization procedure (3-4). A 30 wt.-% solution of **1** or **2** in anhydrous toluene containing 4 mole% of Bis(4-*tert*-butylcyclohexyl)-peroxydicarbonate (Perkadox 16) was deoxygenized by flushing with N$_2$. The mixture was heated at 45°C for 4 hours and after being diluted with toluene, the polymer was prepicitated in methanol. The white solid was washed with methanol and diethyl ether and dried. Yield: 84% (**3**) and 25% (**4**).

Polymers **3** and **4** obtained were characterized by FT-IR, ^{1}H and ^{13}C NMR spectroscopies:

3: FT-IR (KBr): ν = 1800 and 1737 (C=O, anhydride), 1396 and 1372 cm^{-1} (-CH$_3$, tert-butyl). ^{1}H NMR (360 MHz, CDCl$_3$, TMS): δ = 7.70 and 6.50 (2x broad doublet, 4H, phenyl ring, 5.90 and 5.43 (2x broad doublet, 2H, -CH=CH$_2$, pending styryl moieties), 1.20-2.20 (m, 3H, -CH$_2$CH-Phe), 1.59 (s, 9H, -CH$_3$). GPC-analysis: $\overline{M}_w$ = 367,000 $\overline{M}_w$ /$\overline{M}_n$ = 15.26; Tg = 78°C.

4: FT-IR (KBr): ν = 1800 and 1756 (C=O, anhydride), 1395 and 1373 cm^{-1} (-CH$_3$, tert-butyl). ^{1}H NMR (360 MHz, CDCl$_3$, TMS): δ = 6.23 and 5.78 (2x broad doublet, 2H, -CH=CH$_2$, pending methacryl moieties), 1.98 (broad s, 3H, -CH$_3$), 1.00-1.80 (m, 2H, -CH2-), 1.50 (s, 9H, -CH$_3$). GPC-analysis: $\overline{M}_w$ = 44,100 $\overline{M}_n$ = 24,200 $\overline{M}_w$ /$\overline{M}_n$ =1.82; Tg = 57°C.

Preparation of the lithographic printing plate. Onto the lithographic aluminum base was coated a photosensitive composition prepared by mixing 0.4 g of polymer **3** or **4**, 0.05 g of 2,4,6-tris-(trichloromethyl)-s-triazine, 0.025g thioxanton and 0.179 g of a dye dispersion (14 wt.-% HOSTAPERM BLAU B3G01 in methyl ethyl ketone) and 5.4 mL methyl ethyl ketone. This composition was coated to a wet coating thickness of 20 μm resulting in a dry layer thickness of approximately 1.2 g/m^2.

RESULTS AND DISCUSSION

Synthesis of *tert*-butylcarbonic anhydride monomers

According to the early synthetic work described by Tarbell et al. [10], a very useful preparation procedure involves the carbonation of an alkoxide. This pathway is particulary valuable when the corresponding chlorocarbonate is unstable, as with *tert*-butyl alcohol. However, when a ice-chilled solution of potassium t.butoxide in THF was treated with a excess of dry carbondioxide, already after a few minutes, the mixture began to gel. This phenomenon prevented a complete conversion of the carbonation reaction, resulting in unacceptable amounts of *tert*-butyl ester after the addition of the acid chloride. Hence, DMF was used as a cosolvent. Moreover, by introducing lithium as a counterion, an excellent solubility of the corresponding *tert*-butyl carbonate was observed. Subsequential addition of 4-vinylbenzoyl chloride or methacryloyl chloride yielded quantitatively the desired pyrocarbonate monomers within minutes (Fig.1).

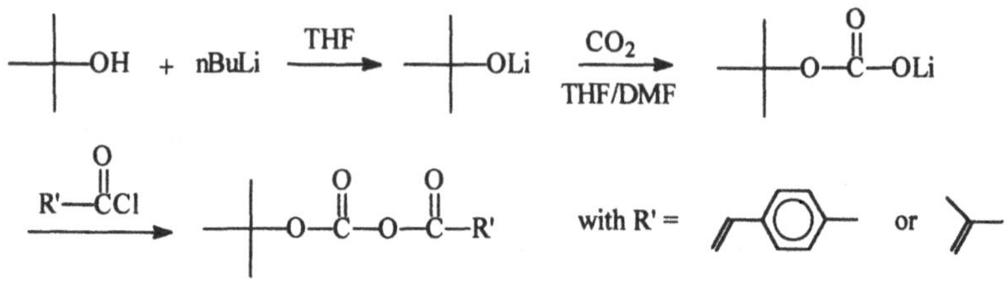

Figure 1: General pathway for the preparation of pyrocarbonate monomers.

Radical polymerization of tert-butylcarbonic anhydride monomers

Initial radical polymerization procedures using 1 mol-% 2,2'-azoisobutyronitrile (AIBN) at 65°C in a 30 wt.-% solution of the monomers **1** or **2** in toluene resulted quickly in a complete crosslinking of the reaction mixture. The reason for this undesired gelation can be found in the thermal decomposition of the *tert*-butyl carbonic anhydride monomers. In addition to the expected concurrent evolution of isobutene and carbon dioxide, thermal decomposition of the monomers **1-2** during polymerization yields bifunctional anhydrides which act subsequently as crosslinking agents (Fig.2).

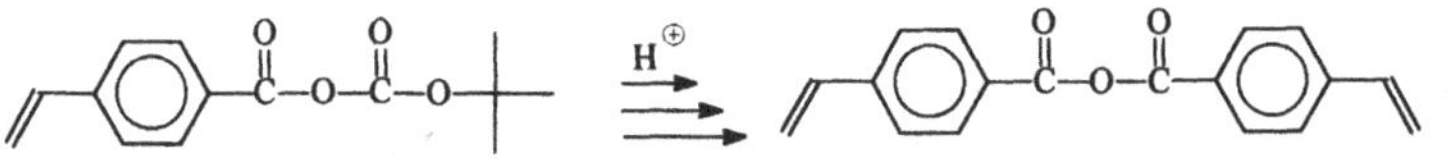

Figure 2: Formation of bifunctional anhydrides.

The formation of 4-vinylbenzoic anhydride was further confirmed by monitoring the thermal decomposition of **1** in solution as a function of time. Following SiO_2-column chromatography, two non volatile degradation products could be isolated. By [^1]H NMR and FT-IR spectroscopic analysis the major product was identified as the symmetrical anhydride, while the minor product was consistent with the *tert*-butyl benzoic ester. This was in good agreement with studies conducted by Tarbell [11] using 4-nitrobenzoic *tert*-butylcarbonic anhydride as a model compound. Similar phenomena were observed with the methacrylic *tert*-butyl carbonic anhydride monomer. Therefore, Perkadox 16 was selected as a suitable radical initiator for the polymerization of monomers **1-2**. Its thermal lability allowed the polymerizations to proceed smoothly at temperatures as low as 35-40°C (Fig.3).

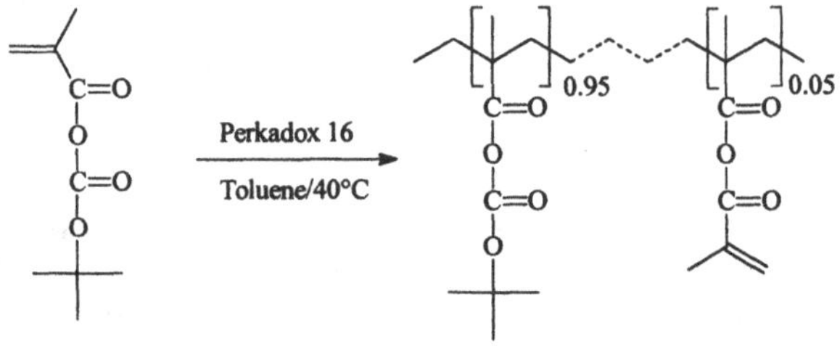

Figure 3: Radical polymerization of methacrylic *tert*-butylcarbonic anhydride at low temperature.

Nevertheless, structural analysis of polymers **3-4** indicated the occurrence (±5 mol-%) of pending styryl and methacrylic vinyl bonds. Moreover, the high molecular weights ($\overline{M}_w$ = 367,000) and the large molecular weight distribution ($\overline{M}_w$ /$\overline{M}_n$ =15.3) of **3** in the case of high conversions clearly resulted from a partial crosslinking reaction during the polymerization, even when using up to 4 mol-% of initiator.

Thermal analysis of polymers 3-4

The thermal decomposition of the prepared polymers was monitored by a thermographic analysis. As shown in Fig.4, the pyrocarbonate structures decompose sharply at 140°C. This thermolysis occurs at a mutch lower temperature as compared with the *tert*-butyl ester (T_d = 253°C) or *tert*-butyl carbonate (T_d = 199°C). The mass loss observed, respectively 36.4% for **3** and 55.7% for **4**, is in good agreement with the expected evolution of isobutene and carbon dioxide.

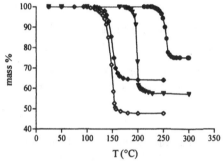

Figure 4: Thermographic analysis of *tert*-butyl protected polymers: -•- Poly[4-(*tert*-butyloxy-carbonyl)styrene], -∇- Poly[4-(*tert*-butyloxycarbonyl)oxystyrene], -♦- Poly (4-vinylbenzoic *tert*-butylcarbonic anhydride), -◊- Poly (methacrylic *tert*-butylcarbonic anhydride).

Lithographic evaluation

Both polymers **3-4** were, in the presence of a photoacid system, evaluated as a positive working photosensitive composition for lithographic printing plates. As UV-sensitive photoacid system, 2,4,6-tris-(trichloromethyl)-s-triazine was used as non-ionic latent Brönsted acid in combination with thioxanton as UV-sensitizer ($\lambda_{max,abs}$ = 370 nm). The lithographic printing plate was placed in face-to-face contact with a test target with a 150 lines per inch screen as well as with fine positive and negative lines. The imaging element was then exposed under vacuum conditions with a UV-exposure unit (DL2000, Agfa-Gevaert, intensity : 10^4 mW/m^2) for 250 s. In the unmasked areas hydrochloric acid is formed as a result of the UV-sensitized decomposition of the 2,4,6-tris-(trichloromethyl)-s-triazine. In a next step, the sample was heated in an oven for 5 min at 120°C. In the exposed areas the thermal decomposition temperature of the pyrocarbonates is reduced to below 120°C due to the presence of the hydrochloric acid formed upon exposure. As a result thermal decomposition occurs only in the exposed areas effecting a dramatic change in polarity. Since this temperature is not reached during the baking step, the resist remains unchanged in the non-exposed areas. Subsequently, the imaging element was processed with Ozasol EP26 to remove the exposed areas resulting in a positive working lithographic printing plate. Due to the formation of polar carboxylic moieties, the exposed areas are easily developed. The plate has a very sharp image resulting in a screen rendering on plate of 1 - 99%.

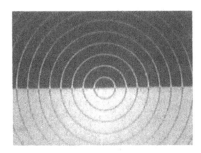

Figure 5: Photograph of a lithographic printing plate consisting of 4μm positive and negative lines.

The obtained image on the lithographic printing plate could be utilized to print on a conventional offset press using a commonly employed ink and fountain. Excellent copies were achieved.

ACKNOWLEDGEMENTS

The authors wish to thank the Flemish Institute for the Promotion of Scientific-Technological Research in Industry (IWT), the Fund for Scientific Research - Flandern (FWO) and the Belgian Ministry of Scientific Programming, IUAP/PAI-IV/11 for their financial support.

REFERENCES

1. H. Ito, C.Willson, and J.Fréchet, *Digest of Technical Papers of 1982 Symposium on VLSI Technology*, 1982, p.86.
2. H.Ito, and C.Willson, *Technical Papers of SPE Regional Technical Conference on Photo-polymers*, 1982, p.331.
3. H.Ito, and C.Willson in *Polymers in Electronics,* edited by T. Davidson (ACS Symposium Series **242**, Washington DC, 1984), p.11.
4. J.Fréchet, E.Eichler, H.Ito, and C. Willson, Polymer **24**, p. 995 (1983).
5. H.Ito, C.Willson, J..Fréchet, M.Farrall, and E.Eichler, Macromolecules **16**, p.510 (1983).
6. H.Ito, and C.Willson, *Proceedings of the ACS Division of Organic Coatings and Plastics Chemistry*, 1983, vol. 48, p.60.
7. R.Allen, G.Wallraff, W.Hinsberg, W.Conley, and R.Kunz, Solid State Technol. **53**, (1993).
8. L.M.Harwood, C.J.Moody, *Experimental Organic Chemistry, Principles and Practice,* Blackwell Scientific Publication, Oxford, 1989, p.588.
9. T.Ishizone, A.Hirao, and S.Nakahama, Macromolecules **22**, 2895 (1989).
10.D.Tarbell, Acc. Chem. Res. **2**, 296 (1969).
11.R.Chow, and D.Tarbell, J. Org. Chem. **32**, 2188 (1967).

Pattern Writing by Implantation in a Large-scale PSII System with Planar Inductively Coupled Plasma Source

Lingling Wu[*], Hongjun Gao[**], Dennis M. Manos[*]
[*]Applied Science Department, College of William and Mary, Williamsburg, VA 23187,
dmanos@as.wm.edu
[**]Solid State Division, Oak Ridge National Laboratory, Oak Ridge, TN 37831. Present Address:
Beijing Laboratory of Vacuum Physics, P. O. Box 27024, Beijing, P. R. China

Abstract

A large-scale plasma source immersion ion implantation (PSII) system with planar coil RFI plasma source has been used to study an inkless, deposition-free, mask-based surface conversion patterning as an alternative to direct writing techniques on large-area substrates by implantation. The apparatus has a 0.61 m ID and 0.51 m tall chamber, with a base pressure in the 10^{-8} Torr range, making it one of the largest PSII presently available. The system uses a 0.43 m ID planar rf antenna to produce dense plasma capable of large-area, uniform materials treatment. Metallic and semiconductor samples have been implanted through masks to produce small geometric patterns of interest for device manufacturing. Si gratings were also implanted to study application to smaller features. Samples are characterized by AES, TEM and variable-angle spectroscopic ellipsometry. Composition depth profiles obtained by AES and VASE are compared. Measured lateral and depth profiles are compared to the mask features to assess lateral diffusion, pattern transfer fidelity, and wall-effects. The paper also presents the results of MAGIC calculations of the flux and angle of ion trajectories through the boundary layer predicting the magnitude of flux as a function of 3-D location on objects in the expanding sheath.

Introduction

High-energy ion implantation is an important surface modification technique, which causes minimal dimensional change, and avoids thermal distortion of the surface profile of the treated object [1, 2, 3]. Plasma source immersion ion implantation (PSII) is advantageous compared to traditional beam line implantation technique for large areas, complicated work piece shapes, and high dose applications [1, 2]. The uniformity of implantation is determined by the spatial plasma density distribution and the plasma sheath distribution. It does not require beam rastering to achieve dose uniformity over large implantation areas. There is also no need for tilting or rotating of the work piece to treat all the surfaces. This greatly simplifies the mechanical construction of the implantation chamber and reduces the cost of building the system.

Different plasma sources for PSII have been used, such as hot filament [2, 3, 4], cylindrical coil RF inductively coupled plasma [5], rf capacitively coupled plasma [2, 6], microwave plasma [7], and glow discharge [2]. To realize PSII technique's potential of treating large work pieces, several large-scale PSII systems have also been built, including PSII systems at Los Alamos National laboratory [6], Hughes Research Laboratories [3], and the University of Wisconsin [2]. None of these large-scale PSII systems takes advantages of the planar coil rf inductively (RFI) coupled plasma source, which is widely used to generate low-pressure, high-density discharge and demonstrates the potential for large area processing and improved spatial uniformity [1, 8]. Since rf power is coupled to the plasma inside the vacuum chamber usually through a quartz window, there are no electrical connections or supporting apparatus inside the chamber to take up extra chamber space to complicate chamber geometry.

Mat. Res. Soc. Symp. Proc. Vol. 624 © 2000 Materials Research Society

Prior operations of the PSII system at the College of William and Mary with hot filament and DC glow discharge plasmas resulted in ion densities estimated to be in the low $10^9 / cm^3$ range [4], in agreement with values previously reported for other large implantation systems [3, 6]. Such low values of n_e made it difficult to produce the desired pulse shape and to control the pulse temporal profile. Although higher densities in the filament driven source were achieved by increasing the filament temperature and bias [4], this led to shortened filament life and the possibility of contamination of work pieces. Similarly, sputtered contaminants made it difficult to increase electron density in a dc glow by raising voltage. The requirement of long mean-free-path for the implanted ions eliminated pressure control of n_e as an option. In addition, both filament and dc glow discharge plasmas demonstrated large non-uniformity over the large radial and vertical dimensions of this chamber (0.61 m ID and 0.51 m tall), which is unacceptable for large area uniform materials processing.

To overcome these difficulties, a large rf inductively coupled (RFI) plasma source has been designed and built for the PSII system. With a 0.43 m in diameter rf antenna sitting on a 0.57 m in diameter quartz window, stable and uniform plasma can be generated in the chamber. More details of the system characterization and operation will be given in a later paper.

In this work, pattern writing has been performed by exposing metal and semiconductor samples through masks, to high ion doses. AFM calibration gratings, made from Si, were also implanted. Implantation depth profiles were characterized by variable-angle spectroscopic ellipsometry (VASE) and AES and compared to TEM results. Measured lateral and depth profiles are compared to the feature morphology to assess lateral diffusion, pattern transfer fidelity, and wall-effects on the depth profile. The paper also presents the results of MAGIC modeling of flux and angle of ion trajectories through the boundary layer predicting the magnitude of flux as a function of 3-D location on objects in the expanding sheath, and to evaluate the fidelity of pattern transfer as a function of feature size.

Experiment

Two kinds of substrates were used for studying pattern transfer: flat substrates and silicon gratings. Flat substrates, including silicon wafers (100 mm in diameter, 0.5 mm thick), polished stainless steel plate (9.12 cm in diameter, 1.2 cm thick), and pure Ti and Ta sheets (25 mm×25 mm, 0.5 mm thick for Ta, 1 mm thick for Ti), were implanted through a 150 μm thick metal mask with 100 μm wide lines and 100 μm ID holes. The mask was weighted to secure it to the stage using a shaped metal ring and a metal cube. The ring is 2.5 cm tall, with a 2.5 cm OD and 2.3 cm ID. The cube is 3.8 cm×3.8 cm×3.8 cm. The ring and the cube were carefully positioned on the 100-mm Si wafer to study wall effect on dose and sheath distributions. Pure silicon gratings with trapezoidal groove cross-sections and of 10 μm groove separation were also implanted.

Research grade nitrogen (99.999%) was used as the gas source. A base pressure of $\sim 10^{-8} Torr$ and operating pressures of $5 \times 10^{-4} Torr$ to $2.5 \times 10^{-3} Torr$ were used. Silicon wafer implants were performed at pulse voltages of 50 kV and repetition rate of 20 Hz for 10 hrs. Stainless steel was implanted at 50 kV at 25 Hz for 9 hrs. The Ti and Ta sheets were implanted through the mask at 40 kV at 25 Hz for 7.5 hrs. Si gratings were implanted at 25 kV at 40 Hz for 5 hrs. Estimated temperature rise for these implantation runs was less than $50^\circ C$.

AES and XPS spectra were taken using a VG Scientific MARKII ESCALAB equipped with EX05 ion gun for depth profiling. Nitrogen concentration depth profiles for high dose nitrogen implanted Si and Ti were obtained by AES. The sputtered depths were calibrated by surface profilometry. The overlapping of nitrogen and titanium peaks and compensation for the presence of oxygen were considered when calculating the nitrogen concentration [9].

A variable-angle spectroscopic ellipsometer (VASE) [10] was used to study the optical properties of the nitrogen-implanted silicon and stainless steel samples. Three incident angles (70, 75, and 80 degrees) were used for each area. The incident optical beam was about 3 mm in diameter. Because the as-received Ti and Ta samples did not have smooth optical surfaces, ellipsometry could not be used to examine them.

The cross-section of nitrogen implanted Si gratings was examined by TEM at Solid State Division of Oak Ridge National Laboratory.

Results and Discussion

The implanted Ti surface showed the gold color characteristic of stoichiometric TiN, with slight pink hue because of the slightly oxidized surface layer. The AES depth profile for high dose nitrogen implantation is shown in Figure 1 for a Ti substrate. The nitrogen concentration fluctuates around the level for stoichiometric TiN (50%) throughout most of the implanted depth, decreasing sharply at ~5000 Å. The native oxide layer of the original substrate was not stripped prior to implantation, accounting for the relatively low nitrogen concentration in this range. Similar nitrogen depth profiles were obtained for the implanted Si wafer, again showing a thick native oxide layer that reduced the nitrogen reaction with silicon, producing low nitrogen concentration. The experimental implantation depth profiles are in good agreement with the high-dose model calculation of Profile Code, which predicts constant concentration for most of the implanted depth and the steep decrease at the end of range [11].

The ellipsometry results for nitrogen implanted Si and stainless steel samples confirmed the

Figure 1. 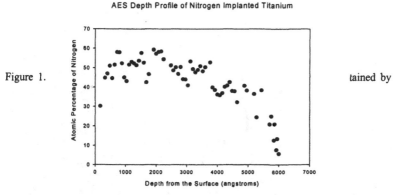 tained by

film-like depth profiles observed by AES and predicted by Profile Code. Table I shows VASE results for 7 different spots on implanted Si wafer. Adding a graded optical layer to model the gradual concentration changes typical of low-dose implantation depth profile, did not improve

Table I. Implanted layer thickness obtained by variable angle spectroscopic ellipsometry (VASE) study of different surface areas on nitrogen implanted Si wafer

Test Position	A	B	C	D	E	F	G
Thickness of Layer (Å)	1770±20	914±3	1055±3	1580±20	851±2	1620±20	1820±20
Color of Spot	Golden	Blue	Light Blue	Golden (lighter)	Blue	Golden (lighter)	Golden

the fits to the data. The optical properties of the implanted layer closely resemble that of a thin film. These optical methods thus provide a distribution for high dose implantation, since the dose is in proportion to converted layer thickness and thickness can be correlated to area color. They are non-destructive and quick compared to other depth profiling techniques, such as AES and SIMS. The color of 100 μm lines and 100 μm I.D. dots produced by implanting through the metal mask were studied using an optical microscope fitted with a CCD color camera and TV monitor. The edges between the implanted and un-implanted area were narrower than the resolution of our camera, indicating pattern transfer contrast considerably sharper than 2 μm. A 5.7 um wide, blue-colored annular zone appears along the circumference of the exposed area. Ellipsometry indicated that this annulus received only $\approx 40\%$ of the maximum dose received at the center of a 100 um hole. The sample stage temperature did not rise significantly during implantation. Analysis of the annular zone just outside the implanted circle confirmed our assumption that lateral thermal diffusion may be ignored.

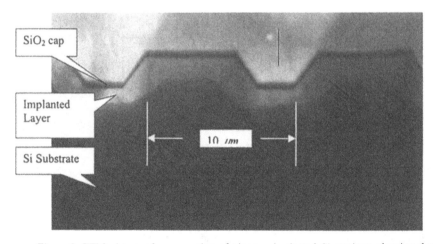

Figure 2. TEM picture of cross-section of nitrogen implanted Si gratings, showing the nitrogen-implanted layer, the Si substrate, and the SiO_2 cap layer.

The AFM calibration gratings were exposed to study the dose distribution in and around narrow features. The trapezoidal groove cross-sections provide virtual mask stripes over the flat Si surface. Our apparatus may be operated in a manner which allows us to co-deposit SiO_2. This mode will be discussed in a future paper. Figure 2 shows TEM picture of Si gratings implanted in this mode. As seen in that figure, the implanted layer is capped with co-deposited SiO₂. The flat tops and bottoms of the grooves show a similar implanted layer thickness. The sloped sidewalls were not uniformly implanted. The converted layer near the corner where the bottom and the sidewalls meet is apparently thinner, while the layer near the corner where the top and the sidewall meet is much thicker.

We modeled the dose and angle distribution for mask-based pattern writing using MAGIC, an electromagnetic particle-in-cell, finite-difference, time-domain code for simulating plasma physics process [12], which we adapted for this purpose. The chamber and sample stage geometry included in the models is assumed to be azimuthally symmetric. Thus only a 2D half-plane cut is necessary for full simulation. We have used this simulation to examine matrix sheath

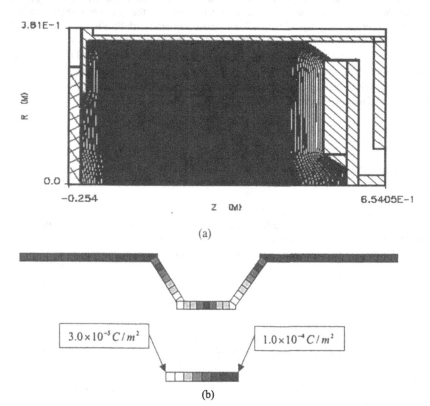

(a)

(b)

Figure 3. MAGIC modeling of PSII process in a trench. (a) Positive nitrogen ion distribution at time $1.2 \times 10^{-7} s$. (b) Ion flux distribution along the trench surface for a $2.5 \times 10^{-7} s$ simulation. (The gray scale is shown below the profile)

formation, sheath propagation, evolving to the steady Child-Law sheath; ion trajectories, ion depletion, and the distribution of ion flux for different mask shapes and aspect ratios. In the simulations shown here, the time step was $5 \times 10^{-10} s$ over a simulated implantation time of

$2.5 \times 10^{-7} s$, representing the initiation phase of a quasi-steady state lasting throughout the $10^{-5} s$ pulses. The modeled ion plasma density was $10^{10} / cm^3$ and applied voltage ~10 kV. The positive ion distribution at $t = 1.2 \times 10^{-7}$ s modeled for a groove with a sidewall to bottom angle of 54.4 degrees is shown in Figure 3. The shallower angle of incidence at the slopes sidewall predicts a non-uniform distribution consistent with TEM observations. The ion flux distribution (Figure 3b) predicts that the bottom of the slope will receive a factor of 2 to 3 less flux, also in agreement with TEM results.

Conclusion

High dose nitrogen implantation into various substrates was performed, yielding film-like depth profiles. A stoichiometric *TiN* layer was produced on the Ti surface. Depth profiles obtained experimentally by AES, VASE and those modeled with Profile Code are in good agreement. Patterns $100 \, um$ in line-width were transferred by implantation, with lateral contrast better than $2 \, \mu m$. The TEM study of nitrogen implanted Si gratings showed implantation can produce sharp patterns as small as $3 \, \mu m$ with good uniformity, in agreement with MAGIC code simulation. With carefully designed masks and good control of implantation parameters, PSII can be used for pattern writing of smaller features.

Reference

[1] M. A. Lieberman, A. J. Lichtenberg, Principles of Plasma Discharges and Materials Processing, John Wiley & Sons, INC., 1994, pp. 387-411.
[2] S. M. Malik, K. Sridharan, R. P. Fetherston, A. Chen, and J. R. Conrad, JVST B, 12 (2), p. 843 (1994).
[3] J. N. Matossian, JVST B, 12 (2), p. 850 (1994).
[4] T. J. Venhaus, Plasma Source Ion Implantation of High Voltage Electrodes, Ph.D dissertation, College of William and Mary, 1999, pp. 25-80.
[5] M. Tuszewski, J. T. Scheuer, I. H. Campbell, and B. K. Laurich, JVST B, 12 (2), p. 973 (1994).
[6] B. P. Wood, I. Henins, R. J. Gribble, W.A. Reass, R. J. Faehl, M. A. Nastasi, and D. J. Rej, JVST B, 12 (2), p. 870 (1994).
[7] S. Qin, C. Chan, JVST B, 12 (2), p. 962 (1994).
[8] J. Hopwood, Plasma Sources Sci. and Technol., 1, p. 109 (1992).
[9] A. Chen, J. Firmiss, and J. R. Conrad, JVST B, 12 (2), p. 918 (1994).
[10] Guide to using WVASE32™, J. A. Woollam Co. Inc., Lincoln, NE 68508, USA.
[11] Profile Code Software Instruction Manual, Version 3.20, Implant Sciences Corporation, Wakefield, Massachusetts.
[12] Bruce Goplen, Larry Ludeking, David Smithe, Magic User's Manual, 1997, Mission Research Corporation, Newington, Virginia.

Lateral Dye Distribution with Ink-Jet Dye Doping of Polymer Organic Light Emitting Diodes

Conor F. Madigan, Thomas R. Hebner, and J. C. Sturm
Department of Electrical Engineering, Princeton University, Princeton, NJ 08544
Richard A. Register, Sandra Troian
Department of Chemical Engineering, Princeton University, Princeton, NJ 08544

ABSTRACT

In this work we investigate the lateral dye distribution resulting from the dye doping of a thin polymer film by ink-jet printing (IJP) for the integration of color organic light emitting diodes (OLED's). The dye is found to segregate into distinct outer rings following rapid droplet evaporation, while slower evaporation rates are found to significantly reduce (or eliminate) this effect. The dye segregation phenomena are found to depend critically on the mechanisms of droplet evaporation. Good dye uniformity was obtained using a low vapor pressure solvent, and integrated, 250 micron red, green, and blue polymer organic light emitting diodes (OLED's) were fabricated with this technique. These devices had good color uniformity over most of the device area and similar electrical properties to comparable spin-coated devices without IJP.

I. INTRODUCTION

Polymer OLEDs are a promising technology for flat panel displays [1]. These devices typically consist of a multi-layer sandwich of a transparent substrate, a transparent anode (in our work, Indium Tin Oxide), a thin film of an organic polymer blend (in our work, the hole-transporting polymer Poly(N-vinyl carbazole) (PVK) doped with an emissive dye), and a reflecting cathode. (The device structure and principle of operation are illustrated in Fig. 1.) When current is driven through the device, holes from the anode and electrons from the cathode combine in the organic film to form excitons, which emit light as they decay. It has been shown that by doping the organic active layer with a small amount of dye one can tune the emission wavelength [2]. Currently, spin coating is the standard method for depositing a polymer

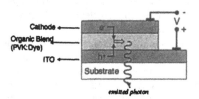

Fig. 1. Basic OLED structure and operation.

blend film, which produces a uniform layer of polymer. However, this does not allow one to integrate multiple colors onto a single substrate, because the film is the same everywhere.

It has been proposed previously [3,4] to locally dope an initially undoped PVK film by depositing droplets of a dye solution onto the film surface and allowing the droplet to evaporate. This task is ideally suited to IJP. This basic procedure is outlined schematically in Fig. 2. To integrate red, green, and blue devices onto a single substrate, solutions of red, green, and blue dyes are locally printed onto the same substrate. Our

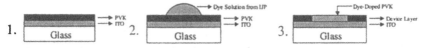

Fig. 2. Procedure for dye-doping by ink jet printing.

objectives in developing this technique are to (a) produce a uniform dye distribution over the device area and film depth and (b) maintain the initial film morphology (so that the electrical device characteristics are not degraded). In employing IJP, this technique should be relatively inexpensive to perform and applicable to large area substrates.

II. IJP DROPLET FORMATION

Our experimental apparatus consists of a piezo-electric type ink jet printer (supplied by MicroDrop GmbH) with a glass print head (which is therefore resistant solvent damage) and x-y-z print head stage motion. In addition, our system has integrated digital imaging equipment, allowing us to view droplet ejection from the print head nozzle directly and to view drying droplets (from above) under high magnification. The print nozzle consists of a 25 μL capillary cavity surrounded by a piezo-electric sleeves which can contract and expand the fluid cavity (see Fig. 3). To drive a droplet out of the nozzle, a first positive then negative pressure pulse is applied to the fluid through the voltages applied

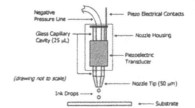

Fig. 3. Schematic of Ink Jet Nozzle. (The negative pressure line is used to balance gravitational forces.)

to the piezo-electric sleeve. The positive pulse drives the fluid down into the nozzle tip (which is 50 μm in diameter), and if sufficient energy is supplied by this pulse, a droplet (with diameter slightly larger than the nozzle diameter) will be ejected. The essential free parameters for controlling droplet ejection are the piezo voltage and the pulse duration.

Several parameters are relevant to understanding droplet formation from an ink jet printer: system geometry (i.e. of capillary cavity and nozzle), properties of the fluid being printed (i.e. viscosity, surface tension, and density), and the relationship between the applied voltage and the resulting pressure pulses. Even each of these parameters is known in detail, a closed-form analysis of the governing Navier-Stokes equations is not possible. However, there is a rapidly growing literature on approximate solutions and numerical simulations of ink-jet flow (e.g. [5-7]), and some important trends are observed.

Droplet formation can be divided up into four regimes, based on the applied voltage. At very low voltages, no droplet is ejected, because the applied pressure pulse has insufficient energy. At higher voltages, single, stable droplet ejection is observed. At still higher voltages, satellite droplets are observed along with the main droplet, and this regime is generally less stable than the single droplet regime. Finally, at yet higher voltages, the ejected fluid will not form into a main droplet and satellites, but break up into an uncontrolled spray, or "jet." These four regimes were clearly demonstrated experimentally on our system with Dimethyl Sulfoxide (DMSO) (see Fig. 4).

 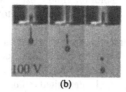 

(a)   (b)   (c)

Fig. 4. Observation of droplet formation regimes with DMSO. No drops were observed for voltages below 60V. (a) Single droplet regime. (b) Satellite droplet regime. (c) "Jet" regime.

We investigated droplet formation with numerous solvents, and found that we could not form droplets (stable or otherwise) at any voltage or pulse duration for solvents with low viscosity and low surface tension on our system. This phenomenon is not reported in the literature. Nevertheless, this introduced a new constraint on solvent selection for our work. The results of our solvent IJP characterization experiments are summarized in Figure 5.

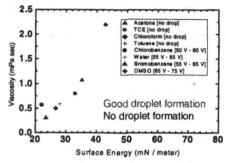

Fig. 5. IJP Solvent Space. For printable solvents, the voltage range for the single droplet regime is given in the legend in brackets.

III. LATERAL DYE DISTRIBUTION

Once a droplet of dye solution is deposited on the polymer film, the actual dye doping occurs over the course of the evaporation of the solvent. Understanding how the dye is laterally distributed in the film, therefore, requires an understanding of how the droplet dries. There are essentially two basic types of droplet drying: unpinned and pinned (see Fig. 6).

It is well known that a droplet of fluid on surface has a characteristic contact angle, θ_c, which is dependent primarily on the fluid-surface interface (and only weakly on the surface-air and fluid-air interfaces). In unpinned evaporation, θ_c remains constant as fluid from the droplet surface evaporations, and the droplet radius shrinks correspondingly. In pinned evaporation, the droplet radius remains constant, and instead θ_c shrinks. Though the phenomenon of pinning is not fundamentally understood, it has been proposed that it is the result of surface roughness (including possible "self-roughening" by solute deposition) [8-10]. In addition, it is believed that for the droplet

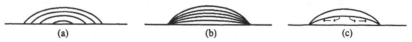

(a)   (b)   (c)

Fig. 6. Evaporation phenomena. (a) Unpinned. (b) Pinned. (c) Flow during pinned evaporation.

radius to remain constant during evaporation, fluid must flow out from the center to the edge of the drop, leading to a possible mass flow of solute towards the edges [8], as illustrated schematically in Fig. 6c.

Our first dye doping experiments were performed on ~100 nm PVK films, using a solution of the green emitting dye Coumarin 6 (C6) in acetone. Since we could not print acetone with our ink-jet printer, in these experiments we used a syringe to deposit individual droplets of ~11μL, which had a deposited radius of ~7mm. We found that for room temperature evaporation, the acetone droplet remained pinned at its initial radius for ~35 s, and then its radius fell rapidly during the remaining 15 s of evaporation (see Fig. 7a). The resulting dye distribution was observed under ultraviolet (UV) photoluminescence (PL) (see Fig. 7b), and revealed distinct rings of high dye concentration corresponding to droplet pinning, while between the rings, the dye concentration was very low. This observation of dye pile-up into rings was further confirmed by X-ray microprobe. These results clearly demonstrated that mass transport of solute from the center of the droplet to the edges (through some mechanism) occurred. Reducing the evaporation rate by cooling the substrate to 4°C greatly improved the dye distribution uniformity over the droplet area; however, substantial segregation of dye towards the edges was still clearly observed.

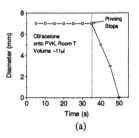

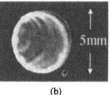

(b)

Fig. 7. Acetone:C6 Deposition. (a) Droplet radius with time during evaporation. (b) Photoluminescence of dye doping, where brightness corresponds to C6 concentration.

Many solvents were considered when determining the best choice for our initial IJP experiments. Essentially, we required a *printable* solvent that dissolved the dye and not the PVK films we would be printing on. In addition, because our work with acetone suggested that slower evaporation times would improve the dye distribution, we desired a solvent with a low vapor pressure. Among the common solvents we tried, DMSO was the only one to meet all these requirements (see Table I).

Table I. Solvents Investigated for IJP Dye Doping Candidacy. (All values for 25°C.)

Name	Formula	Viscosity* (η) (mPa·s)	Surface Energy* (γ) (mN/m)	Vapor Pressure* (p_v) (kPa)	Dissolves Dyes	Dissolves PVK	Prints
Chlorobenzene	C_6H_5Cl	0.753	32.99	1.6	Yes	Yes	Yes
Cyclohexanone	$C_6H_{10}O$	2.02	34.57	0.53	Yes	Yes	Yes
Tetrachloroethane	$C_2H_2Cl_4$	1.84	35.58	1.6	Yes	Yes	Yes
Dimethyl sulfoxide	C_2H_6SO	**2.20**	**42.92**	**0.08**	**Yes**	**No**	**Yes**
Water	H_2O	1.00	71.99	1.00	No	No	Yes
Chloroform	$CHCl_3$	0.58	26.67	26	Yes	Yes	No
Acetone	C_3H_6O	0.30	23.46	31	Yes	No	No

We printed droplets of C6 in DMSO onto PVK films, and observed the drying phenomena. We found that the droplet was pinned for the first 600 s, and for the remaining 400 s evaporation proceeded through a complex sequence of pinning and slipping (see Fig. 8a). The resulting dye distribution was observed using a PL image of the droplet, which revealed a fairly uniform dye distribution, but with a noticeable segregation of dye towards the center (see Fig. 8b). There were no high concentration outer rings observed as in the acetone results. Closer inspection of the PL image shows that instead, a small number of thick, circular bands of uniform dye distribution are present around an essentially uniform central region, with increasing dye concentration towards the center. The edges of these regions correspond well to the pinned radii observed during evaporation, suggesting that in a region of PVK suddenly exposed by a droplet slip, the absorbed solvent evaporates without any dye redistribution. This is consistent with the assumption that the evaporation of the solvent absorbed into the ~100 nm film should occur extremely rapidly and evenly over the exposed film area. (This result also suggests that in the PVK *under* the solvent droplet, the dye concentration is uniform at the time of the slip.)

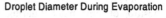

Droplet Diameter During Evaporation

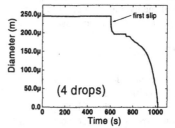

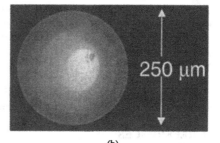

(b)

Fig. 8. DMSO:C6 IJP Deposition.
(a) Droplet radius with time during evaporation.
(b) Photoluminescence of dye doping.

We fabricated integrated red, green, and blue electrical devices on a single substrate, using the dyes Nile Red, C6, and Coumarin 47 to produce each respective color. The resulting 250 μm devices demonstrated similar electrical characteristics to spin-coated devices without IJP, suggesting that the IJP dye doping process did not adversely affect the electrical behavior of the PVK film. The observed electroluminescence (EL) was investigated for uniformity, and an image of a characteristic device is given in Fig. 9. The color uniformity is good over most of the device area, however, there is a distinct dark spot observed in the center of the device, with a slight darkening of the luminance just around this spot. The increased dye concentration in the center of the device (observed in the PL results) could explain a darkening of the luminance (due to dye concentration quenching [11]) in the center, but this effect cannot explain the sharp,

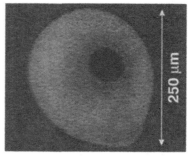

Fig. 9. Electroluminescence of IJP dye doped OLED.

completely dark spot observed. This dark spots appears to be correlated with the surface deposition of some material during the extremely fast drying which occurs in the last stages of the evaporation process over a total region with a radius of about 20 microns, but the details are still unknown.

IV. SUMMARY

We have developed a system to investigate the dye doping by IJP of polymer OLED's, including a highly controllable all-glass ink jet printer and digital imaging equipment for studying IJP droplet formation and droplet drying phenomena. Using a very low vapor pressure solvent, uniform dye distribution over the droplet area was achieved. The dye distribution is critically affected by the dynamics of the drying process, which at present are only qualitatively understood. Integrated 250-micron RGB devices were demonstrated with good color uniformity and with electrical properties comparable to spin-coated devices without IJP.

V. ACKNOWLEDGEMENTS

This work was supported by DARPA/AFOSR, NSF, and NJCST. The authors would also like to thank Joe Goodhouse for his expert assistance in obtaining PL images.

VI. REFERENCES

[1] J.H. Burroughs, D.D.C. Bradley, A.R. Brown, R.N. Marks, K. Mackay, R.H. Friend, P.L. Burns, A.B. Holmes, Nature **347**, 539 (1990).
[2] J. Kido, M. Kohda, K. Okuyama, K. Nagai, Appl. Phys. Lett. **61**, 761 (1992).
[3] T.A. Hebner, J.C. Sturm, APL **73**, 1775 (1998).
[4] C.L. Chang, L.P. Rokhinson, J.C. Sturm, Appl. Phys. Lett. **73**, 3568 (1998).
[5] E.D. Wilkes, S.D. Phillips, O.A. Basaran, Phys. of Fluids **11**, 3577 (1999).
[6] X.G. Zhang, J. Coll. Int. Sci. **212**, 107 (1999).
[7] T.W. Shield, D. B. Bogy, F.E. Talke, IBM J. Res. Dev. **31**, 96 (1987).
[8] R.D. Deegan, Phys. Rev. E **61**, 475 (2000).
[9] E. Adachi, A.S. Dimitrov, K. Nagayama, Langmuir **11**, 1057 (1995).
[10] P.G. de Gennes, Revs. Mod. Phys. **57**, 827 (1985).
[11] C.C. Wu, J.C. Sturm, R.A. Register, J. Tian, E.P. Dana, M.E. Thompson, IEEE Trans. on Elec. Dev. **44**, 1269 (1997).

Printing Methods for
Direct Write Processing

High resolution copper lines by direct imprinting

C.M. Hong, X. Sun, S. Wagner, S.Y. Chou
Department of Electrical Engineering, Princeton University, Princeton, New Jersey 08544

ABSTRACT

One-micrometer wide copper lines are patterned by direct imprinting and thermolysis. First, a layer of plastic copper hexanoate is spun on a substrate and patterned by direct imprint. Then the copper hexanoate line pattern is converted to copper metal lines by thermal and hydrogen anneals. The converted copper film resistivity is ~ 8 $\mu\Omega$cm. The direct imprinting of a fine metal pattern points the way to the direct patterning of device materials at high resolution.

INTRODUCTION

Fine metal lines are becoming ever more important to the technology of integrated circuits (ICs). Interconnecting the devices now is the most costly step in fabricating a silicon IC. The metal lines on a large-area circuit such as active-matrix liquid crystal displays soon could be a mile long. The performance of these circuits depends critically on having metals with high conductivity. The need to reduce cost while raising conductivity has stoked the interest in the direct writing of lines of metal, particularly of copper. Direct writing is a mass production technique that could eliminate many process steps from circuit fabrication. Similar motives have stimulated experiments on the direct printing of other active device materials for transistors [1-4] and organic light emitting diodes [5,6]. These experiments have drawn on conventional printing techniques including screen printing and inkjet printing, whose present line pair resolution however lies considerably above ~ 10 μm. Here we report patterning by direct imprint experiments in which we achieved a copper linewidth of 1 μm by the imprinting of a precursor film followed by its conversion to copper metal. This technique is capable of printing even finer copper lines, because it is closely related to the nanoimprint lithography technique [7,8] that has attained a feature size of 6 nm.

EXPERIMENTS

During experiments on the writing of copper lines by exposing copper hexanoate, $Cu_2(OH_2)_2(O_2CR)_4$ where R = $(CH_2)_4CH_3$, to UV radiation [9,10], we had discovered that this compound can be decomposed and reduced to metallic copper with a simple annealing sequence [3]. Upon heating, the organic ligand and CO_2 or CO are split off and ejected from the surface of the film. The film that remains on the substrate consists of a mixture of copper and copper(I) oxide. Annealing the film in forming gas ($H_2 + N_2$) reduces Cu_2O to copper metal. Copper hexanoate is plastic by virtue of its high aliphatic content and thus is easily molded by imprinting.

Figure 1 shows the steps involved in the patterning of copper lines by imprinting. 1 g of copper hexanoate prepared according to Graddon [11] is dissolved in 5 ml isopropyl alcohol by

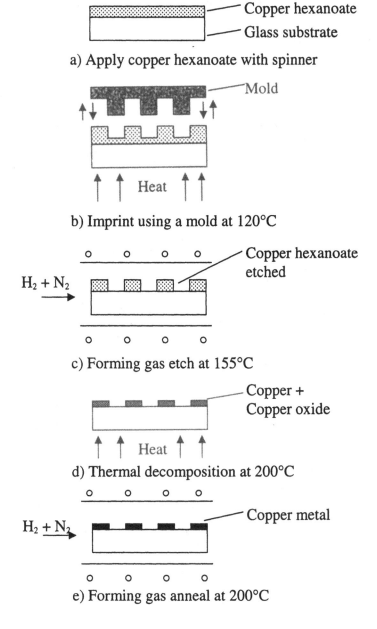

Figure 1. Schematic illustration of the copper imprint process

heating the mixture on a hot plate to 110°C for ten minutes. The surface tension and viscosity of the resulting copper hexanoate solution are 20.9 mN/m and 3.2 mPa·s, respectively. We spin a 1.6-μm thick copper hexanoate film onto a 1.1-mm thick Corning 7059 glass substrate (Figure 1(a)). The substrate is then heated to 120°C and a Si mold was pressed into the copper hexanoate film at a pressure of 600 psi (Figure 1(b)). Subsequently a forming gas (15 vol.% H_2 + 85 vol.% N_2) at 155°C etches away the residual copper hexanoate layer (Figure 1(c)). The remaining copper hexanoate is then converted to copper and copper oxide by heating to 200°C for 30 minutes in a vacuum chamber at a residual pressure of 1.2 Torr (Figure 1(d)). After this thermolysis, the sample is largely converted to copper metal by annealing it in forming gas at atmospheric pressure and 200°C for 40 minutes (Figure 1(e)).

RESULTS AND DISCUSSIONS

Figure 2(a) shows the optical micrograph of a copper hexanoate pattern after the impact printing step of Figure 1(b). Rows of copper hexanoate lines with a width of 2 μm and a separation distance of 2 μm are clearly visible in Figure 2(a). Figure 2(b) shows the atomic force microscopic (AFM) height profile of the copper hexanoate pattern, with a peak-to-valley height of ~ 250 nm. The rms roughness and mean roughness of the film measured by AFM are ~ 5 nm and ~ 4 nm, respectively. An optical micrograph and AFM profiles of the copper film after the thermolysis and forming gas anneal steps at 200°C is shown in Figure 3. The lines of converted copper metal are narrower (~ 1 μm) and thinner than those of the precursor. The rms roughness and mean roughness of the film measured by AFM are ~ 36 nm and ~ 29 nm, respectively. The rougher surface indicates grain growth during the thermal decomposition and annealing steps.

Figure 4 shows the X-ray diffraction spectrum of the copper film, taken with the CuK_α line of 0.154 nm wavelength. The peaks at 43.5°, 50.4°, and 74.2° for the (111), (200), and (220) reflexes are close to the values of 43.3°, 50.4°, 74.1° of the literature [12]. From the line broadening of the (111) peak [13] we obtain an average grain size of ~ 40 nm, which is comparable to the roughness values measued by AFM. The peak intensities indicate that ~ 70 % of the Cu grains have the (111) orientation. The chemical composition of the copper film was studied by Auger electron spectroscopy (AES), using 1.5 keV Ar ion for sputter profiling and a 3 keV primary electron energy. From the AES spectrum, we obtain a film composition of ~ 82 at.% Cu, ~ 12 at.% C, and ~ 6 at.% O. The presence of carbon and oxygen suggest that the conversion of copper hexanoate to metallic copper is incomplete. The electrical resistivity of a contiguous copper film, measured with a four-point probe, is ~ 8 μΩcm, which is 5 times that of pure copper film (1.7 μΩcm). The impurity content but also the granularity will cause an increase in the macroscopic resistivity. Both causes can be addressed by experimentation, the impurity content by more complete conversion and the granularity with an appropriate temperature-time program during the conversion.

CONCLUSIONS

In conclusion, we have demonstrated a direct imprint technique for patterning high resolution copper lines. 1-μm copper lines of ~ 8 μΩcm resistivity have been achieved. The maximum processing temperature of 200°C is compatible with the processing temperature required by large-area electronics technology. The technique is capable of still finer linewidth and lower copper resistivity.

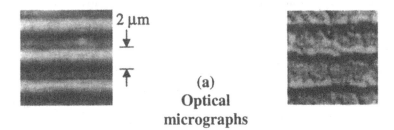

2 μm

(a)
Optical
micrographs

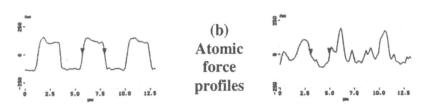

(b)
Atomic
force
profiles

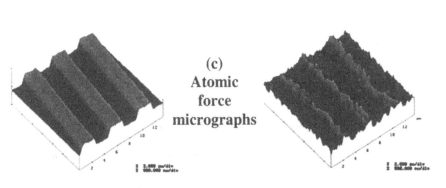

(c)
Atomic
force
micrographs

Figure 2. *Optical micrograph and AFM profiles of copper hexanoate lines on glass made by imprinting.*

Figure 3. *Optical micrograph and AFM profiles of copper lines on glass made by the conversion of imprinted copper hexanoate.*

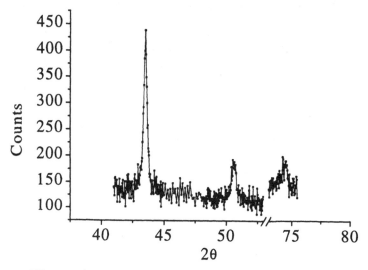

Figure 4. *X-ray spectrum of the copper film.*

ACKNOWLEDGEMENTS

We gratefully acknowledge the support of DARPA under its High Definition Systems and Molecular Level Printing programs.

REFERENCES

1. F. Garnier, R. Hajlaoui, and A. Yassar, Science **265**, 1684 (1994).
2. Z. Bao, J.A. Rogers, H.E. Katz, J. Mat. Chem. **9**, 1895 (1999).
3. C.M. Hong, S. Wagner, Mat. Res. Soc. Symp. Proc. **558**, (1999), to be published.
4. S. Wagner, H. Gleskova, J. C. Sturm, and Z. Suo, in *Technology and Applications of Hydrogenated Amorphous Silicon*, edited by R. A. Street (Springer-Verlag, Berlin, 2000), Springer Series in Materials Science **37**, p. 222.
5. T. R. Hebner, C. C. Wu, D. Marcy, M. H. Lu, and J. C. Sturm, Appl. Phys. Lett. **72**, 519 (1998).
6. S.C. Chang, J. Bharathan and Y. Yang, Appl. Phys. Lett. **73**, 2561 (1998).
7. S.Y. Chou, P.R. Krauss, and P.J. Renstrom, Science **272**, 85 (1996).
8. S.Y. Chou, P. R. Krauss, W. Zhang, L. Guo, and L. Zhuang, J. Vac. Sci. Tech. B. **15**, 2897 (1997).
9. A. A. Avey and R. H. Hill, J. Am. Chem. Soc. **118**, 237 (1996).
10. C. M. Hong, H. Gleskova, and S. Wagner, Mat. Res. Soc. Symp. Proc. **471**, 35 (1997).
11. D. P. Graddon, J. Inorg. Nucl. Chem. **17**, 222 (1961).
12. B.D. Cullity, *Elements of x-ray diffractions*, (Addison-Wesley, Reading, 1967), p. 261.
13. T. Swanson, Natl. Bur. Stand. (U.S.) **539 I**, 15 (1953).

All-Printed Inorganic Logic Elements Fabricated by Liquid Embossing

Colin Bulthaup, Eric Wilhelm, Brian Hubert, Brent Ridley, and Joe Jacobson

Media Lab, Massachusetts Institute of Technology, Cambridge, MA 02139

ABSTRACT

The liquid embossing process, devices made by this process, and their characteristics are presented. Structures fabricated and discussed here include: conductive lines with cross line resistance greater than 100 GΩ and resistivities 4 times that of the bulk material, multilayer structures with etched sacrificial materials, vias that conduct through an embossed insulating layer, photodetectors made with nanocrystal solutions of CdSe, and all printed inorganic field effect transistors.

EXPERIMENT

Process

Liquid embossing[1, 2] is a physical process used to create features in functional materials. Though this process can pattern a wide range of materials this paper is restricted to a discussion of patterning solutions of nanocrystals, spin on glasses (SOG) (Ohka T7), and insulating polymers (Japanese Synthetic Rubber AL 3046). Our paper gives a brief overview of the process, then descriptions of the types of devices produced with liquid embossing and their characteristics. The devices discussed include conducting metal lines, released mechanical structures, and transistors.

A stamp is created by casting an elastomer (PDMS) from a master. The master is typically made of photoresist on a silicon wafer, but can be any surface with raised physical features. As shown in Fig. 1, the stamp is brought into contact with a thin film of functional material in liquid phase. The raised features on the stamp emboss through the thin film to the underlying substrate and remove material. The stamp is then removed while the functional material is still in liquid phase; the material can reflow if embossed again but otherwise will remain patterned. Finally the material is cured, which removes the solvents and sinters or cures the material.

Thin films are created by either a drawdown bar or conventional spin coating. In the draw down process a drop of liquid material is placed on a substrate, and a ground metal cylinder is pulled over the substrate. The only pressure on the substrate is the weight of the cylinder. The thin films obtained are very consistent for a range of liquid material properties including viscosity, vapor pressure, and surface wetting.

225

Figure 1: Schematic of the liquid embossing process: (a) the elastomeric stamp is brought into contact with a thin film of liquid, (b) the stamp contacts the underlying substrate and selectively patterns the liquid film, (c) the stamp is removed and the liquid film remains patterned.

<u>Single-Layer Structures</u>

We have patterned structures with a wide variety of resolutions and geometries. 200 nm line width gratings were fabricated over areas greater than 4 cm with high reliability, (Fig. 2a). Other structures with non-repeating features have been fabricated over the entire surface of a 4" wafer, (Fig. 2b). Typical solid film thickness range from 50 nm – 400 nm.

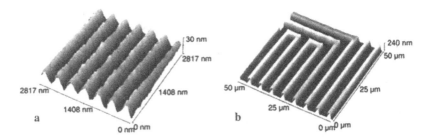

Figure 2: Atomic Force Microscope images of single layers of patterned Au: (a) 400nm grating structure, (b) 3um wide lines in a circuit pattern.

<u>Resistors</u>

Colloidal suspensions of Au nanoparticles in α-terpineol were used as a conducting material. These suspensions were patterned via liquid embossing and then cured at 300° C for 30 minutes to produce a solid metallic film. It was important that the stamp was removed prior to curing in order to guarantee that all of the organic solvent could boil off. Serpentine resistor structures (Fig. 3) were patterned in order to measure the resistivity of the material. These resistors had widths of 1, 3, and 5μm, and lengths of 6, 12, and 30mm. For the narrowest / longest resistor these dimensions corresponded to an aspect ratio of roughly 30,000 to 1. The calculated resistivity for these resistors was 8.32×10^{-6} Ω-cm, with a standard deviation of 5×10^{-7} Ω-cm. The cross-channel resistance between pairs of unconnected gold lines was greater than 100 GΩ.

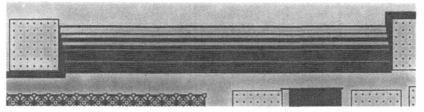

Figure 3: Optical image of a serpentine resistor structure for measuring the resistivity of Au material.

Multi-Layer

Self-planarization and conformal patterning over rough features are very important aspects of liquid embossing. Multilayer structures can be built up by simply embossing on top of a previously patterned and cured layer. Registration between layers is achieved to approximately 5μm of error with a stationary stamp, and a substrate mounted on an x-y positioning stage. Due to the conformability of the PDMS stamp the embossing process can actually clear out material from a previous patterning, as shown in Fig. 4.

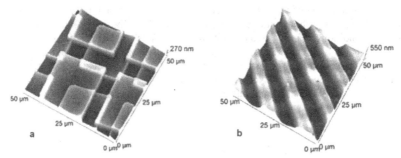

Figure 4: Atomic Force Microscope images of multilayer structures fabricated in Au: (a) two arrays of squares patterned on top of each other, demonstrating conformal patterning, (b) two diffraction gratings patterned on top of each other, demonstrating self-planarization of patterned liquid films.

MEMS

Nanocrystals have already been proven to be able to form heatuators and other MEMS devices by inkjet deposition[3]. Fig. 5 shows a structure we call an aqueduct: triangular profile lines of Au are patterned on top of triangular lines of SOG (spin-on glass). The glass is then etched away leaving only the Au. This method is a route to creating embossed MEMS devices with many layers.

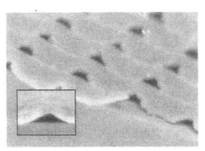

Figure 5: Scanning Electron Microscope image of patterned Au lines with etched away sacrificial lines running perpendicular to the Au lines, (inset) side-view of Au lines.

Transistors & Photodetectors

Recently, it was demonstrated that nanocrystalline films of CdSe could be used as a high mobility semiconductor for thin film transistors (TFTs)[4]. Partially printed TFT structures were created by growing a 300 nm thick thermal oxide on a n+ wafer and then patterning Au source/drain electrodes on top of the oxide by liquid embossing. CdSe was then deposited by pipet in the channel between the electrodes, sintered at 300° C, and encapsulated with a photo-curable adhesive (Norland 73). The device was probed in ambient conditions and showed a mobility of 0.1 $cm^2V^{-1}s^{-1}$ and an on/off ratio of 10^3 (Fig. 6a) which is within an order of magnitude of previously reported results using evaporated Au electrodes. Additionally, photodetectors were created by fabricating similar structures without the Norland encapsulant. These devices showed a strong correlation between relative light intensity and source-drain current (Fig. 6b).

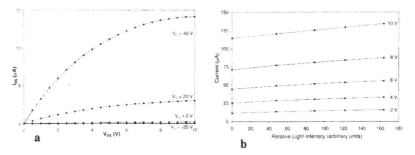

Figure 6: Half-printed devices measured in ambient conditions: (a) Transistor: source/drain current vs. voltage for various gate voltages, (b) Photodetector: source/drain current at a fixed voltage for various different light intensities

Capacitors & Vias

Low-K dielectric capacitors have been made by liquid embossing. A thin film of Au was patterned on top of previously deposited and cured layers of Au and insulator. The insulator was typically SOG or a polyimide. The capacitors, which were fabricated outside of a clean room, were able to insulate between conductive metal layers over areas in excess of 10 cm^2. Combining these low-K dielectric materials with our ability to pattern multiple functional layers allowed us to fabricate vias between electrically isolated metal layers, (Fig. 7).

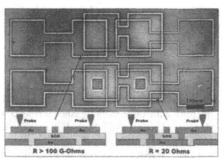

Figure 7: Three layer via structure fabricated entirely by liquid embossing. In the first device there were no vias present and thus no conduction between the two probes. In the second device vias were patterned in the SOG, creating a direct conduction path between the two probes.

All-Printed Transistors

All of the previously mentioned results are necessary constituent elements for fabricating logic. By combining these techniques, we were able to fabricate the first all-printed, all-inorganic transistor. An initial layer of Au was deposited, followed by a 200 nm - 300 nm thick layer of SOG. Source/drain electrodes were then patterned in Au by liquid embossing, and a drop of CdSe was deposited in the channel region and encapsulated with Norland 73. The total fabrication time required was 5 hours, with the vast majority of that time spent curing the SOG. The devices were probed in ambient conditions and showed an on/off ratio of 10^3 (Fig. 8).

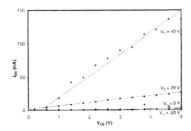

Figure 8: All-printed all-inorganic transistor fabricated by liquid embossing. Source/drain current vs. voltage at various gate voltages, measured in ambient conditions.

CONCLUSION

This paper has discussed some of the structures fabricated by liquid embossing of functional materials. The results are quite exciting since it has been shown that many of the things typically made in clean rooms and billion-dollar silicon fabs can actually be made outside of a clean room for very little cost. Since the process uses inorganic materials, the possibility exists for making devices that have functional characteristics identical to those produced by conventional means. Liquid embossing also presents a possible route to the fabrication of three dimensional logic and memory, and a viable method for creating MEMS with very large numbers of layers.

REFERENCE

1. Ridley, B. A., Nivi, B., Hubert, B. N., Bulthaup, C., Wilhelm, E. J. & Jacobson, J. M. Solution-Processed Inorganic Transistors and Sub-Micron Non-Lithographic Patterning Using Nanoparticle Inks. 1999 MRS Fall Meeting Proc., Nanophase and Nanocomposite Materials III (1999).
2. Jacobson, J. M., Ridley, B. A., Hubert, B. N., Bulthaup, C., Griffith, S. & Fuller, S. B. Printed Micro-Electro-Mechanical Systems and Logic. Invited talk, 1999 MRS Fall Meeting, Boston, Massachusetts (1999).
3. Fuller, S., & Jacobson, J. Ink Jet Fabricated Nanoparticle MEMS, Proceedings of IEEE MEMS 2000 Conference, Miyazaki, Japan (2000)
4. Ridley, B. A., Nivi, B. & Jacobson, J. M. All-Inorganic Field Effect Transistors Fabricated by Printing. *Science* **286**, 746-749 (1999).

RAPID PROTOTYPING OF PATTERNED MULTIFUNCTIONAL NANOSTRUCTURES

Hongyou Fan, Gabriel P. López, and C. Jeffrey Brinker
The University of New Mexico/NSF Center for Micro-Engineered Materials, The Advanced Materials Laboratory, Sandia National Laboratories, Albuquerque, NM

ABSTRACT

The ability to engineer ordered arrays of objects on multiple length scales has potential for applications such as microelectronics, sensors, wave guides, and photonic lattices with tunable band gaps. Since the invention of surfactant templated mesoporous sieves in 1992, great progress has been made in controlling different mesophases in the form of powders, particles, fibers, and films. To date, although there have been several reports of patterned mesostructures, materials prepared have been limited to metal oxides with *no specific functionality*. For many of the envisioned applications of hierarchical materials in micro-systems, sensors, waveguides, photonics, and electronics, it is necessary to define both form and function on several length scales. In addition, the patterning strategies utilized so far require hours or even days for completion. Such *slow* processes are inherently difficult to implement in commercial environments. We present a series of new methods of producing patterns within seconds. Combining sol-gel chemistry, Evaporation-Induced Self-Assembly (*EISA*), and rapid prototyping techniques like pen lithography, ink-jet printing, and dip-coating on micro-contact printed substrates, we form hierarchically organized silica structures that exhibit order and function on multiple scales: on the molecular scale, functional organic moieties are positioned on pore surfaces, on the mesoscale, mono-sized pores are organized into 1-, 2-, or 3-dimensional networks, providing size-selective accessibility from the gas or liquid phase, and on the macroscale, 2-dimensional arrays and fluidic or photonic systems may be defined. These rapid patterning techniques establish for the first time a link between computer-aided design and rapid processing of self-assembled nanostructures.

INTRODUCTION

Living systems exhibit form and function on multiple length scales, and the prospect of imparting life-like qualities to man-made materials has inspired many recent efforts to devise hierarchical materials assembly strategies. For example, Yang et al.[1] grew surfactant-templated mesoporous silica[2] on hydrophobic patterns prepared by micro-contact printing μCP[3]. Trau et al.[4] formed oriented mesoporous silica patterns, using a micro-molding in capillaries *MIMIC* technique[3], and Yang et al.[5] combined *MIMIC*, polystyrene sphere templating[6], and surfactant-templating to create oxides with three levels of structural order. Overall, great progress has been made to date in controlling structure on scales ranging from several nanometers to several micrometers. However, materials prepared have been limited to oxides with no specific functionality, whereas for many of the envisioned applications of hierarchical materials in micro-systems, sensors, waveguides, photonics, and electronics, it is necessary to define both form and function on several length scales. In addition, the patterning strategies employed thus far require hours or even days for completion[1, 4, 5]. Such slow processes are inherently difficult to implement in commercial environments.

We have combined evaporation-induced (silica/surfactant) self-assembly *EISA*[7] with rapid prototyping techniques like pen lithography[8, 9], ink-jet printing[10, 11], and dip-coating on micro-contact printed substrates to form hierarchically organized structures in seconds. In addition, by co-condensation of tetrafunctional silanes $(Si(OR)_4)$ with tri-functional organosilanes $((RO)_3SiR')$[12-14] or bridged silsesquioxanes $(RO)_3Si-R'-Si(OR)_3)$ or by inclusion

231

of organic additives, we have selectively derivatized the silica framework with functional R' ligands or molecules. The resulting materials exhibit form and function on multiple length scales: on the molecular scale, functional organic moieties are positioned on pore surfaces, on the mesoscale, mono-sized pores are organized into 1-, 2-, or 3-dimensional networks, providing size-selective accessibility from the gas or liquid phase, and on the macroscale, 2-dimensional arrays and fluidic or photonic systems may be defined.

EXPERIMENTAL

Precursor solutions used as inks were prepared by addition of surfactants (cationic, CTAB; $CH_3(CH_2)_{15}N^+(CH_3)_3Br^-$ or non-ionic, Brij-56; $CH_3(CH_2)_{15}-(OCH_2CH_2)_{10}-OH$ and Pluronic P123, $HO(CH_2CH_2O)_{20}(CH(CH_3)CH_2O)_{70}-(CH_2CH_2O)_{20}-H)$, organosilanes (R'Si(OR)$_3$, see Table 1), or organic molecules (see Table 1) to an acidic silica sol prepared from TEOS [$Si(OCH_2CH_3)_4$] (A2**). The acid concentration employed in the A2** synthesis procedure was chosen to minimize the siloxane condensation rate, thereby promoting facile self-assembly during printing[20]. In a typical preparation, TEOS, ethanol, water and dilute HCl (mole ratios: 1:3.8:1:5×10^{-5}) were refluxed at 60 °C for 90 min. The sol was diluted with 2 volumes of ethanol followed by further addition of water and HCl. Organosilanes (R'-Si(OR)$_3$, where R' is a non-hydrolyzable organic functional ligand) were added followed by surfactants and (optionally) organic additives (see Table 1). Surfactants were added in requisite amounts to achieve initial surfactant concentrations c_0 ranging from 0.004 to 0.23 M ($c_o \ll cmc$). The final reactant molar ratios were: 1 TEOS : 22 C_2H_5OH : 5 H_2O : 0.093 – 0.31 surfactant : 0.039 – 0.8 organosilanes : 2.6×10^{-5} organic additives. For the ethylene-bridged silsesquioxane, (RO)$_3$Si-R'-Si(OR)$_3$ (R' = CH_2CH_2, R = OC_2H_5), the neat precursor was diluted in ethanol and mixed with 1-8 wt% CTAB or Brij-56 surfactant followed by addition of an aqueous solution of HCl. The final reactant molar ratios were: Si:EtOH:H$_2$O:HCl:surfactant = 1:22:5:0.004:0.054-0.18. Co-hydrolysis of organosilanes with TEOS in the initial A2** sol preparation generally resulted in disordered worm-like mesostructures[21]. After pattern deposition and drying, the surfactant templates were selectively removed by calcination in a nitrogen atmosphere at a temperature sufficient to decompose the surfactant molecules (~350 °C) without degrading the covalently bound organic ligands R' (confirmed by ^{29}Si MAS NMR spectroscopy) or by solvent extraction.

Patterning Procedures

Micropen lithography was performed using a Model 400a micropen instrument purchased from Ohmcraft Inc., Pittsford, NY. The pen orifice was 50 μm and the writing speed was 2.54 cm/s. The pattern was designed using AutoCAD 14 software.

Dip-coating of patterned (hydrophilic/hydrophobic) substrates was performed at a withdrawal speed of 7.6 – 51 cm/min under ambient laboratory conditions. Hydrophilic/hydrophobic patterns were created by microcontact printing of hydrophobic, n-octadecyltrichlorosilane ($CH_3(CH_2)_{17}SiCl_3$) self-assembled monolayers SAMs[15] on hydrophilic silicon substrates (hydroxylated native oxide) or by a technique involving electrochemical desorption of a hydroxyl-terminated SAM prepared from 11-mercaptoundecanol (HO(CH$_2$)$_{11}$SH) from patterned, electrically isolated gold electrodes followed by immersion in a 1 mM ethanolic solution of 1-dodecanethiol, $CH_3(CH_2)_{11}SH$[16].

Ink jet printing was performed using a Model HP DeskJet 1200C printer purchased from Hewlett-Packard Co., San Diego, CA. The pattern was designed using Microsoft PowerPoint 98 software.

Table 1. Functional organosilanes and properties of resultant thin film mesophases

	Functional Silanes[i]/additives[¶] R'-Si(OR)₃	Mesophase	Pore Size[*] (Å)	Surface Area[*] (m²/g)	Properties and Applications
1	F₃C(CF₂)₆CH₂CH₂Si(OC₂H₅)₃ Tridecafluoro-1,1,2,2-tetrahydrooctyltriethoxysilane (TFTS)	3-dH	25	850	Hydrophobic; low k dielectrics
2	HS-(CH₂)₃Si(OCH₃)₃ Mercaptopropyltrimethoxysilane (MPS)	3-dH	25	1060	Coupling of noble metals
3	NH₂-(CH₂)₃Si(OCH₃)₃ Aminopropyltrimethoxysilane (APS)	cubic	22	750	Coupling of noble metals, dye, and bioactive molecules
4	Dye[§]-NH-(CH₂)₃Si(OCH₃)₃	cubic	21	545	pH sensitive
5	O₂N—⟨ ⟩—NH(CH₂)₃Si(OC₂H₅)₃ NO₂	3-dH	22	560	Chromophore; nonlinear optical material (χ²)
6	(H₅C₂O)₃SiCH₂CH₂Si(OC₂H₅)₃	cubic	40	430	low k dielectrics

[*]Pore size and surface area were determined from N₂ sorption isotherms obtained at −196°C, using a surface acoustic wave (SAW) technique. Mass change due to nitrogen sorption was monitored (~80 pg.cm⁻² sensitivity) as a function of nitrogen relative pressure. Pore size and surface area were determined from the isotherms using the BET equation and the BJH algorithm, respectively.

[i] Functional groups are retained through selective surfactant removal during heat treatment in nitrogen. TGA and DTA were used to establish the appropriate temperature window enabling complete surfactant removal without silane decomposition

[¶] Additives investigated include rhodamine-B, cytochrome *c* (from Fluka), oil blue N, disperse yellow 3 (from Aldrich), silver ions and silver nanoparticles.

[§] **4** was prepared by a conjugation reaction between a thin film mesophase containing **3** and the dye molecule (5,6-carboxyfluorecein, succinimidyl ester (5,6-FAM, SE) from Molecular Probes).

RESULTS AND DISCUSSION

Scheme 1

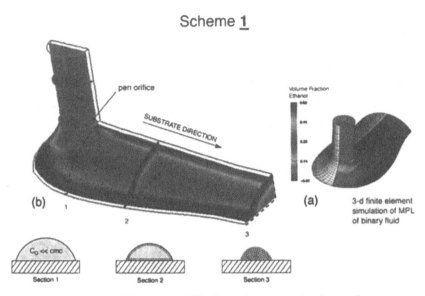

Figure 1. Scheme 1: micro-pen lithography *MPL* of a surfactant-templated mesophase.

Scheme 1 (Figure 1) schematically illustrates direct writing of a mesoscopically ordered nanostructure, using micro-pen lithography *MPL*.[9] (a) shows the simulation of 3-D, binary fluid pattern dispensed on a flat substrate with substrate speed = 2.54 cm/s and fluid injection rate (inlet velocity) = 3.985 cm/s. Color contours represent evaporation-induced, 3-D gradients in alcohol-composition. Fluid was modeled as 54 volume % ethanol and 46 volume % non-volatile phase with Reynolds number = 1.25 and Ca = 0.000833. An ad hoc value of 45° was chosen for the static contact angle. Note that this angle persists at all points on the dynamic contact line because of the dominance of surface tension at this low value of Ca. (b) shows the schematic process: the initially homogeneous sol metered on to the moving substrate experiences preferential evaporation of alcohol creating a complex 3-D (longitudinal and radial) gradient in the concentrations of water and non-volatile surfactant and silicate species. Progressive enrichment of silica and surfactant induces micelle formation and subsequent growth of silica/surfactant mesophases inward from solid-liquid and liquid-vapor interfaces as recently demonstrated for thin films[17] and aerosols[18]. The numerical method utilized for (a) and (b) consisted of a 3D finite element discretization of the Navier Stokes equations augmented with a three dimensional boundary-fitted mesh motion algorithm to track the free surface[19, 20]. Special relations at the 3D dynamic wetting line were also applied. For *ink* we use homogeneous solutions of TEOS, ethanol, water, surfactant, acid, and (optionally) organosilanes and other organic ingredients (see Table 1). These solutions are prepared with an initial surfactant concentration c_o less than the critical surfactant concentration *cmc* and an acid concentration designed to minimize the siloxane condensation rate, thereby enabling facile silica/surfactant self-assembly within the brief time span of the writing operation. As the ink is metered onto the surface, preferential ethanol evaporation causes enrichment of water, surfactant, and silica, establishing a 3-D (longitudinal and radial) gradient in their respective concentrations (Figure 1). Where *cmc* is exceeded, cooperative silica/surfactant self-assembly creates micelles. Further evaporation, of predominantly water, promotes the continuous self-organization of micelles into silica/surfactant liquid crystalline mesophases. As demonstrated previously for *EISA* of films[17] and aerosols[18], liquid crystalline domains are nucleated by incipient surfactant monolayers formed at solid-liquid[21, 22] and liquid-vapor interfaces[23] (at $c < cmc$) and grow inward as evaporation proceeds.

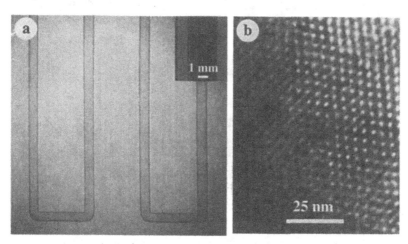

Figure 2. Meandering patterned mesophase created by *MPL*. a) Optical micrograph of patterned rhodamine-B containing silica mesophase deposited on an oxidized [100]-oriented silicon substrate at a speed of 2.54 cm/s. Inset is a fluorescence image of rhodamine-B emission acquired through a 610-nm band pass filter, demonstrating retention of rhodamine-B functionality. b) Representative TEM micrograph of a fragment of the patterned rhodamine-B containing film corresponding to a [110]-oriented cubic mesophase with lattice constant $a =$ 10.3-nm. The sol was prepared by adding 0.01wt% rhodamine-B to a silica/4wt% Brij-56 sol. The TEOS:EtOH:water:HCl:Brij-56:rhodamine-B molar ratio $= 1:22:5.0:0.004:0.075:2.6\times10^{-5}$.

Figure 2a shows a meandering macroscopic pattern formed in several seconds by *MPL* of a rhodamine B-containing solution on a hydrophilic surface (hydroxylated native oxide of <100> silicon). The inset in Figure 2a shows the corresponding fluorescence image of several adjacent stripes, and the TEM micrograph (Figure 2c) reveals the ordered pore structure characteristic of a cubic thin film mesophase. The *MPL* line width can vary from micrometers to millimeters. It depends on such factors as pen dimension[8], wetting characteristics, evaporation rate, capillary number (*Ca* = ink viscosity x substrate speed/surface tension) and ratio of the rates of ink supply and withdrawal (inlet velocity/substrate velocity). The effect of wetting has been demonstrated by performing *MPL* on substrates pre-patterned with hydrophobic, hydrophilic or mixed SAMs. Generally, line widths are reduced by increasing the contact angle and by reducing the inlet/substrate velocity ratio. The conditions providing the minimum stable line width are bounded by a regime of capillary instability – we anticipate that this instability could be exploited to create periodic arrays of dots.

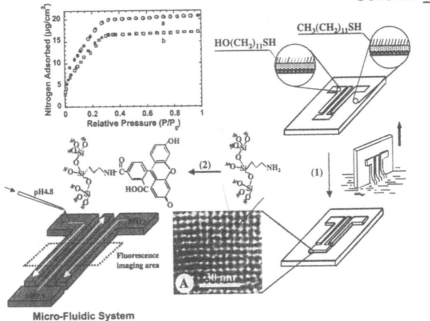

Scheme 2

Micro-Fluidic System

Figure 3. Scheme 2: Patterned functional mesostructure formed by selective de-wetting. The sol was prepared by adding aminopropyltrimethoxysilane (NH$_2$(CH$_2$)$_3$Si(OCH$_3$)$_3$, APS) to a silica/4wt% Brij-56 sol, resulting in a final molar ratio TEOS:APS:EtOH:water:HCl:Brij-56 = 1:0.8:22:5.0:0.011:0.075.

The advantages of *MPL* are that we can use computer-aided design to define any arbitrary 2-D pattern and that we can use any desired combination of surfactant and functional silane as ink to selectively define different functionalities at different locations. However *MPL* is a serial technique. In situations where it is desirable to create an entire pattern with the same functionality, it would be preferable to employ a parallel technique in which the deposition process occurred simultaneously in multiple locations. Scheme 2 (Figure 3) illustrates a rapid, parallel patterning procedure, dip-coating on patterned SAMs. This procedure uses micro-contact printing[24] or electrochemical patterning[16] of hydroxyl- and methyl-terminated SAMs to define hydrophilic and hydrophobic patterns on the substrate surface. Then using homogenous solutions identical to those employed for *MPL*, preferential ethanol evaporation during dip-coating enriches the depositing film in water, causing *selective de-wetting* of the hydrophobic regions. In this fashion, lines, arrays of dots, or other arbitrary shapes can be deposited on hydrophilic patterns in seconds. As described for *MPL*, further evaporation accompanying the dip-coating operation induces self-assembly of silica/surfactant mesophases. Using micro-contact printing or electrochemical desorption techniques, substrates are prepared with patterns of hydrophilic, hydroxyl-terminated SAMs and hydrophobic methyl-terminated SAMs. Preferential ethanol evaporation during dip-coating (1), causes water enrichment and selective de-wetting of the hydrophobic SAMs. Correspondingly film deposition occurs exclusively on the patterned hydrophilic SAMs. Selective de-wetting followed by calcination results in a patterned, amine-functionalized, cubic mesoporous film as is evident from the plan-view TEM micrograph (Inset A), showing a [100]-oriented cubic mesophase with $a = 10.3$-nm and nitrogen adsorption-desorption isotherm (Inset B, curve a) acquired for the thin film specimen using a surface acoustic wave[25] (SAW) technique. The dye conjugation reaction (2) was conducted by immersion in a 0.00002mM solution of 5,6-FAM, SE (Table 1) prepared in dimethylsulfoxide (DMSO) followed by exhaustive, successive washing in DMSO, ethanol, and water. The SAW-based nitrogen adsorption-desorption isotherm of the dye-conjugated mesoporous film is shown in Inset B, curve b, confirming its pore accessibility. BET analyses of the sorption isotherms indicate that the dye conjugation reaction reduces the surface area from 750 to 545 m²/g and the hydraulic radius from 2.2 to 2.1-nm, but pore accessibility is completely retained as evident from combined TEM, SAW, and fluorescent-imaging results (Figure 4a).

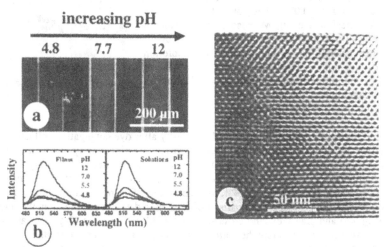

Figure 4. Patterned pH-sensitive fluidic system. a) Fluorescence image of three adjacent 5,6-FAM, SE-conjugated pore channel networks after introduction of aqueous solutions prepared at pH 4.8, 7.7, or 12.0. Patterned dye-conjugated thin film mesophases were prepared according to Scheme 2 (Figure 3). Aqueous solutions of varying pH were introduced on the terminal pads (Figure 3) and transported into the imaging cell by capillary flow. Image was acquired using a Nikon Diaphot 300 inverted microscope and 520-nm band pass filter. b) Fluorescence spectra of 5,6-FAM, SE-conjugated mesoporous films upon exposure to aqueous solutions of pH 4.8, 7.7, and 12.0. Shown for comparison are fluorescence spectra of 0.1 micromolar solutions of 5,6-FAM, SE prepared in aqueous solutions of pH 4.8, 7.7, and 12.0. The similarity of the two sets of spectra confirms the maintenance of dye functionality upon conjugation within the mesoporous channel system. c) Cross-sectional TEM micrograph of the patterned, dye-conjugated thin film mesophase, providing evidence of the 3-D pore channel network.

The patterned dip-coating procedure may be conducted with organic dyes or functional silanes (see Table 1). Scheme 2 illustrates patterned deposition of a propyl-amine derivatized cubic mesophase followed by a conjugation reaction with a pH-sensitive dye, 5,6-carboxyfluorescein, succinimidyl ester (5,6-FAM, SE). The uniform continuous porosity of the amine-derivatized and dye-conjugated films is confirmed by TEM and surface acoustic wave (SAW)-based nitrogen sorption isotherms[25] of the corresponding films deposited on SAW

Figure 5. Patterned dot arrays created by ink-jet printing. a) Optical micrograph of dot array created by ink jet printing of standard ink (from Hewlett-Packard Co., San Diego, CA) on a non-adsorbent surface. b) Optical micrograph of an array of hydrophobic, mesoporous silica dots created by evaporation-induced silica/surfactant self-assembly during *IJP* on an oxidized [100]-oriented silicon substrate followed by calcination. c) Representative TEM micrograph of a dot fragment prepared as in (b). The sol was prepared with molar ratio TEOS:TFT̲S(1):EtOH:water:HCl:Brij-56 = 1:0.05:22.0:5.0:0.004:0.075. The dot pattern used in a) and b) was designed using Microsoft PowerPoint 98 software. The printing rate was approximately 80 dots/s and printer resolution 300 dots/inch. The resolution achieved compared to standard ink and our ability to selectively functionalize the ink suggest applications in display technologies.

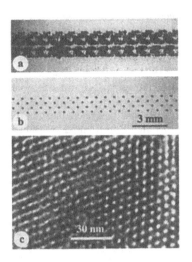

substrates (Figure 4). The reduction in film porosity after dye conjugation reflects the volume occupied by the attached dye moieties. The patterned, functional array can be used to monitor the pH of fluids introduced at arbitrary locations and transported by capillary flow into the imaging cell. Figure 4a shows the fluorescence image of an array contacted with three different aqueous solutions prepared at pH 4.8, 7.7, and 12.0. Figure 4b shows the corresponding emission spectra and provides a comparison with solution data. In combination, the fluorescence image (Fig. 4a) and plan-view and cross-sectional TEM micrographs (Figures 3 and 4c) of the dye-conjugated film demonstrate the uniformity of macro- and mesoscale features achievable by this evaporation-induced, de-wetting and self-assembly route. In comparison, films formed by nucleation and growth of thin film mesophases on patterned SAMs[1] are observed to have non-homogeneous, globular morphologies.

Finally we can create patterned nanostructures by combining *EISA* with a variety of aerosol processing schemes. For example, Figure 5 compares an optical micrograph of a macroscopic array of spots formed by ink jet printing *IJP*[10, 11]on a silicon wafer with *IJP* of standard ink on a non-adsorbent surface. The *IJP* process dispenses the *ink* (prepared as for *MPL*) as monosized, spherical aerosol droplets. Upon impaction the droplets adopt a new shape that balances surface and interfacial energies. Accompanying evaporation creates within each droplet a gradient in surfactant concentration that that drives radially-directed silica/surfactant self-assembly inward from the liquid-vapor interface[18]. The inset in Figure 5 shows a representative TEM micrograph of a hydrophobic, fluoroalkylated silica mesophase formed by *IJP*. The resolution achieved compared to standard ink and our ability to selectively functionalize the ink suggest applications in display technologies. Continuous, nanostructured lines are created by coalescence of overlapping droplets. Patterns may also be created by aerosol deposition through a mask or by aerosol deposition on patterned hydrophilic/hydrophobic surfaces (H. Fan, Y. Lu, and A. Stump, unpublished).

CONCLUSIONS

We have combined evaporation-induced (silica/surfactant) self-assembly *EISA* with rapid prototyping techniques like pen lithography, ink-jet printing, and dip-coating on micro-contact printed substrates to form hierarchically organized structures in seconds. In addition, by co-condensation of tetrafunctional silanes (Si(OR)$_4$) with tri-functional organosilanes ((RO)$_3$SiR') or by inclusion of organic additives, we have selectively derivatized the silica framework with functional R' ligands or molecules. The resulting materials exhibit form and function on multiple length scales and multiple locations.

ACKNOWLDGEMENTS:

This work was supported by the US Department of Energy Basic Energy Sciences Program, the Sandia National Laboratories Laboratory-Directed Research and Development Program, and the Defense Advanced Research Projects Agency Bio-Weapons Defense Program. The authors also thank Mr. Pin Yang, Mr. Scott Reed for technical assistance with micro-pen lithography, and Dr. Tom Baer and Dr. Randy Schunk for 3D simulation of micropen. TEM investigations were performed in the Department of Earth and Planetary Sciences at the University of New Mexico. Sandia is a multiprogram laboratory operated by Sandia Corporation, a Lockheed-Martin Company, for the U.S. DOE under Contract DE-AC04-94AL85000.

References

[1] H. Yang, N. Coombs, G. A. Ozin, *Adv. Mater.* **1997**, *9*, 811.
[2] C. Kresge, M. Leonowicz, W. Roth, C. Vartuli, J. Beck, *Nature* **1992**, *359*, 710.
[3] Y. Xia, G. M. Whitesides, *Angew. Chem. Int. Ed.* **1998**, *37*, 550.
[4] M. Trau, N. Uao, E. Lim, Y. Xia, G. M. Whitesides, I. A. Aksay, *Nature* **1997**, *390*, 674.
[5] P. Yang, T. Deng, D. Zhao, P. Feng, D. Pine, B. F. Chmelka, G. M. Whitesides, G. D. Stucky, *Science* **1998**, *282*, 2244.
[6] M. Antonietti, B. Berton, C. Goltner, H. P. Hentze, *Adv. Mater.* **1998**, *10*, 154.
[7] c. J. Brinker, Y. Lu, A. Sellinger, H. Fan, *Adv. Mater.* **1999**, *11*, 579.
[8] R. D. Piner, J. Zhu, F. Xu, S. Hong, C. A. Mirkin, *Science* **1999**, *283*, 661.
[9] P. Yang, D. Dimos, M. A. Rodriguez, R. F. Huang, S. Dai, D. Wilcox, *Mat. Res. Soc. Symp. Proc.* **1999**, *542*, 159.
[10] S.-C. Chang, J. Liu, J. Bharathan, Y. Yang, J. Onohara, J. Kido, *Adv. Mater.* **1999**, *11*, 734.
[11] D. Pede, G. Serra, D. De Rossi, *Mat. Sci. and Eng.* **1998**, *C5*, 289.
[12] S. L. Burkett, S. D. Sims, S. Mann, *Chem. Commun.* **1996**, 1367.
[13] C. E. Fowler, S. L. Burkett, S. Mann, *Chem. Commun.* **1997**, 1769.
[14] M. H. Lim, C. F. Blanford, A. Stein, *J. AM. Chem. Soc.* **1997**, *119*, 4090.
[15] C. D. Bain, E. B. Troughton, Y.-T. Tao, J. Evall, G. M. Whitesides, R. G. Nuzzo, *J. Am. Chem. Soc.* **1989**, *111*, 321.
[16] L. M. Tender, W. R.L., H. Fan, G. P. Lopez, *Langmuir* **1996**, *12*, 5515.
[17] Y. Lu, R. Ganguli, C. A. Drewien, M. T. Anderson, C. J. Brinker, W. L. Gong, Y. X. Guo, H. Soyez, B. Dunn, M. H. Huang, J. I. Zink, *Nature* **1997**, *389*, 364.
[18] Y. Lu, H. Fan, A. Stump, T. L. Ward, T. Reiker, C. J. Brinker, *Nature* **1999**, *398*, 223.
[19] R. A. Cairncross, P. R. Schunk, T. A. Baer, R. R. Rao, P. A. Sackinger, *Int. J. Numer. Meth. Fluids* **to appear 2000**.
[20] T. A. Baer, R. A. Cairncross, P. r. Schunk, R. R. Rao, P. A. Sackinger, *Int. J. Numer. Meth. Fluids* **to appear 2000**.
[21] I. A. Aksay, *Science* **1996**, *273*, 892.
[22] H. Yang, A. Kuperman, N. Coombs, S. Mamiche-Afara, G. A. Ozin, *Nature* **1996**, *379*, 703.
[23] H. Yang, N. Coombs, I. Sokolov, G. A. Ozin, *Nature* **1996**, *381*, 589.
[24] J. L. Wilbur, A. Kumar, H. A. Biebuyck, E. Kim, G. M. Whitesides, *Nanotechnology* **1996**, *7*, 452.

[25] G. C. Frye, A. J. Ricco, S. J. Martin, C. J. Brinker, in *Better Ceramics Through Chemistry III, Vol. 121* (Eds.: C. J. Brinker, D. E. Clark, D. R. Ulrich), Mat. Res. Soc., Reno, Nevada, **1988**, pp. 349.

Considerations for
Direct Write Processing

NUMERICAL SIMULATION OF LASER INDUCED SUBSTRATE HEATING FOR DIRECT WRITE OF MESOSCOPIC INTEGRATED CONFORMAL ELECTRONICS (MICE)

S. Lowry[*] and J. C. Sheu
CFD Research Corporation, 215 Wynn Drive, Huntsville, AL 35805, USA
sal@cfdrc.com and jcs@cfdrc.com

Robert Stewart and Robert Parkhill
CMS Technetronics Inc., Stillwater, Oklahoma.
5202-2 North Richman Hill Road

ABSTRACT: A numerical tool is developed to simulate the optical and thermal interactions of selected lasers with precursors and substrates in support of the emerging technology for the direct write of Mesoscopic Integrated Conformal Electronics (MICE). The code couples the Discrete Ordinate Method (DOM) radiation model with the multi-physics computation fluid dynamics code CFD-ACE to predict the conductive and radiative heat transport in the process.

This paper provides a brief overview of the numerical model. Selected simulations are presented including comparison with empirical data. The capabilities, limitations, and potential applications of the model with respect to MICE are discussed. Future model enhancements are proposed.

INTRODUCTION

Under DARPA sponsorship, advanced techniques are being developed to allow the direct write of Mesoscopic Integrated Conformal Electronics (MICE). Laser processing of selected precursors is one such technique being developed by CMS Technetronics to enable the direct deposition of electronic components onto diverse substrate materials. For complex circuits, controlling the temperature of the deposition process is critical and potentially difficult when a range of thermal and optical material properties is involved. Temperature control is critical because the temperature dictates the rate, extent and quality of the process. Simulation tools that can predict the temperature history of the deposited substances will help optimize this technology by preventing underheating or overheating of the materials.

THEORETICAL FORMULATION

Conductive Heat Transport The foundation of the thermal simulation is the CFD code CFD-ACE which can solve for the conductive and convective heat transfer through disparate solids and fluids. The basic energy equation to solve is

Energy: $$\frac{\partial \rho h}{\partial t} + \nabla \cdot \rho \mathbf{u} h = \nabla \cdot \mathbf{q} + \tau : \nabla \mathbf{u} + \frac{dp}{dt} \qquad (1)$$

[*] Point of Contact: Sam Lowry (256) 726-4853

Mat. Res. Soc. Symp. Proc. Vol. 624 © 2000 Materials Research Society

where ρ, **u**, p,**q**, τ, and h are density, velocity, pressure, heat flux, stress tensor, and enthalpy, respectively.

Radiation Radiative heat transport in the laser process is currently modeled using the Discrete Ordinate Method (DOM) [1,2]. The DOM method is capable of modeling the transmittance and absorption of non-gray radiation (i.e., with spectral effects) through semi-transparent materials. It also accounts for the reflection and absorption of radiation at opaque surfaces. The limitation of the DOM method for modeling laser radiation is that it is relatively dispersive as compared to the non-dispersive beam of a laser. This effect is minimized by introducing the laser into the solution domain within less than 100 microns of the substrate to reduce any numerically induced beam divergence prior to striking the surface.

Laser Intensity The laser intensity distribution is introduced at the upper boundary of the solution domain assuming either a Gaussian or "Tophat" (square wave) distribution. The duration, pulse frequency, and power of the laser may also be varied. The path and translation speed of the laser are specified via an input text.

Properties Determining the effective properties of the materials to be simulated is critical for accurate model prediction. These properties include density, thermal conductivity, specific heat, and optical properties and typically depend on the temperature of the material. Furthermore, the precursors used in the direct write process are often mixtures of materials with complex geometrical structures, such as powders with binders, making the determination of the effective mixture properties difficult. Selected properties used in the current simulations are provided in Table 1. The properties of the resistor paste modeled in this paper are currently very preliminary approximations and remain to be refined based on future analysis.

Table 1. Material Properties for Laser heating simulation

Material	Conductivity (W/m-K)	Specific Heat (J/kg-K)	Density (kg/m³)	Emissivity
Aerogel	10.0	981	221	
Alumina	30.27	800	4000	0.35 (opaque)
Resistor	0.5	250	4000	0.4 (opaque)
Silver	f_1 (T)	235	10,500	0.05 (opaque)
Silicon	f_2 (T)	702	2,330	

$$f_1(T) = 425 + 0.07 \cdot T - 0.0002 \cdot T^2 + 1.03 \times 10^{-7} \cdot T^3 + 1.03 \times 10^{-11} \cdot T^4 - 1.72 \times 10^{-14} \cdot T^5$$

$$f_2(T) = 445 - 1.65 \cdot T + 0.0028 \cdot T^2 - 2.4 \times 10^{-6} \cdot T^3 + 1.0 \times 10^{-9} \cdot T^4 - 1.37 \times 10^{-13} \cdot T^5$$

SIMULATION RESULTS

Validations A preliminary validation case was performed based on empirical measurements of laser heating of two crosslines by Toivo et al. [3]. Figure 1 shows the basic configuration with the materials and properties. In this experiment, a stationary 0.5 watt laser is focused on the intersection of nickel and gold strips overlaid on a silicon dioxide substrate. Figure 2 indicates the good comparison between the model predictions and the measured temperature distribution.

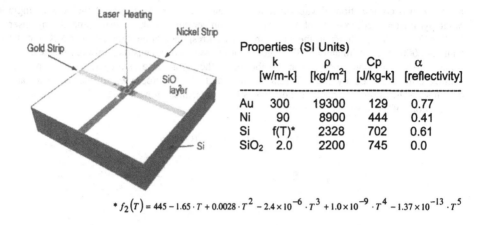

Properties (SI Units)

	k [w/m-k]	ρ [kg/m²]	Cp [J/kg-k]	α [reflectivity]
Au	300	19300	129	0.77
Ni	90	8900	444	0.41
Si	f(T)*	2328	702	0.61
SiO₂	2.0	2200	745	0.0

$$* f_2(T) = 445 - 1.65 \cdot T + 0.0028 \cdot T^2 - 2.4 \times 10^{-6} \cdot T^3 + 1.0 \times 10^{-9} \cdot T^4 - 1.37 \times 10^{-13} \cdot T^5$$

Figure 1. Experimental Configuration for Laser Heating of Nickel and Gold Crosslines

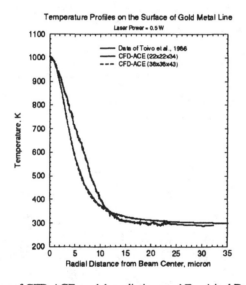

Figure 2. Comparison of CFD-ACE model predictions and Empirical Data for Laser Heating

Additional validation efforts are currently underway using infrared measurements of the CMS laser process. Figure 3 shows a photograph of a series of parallel silver lines deposited on an alumina substrate, crossed by a single line of resistor material. Prior to processing, the resistor line is a combination of glass powder : ~55 % wt, Ruthenium 2-ethyl Hexanoate ~38% wt and Terpineol ~7% wt. The width of the lines is on the order of 1000 microns and the height is approximately 75 microns. For the case shown, a CO2 CW laser with a Gaussian beam 300 microns in diameter at a power of 2.2 watts is subsequently traversed along the resistor line resulting in a combination of vaporization, reaction and sintering. Figure 3 includes a single point pyrometer measurement of the maximum temperature of the resistor line during laser heating. The value of 4096 on the scale provided corresponds approximately to 800K and the value of 2000 to approximately 590K. Figure 4 provides a two-dimensional infrared image of the same configuration and process. The distribution of the color contours indicates the anisotropic heating effects caused by the different thermal and optical properties of the substrate and crosslines. The more thermally conductive silver crosslines provide a cooling effect at the intersections of the silver and resistor.

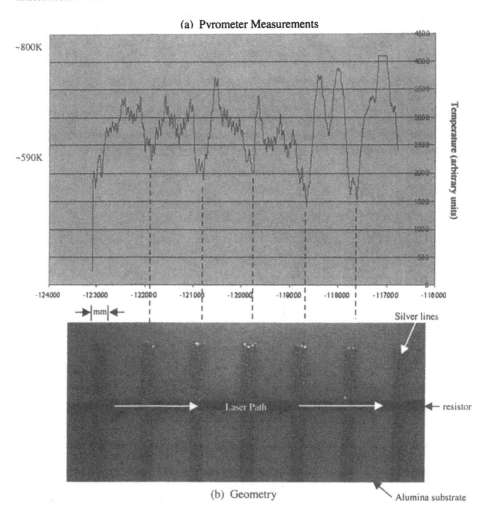

Figure 3. CMS Crossline Laser Processing Experiment #7

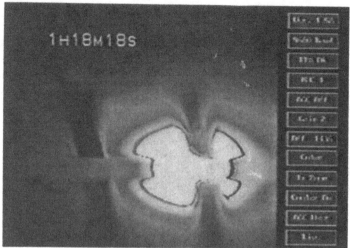

Figure 4. CMS Crossline Laser Processing Experiment #7, 2-D Pyrometer Image

The vertical darker blue lines in Figure 4 indicate cooler temperatures corresponding to the silver crosslines. This is due, in part, to the high reflectivity of the silver, preventing the absorption of laser radiation. The sharp drop in temperature between the silver and surrounding materials may also indicate poor thermal contact with the overlaying resistor paste and the adjacent alumina. Another phenomenon indicated by the experimental image is that the effects of the laser beam extend well beyond the reported 150 micron Gaussian diameter of the beam. The silver lines in the Figure 4 are 1000 microns wide, and yet the effects of the radiatve heating of the alumina substrate extend well beyond this radius.

Figure 5 shows a preliminary numerical simulation of the above case. In an attempt to get a better match with the experimental image, the boundary conditions and material properties listed in Table 1 were varied to determine the sensitivity of the solution to these parameters. In the prediction shown, the thermal conductivity of the alumina was reduced an order of magnitude and its emissivity increased to 0.9, as compared with the published value listed in Table 1. The diameter of the laser beam was also increased 4 fold in the simulation based on the extended heating indicated by the experimental image. This was done in an attempt to get more heating of the substrate, as compared to the silver and resistor line. Even with these modifications, the temperature of the resistor is still too hot, as compared with the alumina, resulting in a qualitatively different result that the cloverleaf pattern seen in the infrared image. In addtion, the relative magnitude of the maximum resistor temperature in the simulation cycled between 1000K and 1500K which is clearly higher that indicated in Figure 3. These discrepancies may be due to the energy being absorbed by the unprocessed paste as it converts to processed resistor material.

As more quantitative data become available, the source of these discrepancies will be isolated and the effective properties of the materials will be calibrated. The influence of boundary conditions will also be investigated. The end objective is to validate the model such that it can be used as a predictive tool for arbitrary configurations. Future improvements to the model include incorporating the effects of semi-transparent materials and the effects of reflection and absorption as functions of both temperature and wavelength. The next step will be to add reaction models such that the effects of the chemistry/sintering in the lines can be simulated.

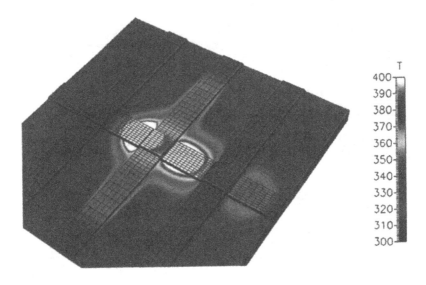

Figure 5. CFD-ACE prediction of resistor heating with crosslines.

CONCLUSIONS

A preliminary heat transfer model of laser heating of a substrate has been developed and validated for crossline geometries with well know material properties. The transfer of energy from the laser to the surface is simulated using the Discrete Ordinate Method. The optical properties of the material surfaces are taken into account. Work is currently in progress to calibrate and test the model for direct write applications in collaboration with CMS Technetronics. This direct comparison with test runs for the MICE process has revealed discrepancies between predictions and measurements that are not unexpected for a first attempt. Qualitatively, the preliminary results are meaningful and a better understanding and improvement in the correlation is expected to evolve rapidly as the validation effort continues.

ACKNOWLEDGEMENTS

This work is being supported under DARPA contracts: MDA972-99-C-0017. The authors are grateful to Dr. Bill Warren, DARPA program manager, for his comments, suggestions and his enthusiastic support of this work.

REFERENCES

1) M.G. Giridharan, S.A. Lowry, and A. Krishnan, "A Multi-Block BFC Radiation Model for Complex Geometries," 30th AIAA Thermophysics Conference, San Diego, CA, 19-22 June, 1995.
2) M. Kannapel, S.A. Lowry, A. Krishnan, I. Clark, P. Hyer, and E. Johnson, "Preliminary Study of the Influence of Grashof and Reynolds Numbers on the Flow and Heat Transfer in an MOCVD Reactor," CFDRC No. PA-97-01, SPIE International Symposium, San Diego, CA, July 27 – Aug.1, 1997.
3) Toivo Kodas et al., "J. Appl. Phys. 61 (8), pp. 2749-2753, April 1987.

THERMAL STABILITY AND ANALYSIS OF LASER DEPOSITED PLATINUM FILMS

G.J. Berry[†], J.A. Cairns[†], M.R. Davidson[†], Y.C. Fan[†], A.G. Fitzgerald[†]
A.H. Fzea[*], J. Lobban[*], P. McGivern[*], J. Thomson[*] and W. Shaikh[§]

[†] Department of Applied Physics and Electronic & Mechanical Engineering, University of Dundee, Dundee DD1 4HN, UK
[*] Department of Chemistry, University of Dundee, Dundee DD1 4HN, UK
[§] Central Laser Facility, Rutherford Appleton Laboratory, Chilton, Didcot, Oxfordshire OX11 0QX, UK

ABSTRACT

As the trend towards device miniaturisation continues, surface effects and the thermal stability of metal deposits becomes increasingly important. We present here a study of the morphology and composition of platinum films, produced by the UV-induced decomposition of organometallic materials, under various annealing conditions. The surface composition of the metal deposits was studied by X-ray photoelectron spectroscopy, both as-deposited and following thermal treatment. In addition, the morphology of the surface was studied by atomic force microscopy which enabled the investigation of film restructuring. These studies were performed over a range of temperatures up to 1000°C in air and up to 600°C in reducing environments. Complementary information regarding the film morphology has been obtained from transmission electron microscopy. The data has been used to provide an insight into the effects of elevated temperatures on metal films deposited by a direct write method.

INTRODUCTION

Many studies have been made into the photolytic and pyrolytic decomposition of organometallic molecules in the gaseous [1], liquid [2] or solid phase [3-7]. We have developed a range of organometallic compounds which can be deposited as thin films by thermal evaporation [8,9]. This method of producing thin films has many advantages over alternative deposition methods such as spin coating. The absence of a solvent means that a film of high purity organometallic can be deposited onto a wide range of substrates including non-wettable or fragile membranes such as those used in X-ray lithography. The organometallic films exhibit excellent adhesion as a result of the absence of solvent molecules competing with the organometallic for bonding sites [10]. The films are then patterned through a standard chromium-on-quartz photomask using UV from an excimer laser. We present here a study of the composition and morphology of platinum films produced by the UV induced decomposition of organometallic films. UV deposited films were studied using Atomic Force Microscopy (AFM), Transmission Electron Microscopy (TEM) and X-Ray Photoelectron Spectroscopy (XPS) following various annealing regimes which involved heating the films in a range of reducing and oxidising environments at temperatures of up to 1000°C. Patterned organometallic films were also studied by Scanning Electron Microscopy (SEM) to investigate any changes in linewidth after heating the samples in both oxidising and reducing environments.

Mat. Res. Soc. Symp. Proc. Vol. 624 © 2000 Materials Research Society

EXPERIMENTAL DETAILS

The organometallic compound used throughout this study was cis-Dichlorobis (triphenylphosphine) platinum(II). The substrates used included <100> silicon wafers for SEM, AFM and XPS analyses and silicon nitride membranes for TEM analysis. As mentioned above, the organometallic was deposited by thermal evaporation. This was done *in vacuo* at a base pressure of typically 1×10^{-5} Torr with the temperature of the boat being typically of the order of 200°C. The substrate was held at room temperature. The thickness of the deposited films was measured by Atomic Force Microscopy and found to be typically 700nm.

Following deposition, the films were exposed to ultraviolet radiation from a Lambda Physik 210iF excimer laser. This laser had a pulse length of approximately 20ns with a maximum output power of 150mJcm^{-2} per pulse. The beam intensity was varied by the insertion of suitable attenuators in the form of thin quartz plates, and was measured using a large area photodiode (Gentec ED500) and joulemeter (Sun EM1). Exposures were performed by opening a shutter for the desired length of time. In this study the pulse energy was varied from 25 to 52mJcm^{-2} per pulse and the total exposure dose ranged from 0.03 to 60Jcm^{-2}. Patterned samples for linewidth analysis were made by exposing the organometallic through a chromium-on-quartz photomask (Compugraphics International). After exposure to UV the samples were rinsed with acetone to remove any unexposed organometallic. The films were then studied as deposited and following thermal treatment. The topography of the thin films was viewed in an AFM, and the surface composition of the irradiated film was studied by XPS in a VG HB 100 Multilab instrument.

Linewidth analysis was performed by taking SEM micrographs of the features, which were then digitised and measured using image processing software. (NIH Image).

RESULTS AND DISCUSSION

A film of dichlorobis(triphenylphosphine)platinum(II) was prepared for laser irradiation by thermal evaporation. This film was irradiated with 248nm UV with various doses ranging from 0.3Jcm^{-2} to over 30Jcm^{-2}. The morphology of the films was then studied by atomic force microscopy. Figure 1 shows a comparison between an unirradiated film and films exposed to increasing doses of UV. Table 1 shows the effect of increasing exposure dose on the RMS roughness of the irradiated organometallic film.

Table 1 - Effect of exposure dose on the surface roughness of irradiated organometallic films.

Dose (Jcm^{-2})	RMS Roughness (nm)
0	0.40
0.323	12.33
1.615	12.91
3.23	19.85
16.15	80.33
32.3	83.01

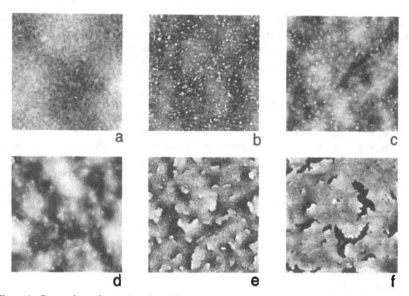

Figure 1. Comparison of an unirradiated film, (a), with films exposed to (b)0.323, (c)1.615, (d)3.23, (e)16.15, and (f)32.3Jcm⁻² 248nm UV irradiation N.B. Each image has a scan size of 10×10μm.

Figure 1a shows an AFM image of an unirradiated film of dichlorobis(triphenylphosphine) platinum(II). Figures 1b to 1f illustrate how a film of organometallic is affected by increasing doses of UV irradiation.

Table 2: Summary of XPS results.

Sample	Total Dose (Jcm^{-2})	Binding Energy of Pt 4f $_{7/2}$ (eV)
Unirradiated	0	73.04
A1	0.323	72.88
A2	1.615	72.77
A3	3.23	72.78
A4	16.15	71.22
A5	32.3	71.28
B1	0.55	71.81
B2	2.75	71.31
B3	5.5	71.54
B4	27.5	71.41
B5	55	71.31

Table 2 shows XPS results comparing the position of the platinum 4f $_{7/2}$ peak for two sets of samples. Samples A1 to A5 were irradiated with a fluence of 25mJcm⁻² per pulse and samples B1 to B5 were exposed with a fluence of 52mJcm⁻² per pulse. All samples were of the same initial film thickness. The position of the Pt 4f $_{7/2}$ peak for the unirradiated film at 73.04eV corresponds well with the published value for this compound at 73.3eV [11]. Looking initially at samples A1 to A5, at low doses the position of the Pt 4f $_{7/2}$ peak shows little change from the value obtained from the unirradiated sample. At higher doses (A4 and A5) the binding energy drops to a value which is consistent with the published value of

71.2eV for metallic platinum [12]. In contrast, for organometallic films exposed to higher fluences (samples B1 to B5), the Pt 4f $_{7/2}$ peak shifts to lower binding energy after a very low dose. In both sets of samples, this reduction to metallic platinum is accompanied by the disappearance of the chlorine 2p peak from the spectra. The phosphorous 2p peak also reduces significantly in intensity at this point but does not disappear entirely.

Correction has been made for the shift in binding energy resulting from surface charging. Carbon 1s photoelectrons are those most generally adopted for referencing purposes. However, this is not a simple exercise when carbon is a major constituent of the film under study, as is the case in this work. The carbon is initially bound in a phenyl group, but subsequently reduces to graphitic carbon. Deconvolution of the carbon 1s spectrum obtained from the unirradiated film reveals a single peak corresponding to carbon atoms bound in a phenyl group. The published value for this peak is 284.9eV [13]. The carbon 1s signal in the spectra obtained from samples A4, A5 and B1-B5 consists of a single peak with a shoulder at a higher energy. Deconvolution of this peak demonstrates that the C1s XPS signal is now dominated by graphitic carbon, which has a binding energy of 284.6eV [13]. Charge referencing was carried out using the appropriate carbon 1s peak.

The samples were then heated in air to 1000°C in 250°C steps. The films were heated at 5°C per minute, held at the annealing temperature for 5 minutes, then cooled at the same rate. The XPS spectra of the Pt4f and the C1s signals are shown in figure 2 before and after heating to 500°C.

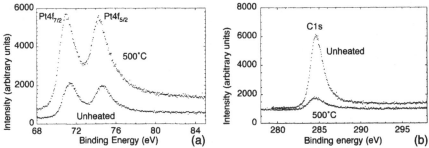

Figure 2 (a) Platinum 4f signal and (b) carbon 1s signal from sample B5 as deposited and after heating to 500°C in air.

Figure 2a shows the platinum 4f peaks from sample B5 for the film in its unheated state and after annealing at 500°C. Figure 2b shows the carbon 1s signal from this sample under the same conditions. It can be seen from these spectra that the carbon content of the sample has been dramatically reduced upon heating to 500°C, which is accompanied by a similarly large increase in the platinum signal. The spectra obtained from the samples after heating to 250°C showed negligible change from the unheated case. XPS data from samples heated to 750°C and above had high levels of contamination. This was subsequently traced to the furnace used for the annealing of the samples. Argon ion etching was not used to remove any probable surface contamination because of its tendency to induce changes in the elemental composition of the surface as a result of differential sputtering. Quantitative compositional analysis of the films could not be performed for this reason.

Patterned samples were made by exposing films of organometallic through a chromium-on-quartz photomask with a dose of 12Jcm^{-2} at a fluence of 60mJcm^{-2} per pulse. The pattern was a simple line-space design, having a 1:1 mark-space ratio and a period of 5μm (i.e. linewidth of 2.5μm). These samples were heated in air and hydrogen, in a similar manner to the XPS samples, and the linewidths measured at each stage. Figure 3 shows the variation in linewidth as a function of annealing temperature in air and hydrogen.

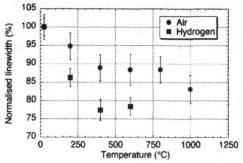

Figure 3 - Linewidth reduction as samples are heated in air and hydrogen.

An example of the lines produced before and after annealing is shown below in Figure 4.

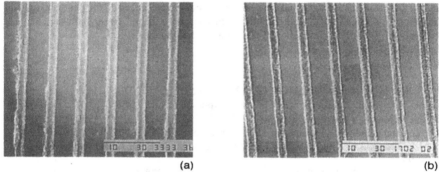

| (a) | (b) |

Figure 4 - Unheated Pt pattern (a) and the same heated to 600°C in hydrogen (b). Scale mark represents 10μm.

The sample heated in air shows a decrease in linewidth to around 89% of the original value at 400°C. Above this temperature, little change is observed until the sample is heated to 1000°C, at which point a decrease to approximately 83% is observed. The sample heated in hydrogen exhibits similar behaviour, but falls to a lower value of 77% of initial width. This is believed to be a result of the ability of hydrogen to reduce platinum efficiently. The samples could not be heated above 600°C in a hydrogen containing atmosphere due to safety limits on the equipment used.

Selected area electron diffraction has been used to study the crystallinity of UV deposited platinum films on silicon nitride substrates. These studies yield diffraction patterns consistent with the presence of polycrystalline metal.

CONCLUSIONS

The deposition of platinum features produced via the UV laser induced decomposition of organometallic films has been demonstrated. The topography and surface composition of these films has been studied. We have found that lower doses of UV produce better quality features which have a low surface roughness. XPS was used to study changes in surface chemistry. It was found that the platinum was reduced more efficiently at higher fluences. However, a balance must be met since higher fluences tend to result in undesirable ablation of the film. In common with the majority of metal films produced by a direct write method, our deposits were found to contain carbonaceous impurities. Annealing the deposits was found to cause some shrinkage of small metal features but impurity levels were dramatically reduced.

ACKNOWLEDGEMENTS

This work was funded by an Engineering and Physical Sciences Research Council grant, number GR/M41223. The project was also part funded by the European Community, European Regional Development Fund through the Eastern Scotland Objective 2 Programme, and further financial assistance was provided by Scottish Enterprise and the Scottish Higher Education Funding Council. The authors would also like to thank Compugraphics International Ltd for the production of photomasks used in this work.

REFERENCES

1. S. Maeda, K. Minami and M Esashi, J. Micromech. Eng., **5**, 237-242, (1995).
2. H. Yokoyama, S. Kishida, and K. Washio, Appl. Phys. Lett., **44**(8), 755-757, (1984).
3. A. M. Mance, Appl. Phys. Lett., **60**(19), 2350-2352, (1991).
4. Y. Yutang and R. G. Hunsperger, Appl. Phys. Lett., **51**(25), 2136-2138, (1987).
5. J.Y. Zhang, S.L. King, I.W. Boyd and Q. Fang, Appl. Surf. Sci., **109/110**, 487-492, (1997).
6. A.A. Avey and R.H. Hill, J. Am. Chem. Soc., **118**, 237-238, (1996).
7. J.Y. Zhang, H. Esrom and I.W. Boyd, Appl. Surf. Sci., **96-98**, 399-404, (1996).
8. G.J. Berry, J.A. Cairns and J. Thomson, J. Mater. Sci., **14**, 844-846, (1995).
9. G.J. Berry, J.A. Cairns, M.R. Davidson, Y.C. Fan, A.G. Fitzgerald, J. Thomson and W. Shaikh, Appl. Surf. Sci., in press.
10. W.M. Moreau, Semiconductor Lithography, Plenum Press, (1988), pp. 304.
11. G.M. Bancroft, T. Chan, R.J. Puddenphatt and M.P. Brown, Inorg. Chim. Acta, **53**, L119-L120, (1981).
12. Practical Surface Analysis, 2nd Edition, Vol 1, Edited by D. Briggs and M.P. Seah, John Wiley, 1996 pp 621.
13. Ibid, pp 599.

CFD MODELING OF LASER GUIDED DEPOSITION FOR DIRECT-WRITE FABRICATION

J. C. Sheu, M. G. Giridharan, and S. Lowry
CFD Research Corporation, 215 Wynn Drive, Huntsville, AL 35805, USA
jcs@cfdrc.com and mgg@cfdrc.com

ABSTRACT

A computational model for simulating laser-guided particle deposition process is described. This model solves for the transport of gas and particle phases in a fully coupled manner. The optical forces on the particles are evaluated from Mie theory based on local laser intensity. Simulations were performed for different operating conditions of laser power, ambient pressure, and substrate traverse speed. Simulations revealed potential problems such as particle deflected away from substrate due to ambient air current disturbances, substrate overheating, and optical fiber clogging. Possible solutions for these problems are discussed.

INTRODUCTION

Direct fabrication processes to produce 3-dimensional metal and plastic parts have been used widely in the industry. These rapid fabrication techniques do not require a master pattern or other intermediate steps to produce the final product. A variety of direct fabrication techniques are commercially available now. These include Stereolithography (SLA) [1], Laminated Object Manufacturing (LOM) [2], Selective Laser Sintering (SLS) [3], Fused Deposition Modeling [4], Inkjet Printing or Droplet Deposition [5], and Laser Engineered Net Shaping (LENS) [6], etc. More recently, a new laser-guided particle deposition (LPD) technique for direct-write fabrication of electronic components was reported [7] and currently being developed by Optomec, Inc. A schematic of a LPD system is shown in Figure 1. In this LPD technique, particles in a laser beam are confined and propelled into a hollow optical fiber. The optical forces that cause confinement and propulsion of particles are generated by absorption and scattering of laser energy. The particles after traversing through the fiber get deposited on a substrate. This technique is being developed for fabrication of wireless and GPS circuits. Successful commercialization of the LPD technique for these high volume applications depends on the maximum rate at which particles can be deposited on the substrate. The maximum particle flux rate is limited by parameters such as laser power, particle capturing efficiency, fiber size, alignment, etc. Achieving high flux rates without clogging the fiber is a challenging task and optimizing the parameters based on physical testing will be expensive and time consuming. Computational modeling complements the experiments in identifying the optimum operating parameters in a shorter period of time.

This paper describes a computational model to simulate the LPD process. The model requires solution algorithms for: optical forces; particle tracking; and gas phase transport. The latter two are available in the commercial CFD code, CFD-ACE+ [8]. The optical force can be divided into two components: propulsion force acting in the direction of the incident light; and gradient force acting in the direction of the intensity gradient. Equations for evaluating these forces on a Rayleigh particle (particle size much smaller than the wavelength of incident light) are available

255

[9]. But for the LPD process, since the particle sizes are comparable to the wavelength, Mie scattering theory must be used. An approach for evaluating the optical forces is given in the following section.

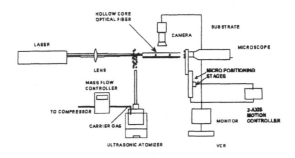

Figure 1. Schematic of the LPD System

THEORETICAL FORMULATION

Optical Force

The propulsion force acting along the direction of incident light can be expressed in terms of particle size as [11]:

$$F_{prop} = Q_{pr} \frac{\pi d_p^2}{4} \frac{I}{C}$$

where Q_{pr} is the pressure efficiency, I the laser intensity, and C the speed of light. The pressure efficiency is a function of laser wavelength, particle size, and complex index of refraction of the particle material. The pressure efficiency is calculated from Mie scattering theory, Wiscombe [12-14]. The gradient force confines the particles to the center of laser light. This force is conservative and attractive toward high intensity for positive polarizability of particles. The calculation of gradient force is based on the following formula:

$$F_{grad} = \nabla_r U$$

where U is the potential energy of the particles. In general, the potential energy is a function of particle size and accurate calculation of potential energy is complex and difficult. Here, the potential energy is assumed to be constant (5×10^{-18} J) and hence the gradient force is independent of particle size. Future calculations, however, would include the particle size dependence in the gradient force calculation.

Laser Intensity

The laser intensity distribution in regions outside of the fiber can be assumed to follow the Gaussian decay characteristics. This Gaussian distribution is derived from the theory of wave optics [15] and is sufficiently accurate under the condition that the medium (air) is essentially transparent. Inside the fiber, the laser light is coupled with a low-order grazing incidence mode. From the modal analysis of Marcatile and Schmeltzer [16], the zeroth order Bessel beam gives the lowest mode. Therefore, the intensity distribution inside the fiber is assumed to follow the zeroth order Bessel function.

Particle Dynamics

The equation governing the motion of particles is momentum conservation equation written as

$$\frac{du_p}{dt} = \frac{3}{4} \frac{C_D \mu_g R_e}{\rho_p d_p^2} \left(u_g - u_p \right) + g + S_f$$

where μ is gas viscosity, g is acceleration of gravity, R_e the Reynolds number based on relative velocity, and subscripts p and g denote the properties of particle and gas, respectively. The first term on the right hand side represents the drag force, the second term represents the gravity force, and the third term represents the optical forces generated by the laser. Integration of the above equation over time from specified initial conditions gives the instantaneous positions of particles.

From Crowe et al. [9], the drag coefficient C_D is estimated from

$$C_D = \frac{C_{D,noslip}}{1 + Kn \left[2.49 + 0.84 \exp\left(-\frac{1.74}{Kn} \right) \right]}$$

where $C_{D, noslip}$ is the drag coefficient calculated based on no-slip flow and the denominator represents correction due to slip effect in transition flows where the Knudsen number (Kn) defined as the ratio of mean free path of gas molecules (λ) to the particle diameter is above 0.01. It has been found that the slip effect may become significant for particle sizes in the range of 0.5 < d_p < 2 micrometers and it is included in all the calculations discussed below.

RESULTS

A sketch of a simplified, two-dimensional geometry of a LPD system (designed by Optomec Design Company [17]) is illustrated in Figure 2. The geometric and operating conditions are given in Table 1. The laser is focused at the fiber inlet. The velocity of atomized particles near the atomizer (size ranging from 0.5 to 2.0 micrometers) is assumed to be the same as that of the carrier gas. Though heat transfer to particles and substrate is included, phase change of particles is not included in the simulations discussed below.

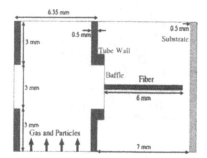

Hollow core inner diameter	20 microns
Hollow core outer diameter	360 microns
Laser power	300 mW
Laser wavelength	532 nm
Laser spot size	14 microns
Particle size	05.-2 microns
Gas inlet flow rate	10 ml/min

Table 1. Simulation Conditions for LPD System

Figure 2. Computational domain for the 2-D, LPD System

For the conditions described above, Figure 3 shows the predicted particle trajectories. Before interacting with the laser beam, all the particles travel along the direction of the carrier gas as the particle relaxation time is small for micron-sized particles. Inside the laser beam, however, the particles experience the optical forces causing the particle trajectories to shift along the laser beam direction. This shift in position traps some particles (close to the fiber entrance) into the fiber. It is found that only a small fraction of the particles (< 1%) are captured into the fiber. For a given size range of the particles, the capturing efficiency can be improved either by increasing the laser power or by making the carrier gas flow towards the fiber entrance. The latter option is being explored. Since the fiber acts as a waveguide, the axial decay of laser intensity is small enough that optical forces continue to propel the particles through the fiber and onto the substrate. The maximum velocity of particles was found to be about 1 cm/s. The Stokes number (ratio of particle relaxation time to flow time scale) for this case is found to be so small (< 0.002) that the particles would just follow the flow streamlines.

Next simulations are performed to predict the substrate temperature at the laser spot. The substrate materials modeled include silicon, glass, and PVC plastic. The results indicate that the temperature rise at the laser spot for silicon and glass substrates was less than 10 degrees. Though silicon is semitransparent, it has high thermal conductivity to distribute the laser energy. The PVC substrate, on the other hand, is opaque with low thermal conductivity and the laser energy heats the substrate close to melting temperature. Simulations revealed that the high temperature at the laser spot caused buoyancy driven flow towards the hot spot (see figure 4). At high temperatures, this flow was strong enough to deflect the particles away from the laser spot. The buoyancy flow could be avoided by orienting the system in such a way that laser from below shines on the substrate at top. Clearly, the maximum laser power that can be used depends on the substrate material. Pulsing the laser or increasing the substrate traverse speed could be used to control substrate overheating. But high substrate traverse speeds again deflect the particles as they approach the substrate due to the boundary layer effect. Figure 5 shows the distance where the particles land on the substrate from the laser beam center as a function of substrate traverse speed. As can be seen, the maximum substrate speed is limited which in turn limits the writing speed. The substrate heating which could produce significant buoyancy flow and the boundary layer flow limits the operating window for laser power and traverse speed which depend on the particle as well as substrate material.

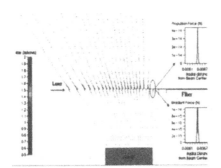

Figure 3. Particle Trajectories Influenced by Optical Forces.

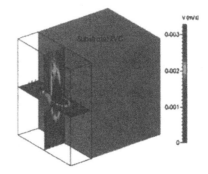

Figure 4. Buoyancy Driven Flow Caused by Substrate Overheating.

From the above-mentioned simulations, it is clear that the momentum imparted to the particles by the optical forces was not high enough to overcome ambient air current disturbances. It was felt that convective flow inside the fiber would increase the particle momentum in addition to that imparted by optical forces. Air flow inside the fiber could be introduced by maintaining a lower pressure at the exit of the fiber. To investigate this idea, simulations are performed with different fiber downstream pressures: 750 and 1 torr, while the upstream pressure is maintained constant at 760 torr. When the downstream pressure is 750 torr, the capturing efficiency increased to 20% from 1% for the no pressure gradient case. However, very few particles hit the substrate because the particles are carried away by the boundary layer flow (substrate traverse speed is 1 cm/s) and ambient air currents. For the 1 torr case, the capturing efficiency increased to 100%, but most of the particles unable to make a sharp 90° turn into the fiber and hence they hit the front face of the fiber. This would certainly lead to clogging in the fiber. Also, the air velocity inside the fiber was so high (about 100 m/s) that the drag force on particles is several orders of magnitude higher than the optical forces. Clearly, there is an optimum downstream pressure at which the capturing efficiency can be increased and at the same time both optical and drag forces on particles impart just enough momentum to overcome ambient air current disturbances.

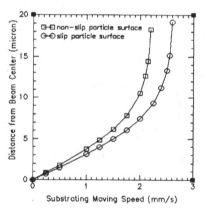

Figure 5. Particle Deflection Distance as a Function of Substrate Traverse Speed.

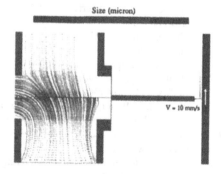

Figure 6. Particle Trajectories Predicted for Axial Pressure Gradient in Fiber (downstream pressure = 750 torr).

CONCLUSIONS

Simulations of the LPD technique for direct-write fabrication of electronic components have been presented. Simulations have shown that the temperature of the substrate at the laser spot would depend on the optical properties and the thermal conductivity of the substrate. Opaque and low conductivity substrate was shown to produce buoyancy driven air currents that are strong enough to deflect the particles away from substrates. To avoid substrate overheating, the laser power can be decreased or the substrate relative velocity can be increased. But both of these solutions severely limits the writing speed because: decreasing the laser power decreases the propulsion velocity of the particles towards the substrate; and increasing the substrate relative velocity causes the particles to drift away by the boundary layer flow on the substrate. In order to increase the particle flux, an axial pressure gradient in the fiber can be introduced by maintaining suction pressures downstream of the fiber. Simulations have indicated the encouraging result that maintaining reasonable pressure differential in the fiber could increase the particle flux without clogging the fiber.

ACKNOWLEDGEMENTS

This work is being supported by a DARPA contract: DAAH01-99-C-R015. The authors are grateful to Dr. Bill Warren, DARPA program manager, for his comments, suggestions and his enthusiastic support of this work. The authors would like to thank Drs. Mike Renn, Marcelino Essien, Bruce King and Doyle Miller of Optomec, Inc. for their inputs and providing the geometry and operating conditions for the simulations.

REFERENCES

1. Ohr, S., *Electronic Engineering Times*, June 15, 1998.
2. Lauwers, B., Meyvaert, I., and Kruth, J. P., *Prototyping Technology International 97* (1997) p. 139.
3. Chartoff, R. P., Ullett, J. S., and Klosterman, D. A., *Prototyping Technology International 97*, (1997) p. 191.
4. Jepson, L. R. and Beaman, J. J., *Prototyping Technology International 97*, (1997) p. 166.
5. Stanley, M., *Prototyping Technology International 97*, (1997). p. 207.
6. CAD/CAM Publishing, Inc., *Rapid Prototyping Report*, (March 1998) p. 3
7. Renn, M. J., Pastel, R., and Lewandowski, Physical Review Letters (in press).
8. CFD-ACE+: Theory Manual, Version 6, CFD Research Corporation, (2000).
9. Kerker, M., *The Scattering of Light and Other Electromagnetic Radiation*, (Academic Press, New York, 1969).
10. C. Crowe, M. Sommerfeld, and Y. Tsuji, *Multiphase Flows with Droplets and Particles*, (CRC Press, 1998).
11. H. C. van de Hulst, *Light Scattering by Small Particles*, (Dover Publications, Inc., New York, 1981).
12. Wiscomb, W. J., *Mie Scattering Calculations: Advanced in Technique and* Fast, Vector-Speed Computer Codes, NCAR/TN-140+STR, NCAR Technical Note, July 1979.
13. Wiscomb, W. J., Applied Optics, **19**, 1505, (1980).
14. Nussenzveig, H. M., and Wiscomb, W. J., Physical Review A, **43**, 2093, (1991).
15. B. E. Saleh, and M. Teich, *Fundamentals of Photonics*, (John and Wiley & Sons Inc., New York, 1991).
16. Marcatile, E. A., and Schmeltzer, A. A., Bell System Tech J., July, 1783, (1964).
17. Optomec, Inc. dba Optomec Design Company, (private communication).

COMPUTATIONAL MODELING OF DIRECT PRINT MICROLITHOGRAPHY

A. A. DARHUBER*, S. M. MILLER*, S. M. TROIAN*, S. WAGNER+

*Interfacial Science Laboratory, Dept. of Chemical Engineering, Princeton University
+Dept. of Electrical Engineering, Princeton University, Princeton, New Jersey 08544, USA

ABSTRACT

Using a combination of experiment and simulations, we have studied the equilibrium shapes of liquid microstructures on flat but chemically heterogeneous substrates. The surface patterns, which define regions of different surface energy, induce deformations of the liquid-solid contact line, which in turn can either promote or impede capillary break-up and bulge formation. We study numerically the influence of the adhesion energies on the hydrophilic and hydrophobic surface areas, the pattern geometry and the deposited fluid volume on the liquid surface profiles.

INTRODUCTION

In the last decade many efforts have been made both to reduce the minimum feature size of electronic devices and to increase the throughput and reduce fabrication costs. Several groups have explored printing techniques such as gravure offset printing, screen printing, inkjet-printing and micro-contact printing for lithography or direct deposition of semiconductor- or polymer-based thin film transistors and light emitting diodes [1-11]. We are investigating wet printing techniques for the transfer of liquid inks from a chemically patterned surface onto an unpatterned target substrate. There are five major technological challenges involved in this process: (a) the fabrication of the printing plates, (b) the deposition and distribution of the ink on these stamps, (c) the control of the behavior of the ink patterns between ink deposition and printing, (d) the printing and (e) the stability of the printed ink patterns. In this article we focus on aspects of the behavior of liquid microstructures on chemically heterogeneous surfaces and the transfer process during printing.

PRINTING PLATE FABRICATION

The selective distribution of the ink on the imaging areas of the printing plate is achieved with the aid of a hydrophobic self-assembled monolayer (SAM) [12], which repels the ink from the non-imaging areas. There are several approaches to stamp fabrication. For example one can homogeneously coat a substrate like a silicon wafer with an SAM of octadecyltrichlorosilane (OTS) and subsequently remove the monolayer from the imaging areas by exposure to deep-UV light (λ = 193 nm) through a chromium mask [13]. Alternatively, conventional photolithography can be used to pattern the SAM by first spin-coating photoresist on the SAM, exposing it to UV light through a chromium mask, developing the resist, removing the monolayer by oxygen plasma treatment and lastly removing the remaining photoresist. Since the mechanical resistance of SAMs to scratches is not very high and because scratches render the stamp surface hydrophilic in undesired locations, scratched stamps have to be reprocessed from the beginning or disposed. This problem can be partially overcome by depositing and patterning a 50 nm thick gold layer on the silicon surface by wet chemical etching and coating the gold with an SAM of hexadecanethiol (HDT). HDT forms a chemical bond with gold but does not adhere to the bare silicon regions. In case of surface scratches, which only damage the HDT layer, the sample only needs to be immersed in an HDT solution again, but no lithography step is required since the pattern is already defined in the gold layer. Kumar and Whitesides have developed another stamping technique called micro-contact printing, where a thiol-SAM is deposited on a flat gold surface by selective deposition with a polymer stamp [14].

NUMERICAL PROCESS SIMULATIONS

Numerical simulations of the equilibrium shapes of liquid microstructures have been performed with Surface Evolver [15], taking into account the surface tension γ_{lv} of the liquid and the contact angles on both the hydrophilic and hydrophobic parts of the surface. An example of a droplet on a hydrophilic channel, which spread over the channel boundaries, is presented in Fig. 1 along with a corresponding calculated droplet shape. A more extensive discussion of the simulation technique can be found in Ref. [16].

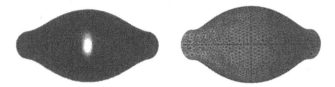

Figure 1: A droplet of glycerol (left, width 47 μm) on a hydrophilic channel defined with an OTS monolayer on a silicon wafer and a simulated shape (right).

Figure 2 shows top and side view profiles of ink lines at the junction of two hydrophilic channels, which meet at an angle of 60°. A corresponding volume filling factor f may be defined as the ratio of the total ink volume and the product of the total hydrophilic surface area and the channel width. If the junction of the lines has an acute outer corner [Fig. 2(a),(b),(e),(f)], the liquid forms a bulge in the corner. Such an uneven height profile would be unfavorable for printing. If the outer corner is cut and the junction width is made narrower than the channel width [Fig. 2(c),(g)], the height profile is more or less even for a filling factor of $f = 0.2$. However, if the volume is reduced by 50 percent [Fig. 2(d),(h)] the height of the liquid is reduced in the junction region. Upon further reduction of the volume the continuous liquid surface would pinch off at the corner. Thus, maintaining an even height profile requires both the proper design of the surface pattern and the control of the deposited liquid volume.

An explanation for the bulge formation is sketched in Fig. 3(a). Because of surface tension, the liquid prefers to reside in regions where it can minimize its surface curvature. Since the radius of a circle, which can be inscribed within the boundaries of the hydrophilic region, is

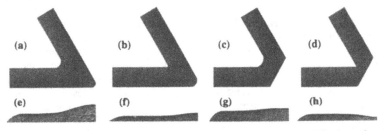

Figure 2: Top and side-view profiles of ink lines at a junction of two hydrophilic channels, which meet at an angle of 60°. (a)-(d) are top views, (e)-(h) are side views along the plane of symmetry. Only one half of the surface profile is shown in the side views because of mirror symmetry. The volume filling factors are $f = 0.2$ for (a), (c), (e) and (g) and $f = 0.1$ for (b), (d), (f) and (h).

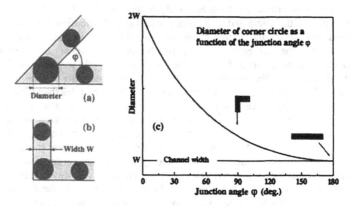

Figure 3: Qualitative explanation for the bulge formation in the junction region of two hydrophilic lines. (a,b) Sketches of the junction geometry for a junction angle φ of (a) 60° and (b) 90°. (c) The radius of the largest circle inscribed in the corner as a function of φ.

maximal in the junction region, the liquid will preferably accumulate there. This explanation is qualitative because the liquid can cross these boundaries and enter the interior hydrophobic region of the junction as shown in Figs. 2(a) and (c). Fig. 3(c) plots the radius of a circle inscribed in the corner as a function of the junction angle φ. As evident, the effect and thus the incentive for bulge formation becomes more pronounced for smaller values of φ.

In view of the trends shown in Figs. 2 and 3, the optimal channel geometry seems to be one in which the radius of an inscribed circle is constant. However, if the deposited liquid volume is very high, bulges will nonetheless form in the corner regions [17]. Thus, acute angles and small radii of curvature of the boundary lines between hydrophilic and hydrophobic regions should be avoided if pattern fidelity is to be maintained.

When liquid is transferred to a non-porous substrate and the separation of the plates becomes small, the liquid will be squeezed between the plates beyond the boundaries of the hydrophilic regions. Therefore, the spacing of the printing plates must be controlled and maintained above a certain minimum value. A suitable solution is to place rigid spacer elements on the printing plate, which mechanically impede too close a contact between stamp and target surface. These spacers must be hydrophobic or they will attract ink during the deposition process. The spacer thickness must be tuned such that the contact line on the target substrate

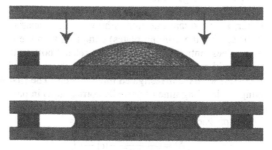

Figure 4: The use of spacer elements (black boxes) prevents the squeezing of the ink beyond the designed pattern boundaries.

matches the designed pattern on the stamp as closely as possible.

Since contact of the stamp and target plates leads to a redistribution of liquid, the required spacer thickness depends on the pattern geometry. Two limiting cases are straight, long lines, and circular pads. For the line geometry the redistribution of ink occurs only in the direction transverse to the channels, while for the circular pads, the ink spreads radially in-plane. Assuming identical ink profile heights for circles and lines, the plate separation required to maintain registry of designed and printed dimensions is smaller for circles than for lines.

Figure 5 shows the spacer thickness s as a function of the apparent contact angle for infinite lines and circular pads. s is given in units of the feature width, which corresponds to the diameter of the circular pads or the width of the lines. The apparent contact angle increases with the volume deposited on the surface pattern. As outlined above, the spacer thickness is smaller for circular pads than for long lines. Moreover, the spacer thickness s depends on the contact angle on the target substrate θ_{target}; the smaller θ_{target}, the larger is s. However, this dependence is weak for apparent contact angles below 40° on the stamp. Therefore, the dimensions of the designed pattern may need to be corrected, if it contains both elongated and compact shapes.

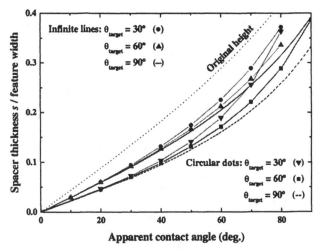

Figure 5: The required spacer thickness in units of the feature width as a function of the apparent contact angle for infinite lines and circular pads. The dotted line indicates the height of the lines and dots prior to contact with the target surface.

For the actual ink transfer, an important question is how the time-scale of the printing process (e.g. the time required to separate the printing plates) compares with the intrinsic time-scales of the liquid microstructures. These intrinsic time-scales are related both to the capillary relaxation time and the speed of the liquid-solid contact line motion. To obtain an estimate for the speed of capillary relaxation processes, we have investigated three models. The first was developed by S. E. Orchard for calculating the leveling time of periodic corrugations in paint films applied with a brush [18]. Assuming that the wavelength λ of the corrugations is much larger than the average film thickness h, the final expression for the leveling time Δt is

$$\Delta t = \frac{3}{16\pi^4} \frac{\eta \lambda^4}{h^3 \gamma_{lv}} \ln\left(\frac{a_0}{a}\right), \tag{1}$$

264

where η is the viscosity of the liquid, a_0 the initial amplitude of the corrugation and a its amplitude after time Δt. Inserting $\lambda = 4$ μm, $h = 0.5$ μm, $\gamma = 0.07$ N/m, $\eta = 0.001$ Pa s and $a_0/a = 10$ results in $\Delta t = 1.3 \cdot 10^{-7}$ s. The second model describes the spreading of a droplet on a smooth, flat and partially wetting surface [19]. It takes into account the friction processes associated with the contact line motion and gives similar results. The third model [20] describes the free oscillation of a spherical droplet, whose frequencies are given by

$$\omega^2 = n(n-1)(n+2)\, \gamma/(\rho R^2). \qquad (2)$$

Assuming a radius of $R = 1$ μm and a density of $\rho = 1000$ kg/m^3, gives a period of $T = 2.6 \cdot 10^{-7}$ s.

Thus, for identical material parameters and similar feature sizes, all three models yield Δt of order 10^{-7} s for the capillary relaxation times. According to Eq. (1) Δt is proportional to η and inversely proportional to γ_{lv}. The viscosity of glycerol, which we use as a model ink because of its low vapor pressure, is about 2000 times higher and its surface tension is 0.063 N/m. The maximum separation velocity of our laboratory printing press is 500 μm/s, thus the time required to separate the plates by 2 μm is longer than $4 \cdot 10^{-3}$ s. Therefore, the printing process for micron-sized glycerol lines and droplets is to be considered slow and determined by the chemical surface properties of the stamp and the target [21]. If the separation velocity, the ink viscosity or the feature size is significantly higher, the chemistry no longer determines the transfer process. We then enter the viscosity-controlled regime, in which the snap-off velocity of the ink meniscus is much higher than the liquid-solid contact line velocity on the plates and the transfer ratio of the ink is expected to become 50% irrespective of surface properties [22].

SUMMARY

We have presented numerical simulations of the equilibrium shapes of liquids on chemically patterned surfaces. From these model calculations we derived design rules for the plate fabrication, which include that connected regions of different width and acute junction angles of straight channels should be avoided. High curvatures of the boundary lines between hydrophilic and hydrophobic regions tend to become smoothened, particularly for higher deposited ink volumes. An estimate of the time-scales of the printing process allowed to identify and quantify the regimes where the ink transfer is governed by the chemical surface properties or by the dynamical material and process parameters.

ACKNOWLEDGMENTS

This work is supported by the Defense Advanced Research Projects Agency under the Molecular Level Printing Program. We are grateful to the Austrian Fonds zur Förderung der Wissenschaftlichen Forschung for a post-graduate fellowship (AAD) and the Eastman Kodak Corporation for a graduate fellowship (SMM). Surface Evolver was developed by Kenneth Brakke, Susquehanna University, Selinsgrove PA.

REFERENCES

[1] E. Kaneko, Electrochem. Soc. Proc. **96-23**, 8 (1996).

[2] Y. Mikami et al., IEEE Transact. Electr. Dev. **41**, 306 (1994).

[3] H. Asada, H. Hayama, Y. Nagae, S. Okazaki, Y. Akimoto, T. Saito, Conference Record of the 1991 Int. Display Research Conference, p. 227 (1991)

[4] F. Garnier, R. Hajlaoui, A. Yassar, P. Srivastava, Science **265**, 1684 (1994).

[5] B. A. Ridley, B. Nivi, J. M. Jacobson, Science **286**, 746 (1999).

[6] T.-X. Liang, W. Z. Sun, L.-D. Wang, Y. H. Wang, H.-D. Li, IEEE Transact. Components,

Packaging and Manufacturing Technology B19, 423 (1996).

[7] Z. Bao, Y. Feng, A. Dodabalapur, V. R. Raju, A. Lovinger, Chem. Mater. **9**, 1299 (1999).

[8] J. A. Rogers, Z. Bao, A. Makhija, P. Braun, Adv. Mater. **11**, 741 (1999).

[9] Z. Bao, Adv. Mater. **12**, 227 (2000).

[10] T. R. Hebner, C. C. Wu, D. Marcy, M. H. Lu, J. C. Sturm, Appl. Phys. Lett. **72**, 519 (1998).

[11] J. Bharathan, Y. Yang, Appl. Phys. Lett. **72**, 2660 (1998).

[12] J. D. Swalen, D. L. Allara, J. D. Andrade, E. A. Chandross, S. Garoff, J. Israelachvili, T. J. McCarthy, R. Murray, R. F. Pease, J. F. Rabolt, K. J. Wynne, H. Yu, Langmuir **3**, 932 (1987).

[13] C. S. Dulcey, J. H. Georger, V. Krauthamer, D. A. Stenger, T. L. Fare and J. M. Calvert, Science **252**, 551 (1991).

[14] A. Kumar, G. M. Whitesides, Appl. Phys. Lett. **63**, 2002 (1993).

[15] K. Brakke, Experimental Mathematics **1**, 141 (1992).

[16] A. A. Darhuber, S. M. Troian, S. M. Miller, S. Wagner, J. Appl. Phys. (2000), in press.

[17] H. Gau, S. Herminghaus, P. Lenz and R. Lipowsky, Science **283**, 46 (1999).

[18] S. E. Orchard, Appl. Sci. Res. A**11**, 451 (1962).

[19] M. J. de Ruijter, M. Charlot, M. Voue, J. De Coninck, Langmuir **16**, 2363 (2000); M. J. de Ruijter, J. De Coninck, G. Oshanin, Langmuir 15, 2209 (1999).

[20] H. Lamb, *Hydrodynamics*, Dover Publications (New York, 1945).

[21] A. A. Darhuber, S. M. Troian, S. M. Miller, S. Wagner, Proc. of the 3rd Int. Conf. on Modeling and Simulation of Microsystems (San Diego, March 2000), Computational Publications (Cambridge, 2000).

[22] E. D. Yakhnin, A. V. Chadov, Kolloidnyi Zhurnal **45**, 1183 (1983).

USING CONVECTIVE FLOW SPLITTING FOR THE DIRECT PRINTING OF FINE COPPER LINES

T. CUK*, S.M. TROIAN**, C.M. HONG and S. WAGNER***
*ELE Department, Princeton University, Princeton, NJ 08544, tanjacuk@princeton.edu
**CHE Department, Princeton University, Princeton, NJ 08544
***ELE Department, Princeton University, Princeton, NJ 08544

ABSTRACT

We have developed a technique for the printing of copper lines using solutions of a metal organic precursor, copper hexanoate. A 500-μm written liquid line is observed to split into two 100-μm wide lines. We observe further splitting into four parallel lines in experiments with written lines of copper hexanoate solution in chloroform. Surface profiles indicate that the thickness, width and number of lines formed are strongly dependent on the solution viscosity and volume per unit length deposited. From particle tracking visualization and surface profiling, we have found that evaporative cooling produces Marangoni convection patterns that accrete the solute along two key boundaries of flow.

INTRODUCTION

Recently, many studies have been made on the self-organized ring formation in drying drops[1-5]. One can observe this ring formation in the everyday occurrence of a coffee-stain—a stain that is much darker at the edges than in the center of the drop. Applications include high-resolution printing, nanoparticle arrays for light emitting and light processing devices, and colloidal self-organization for photonic crystals. This study specifically demonstrates the use of solute segregation in the printing of fine metallic lines.

Past studies targeted the "pinning" of the "contact line" as the means for solute accretion in drying drops. As the suspension evaporates, the "contact line"—the triple phase junction of the drying drop—moves inward; its motion can be halted due to either chemical or geometrical surface roughness. In its halted state, the contact line is "pinned"; when the contact line resumes its motion, it "depins"[6-7]. The pinning of the contact line can cause ring formation in two ways—capillary motion from center to edge[1] and meniscus pinning of the particle array[5].

In this study, we utilize a new geometry—that of a ribbon—to form multiple pairs of parallel lines, rather than a drop to form rings. We also demonstrate the importance of thermocapillary forces, in addition to contact line pinning, in directing solute accretion.

Figure 1: Optical micrograph of two copper lines, each 100 μm in width, formed by capillary writing a 544 μm wide liquid ribbon of 0.10 g CuHex/ ml CHCl₃ onto a glass slide.

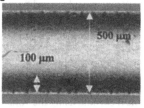

500 μm
100 μm

EXPERIMENT
Ribbon Deposition

Liquid ribbons of copper hexanoate, $Cu_2(OH_2)_2(O_2CR)_4$ with R= $(CH_2)_4CH_3$, in chloroform, were written with glass capillaries of inner diameter approximately 100μm. The precursor copper hexanoate can be reduced to pure metallic form by annealing.[8] We have

267

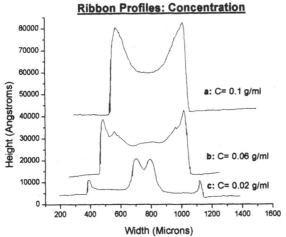

Figure 2: Surface profiles of dried CuHex from precursor liquid ribbons with C= 0.10, 0.06, and 0.02 g/ml. Other relevant parameter values are listed in entries 1a-1c of Table 1.

observed that the copper hexanoate (CuHex) accretes into pairs of parallel lines narrower in width than the original liquid ribbon. In fig. 1 is shown one pair of converted copper lines (the black in the figures is pure metallic copper). In his study, Cheong-Min Hong showed that the same could be achieved by ink-jetting a solution of CuHex in isopropanol.[8]

We studied the drying behavior of three concentrations, C= 0.02, 0.06, and 0.10 g/ml of CuHex / CHCl₃, prepared by dissolving CuHex in chloroform (CHCl₃) and for two minutes at 63°C. Upon heating deposition, the chloroform rapidly evaporated and left a characteristic line pattern on the leveled glass surface. A Dektak profilometer was used to obtain cross-sectional scans of the dried solid lines. Fig. 2 depicts typical transverse scans—it shows that, for lower solute concentrations, solute accretes well inside the ribbon as well as at the contact line. Thus, two pairs of lines are formed from the original ribbon, rather than the expected one pair shown in the transmission photos of Fig.1.

We measured several parameters intrinsic to the three solutions—including viscosity (by a visconometer), density (by a pycnometer), evaporation rate (by a balance in ambient air), and surface tension (by a Wihlemy plate). We also calculated two parameters of the ribbons themselves: volume/length deposited and the contact angle. The vol/length (ml/cm) was estimated from the ratio $\rho A_{CuHex} / C$, where $\rho = 1.1$ g/ml is the density of CuHext, A_{CuHex} the integrated cross-sectional area of dried CuHex as measured by the profilometer, and C the initial solute concentration. By approximating the cross-section of a liquid ribbon as that of a perfect cylindrical cap and using the volume/length, one can determine by applying simple trigonometry the angle the liquid makes with the glass slide.

Flow Visualization

In the results section and in Fig 2, we show that, under certain conditions, solute accretes well inside the ribbon as well as at the contact line. The mechanisms that ascribe solute accretion to contact line pinning cannot explain simultaneous solute accretion at the center of the ribbon and at the contact line. By flow visualization, we have correlated this simultaneous solute accretion with studies done by Zhang, Yang, and Duh (in the years between 1982-1989) on convection in evaporating drops.

Convection exists in evaporating drops because evaporation sets up a temperature differential between the substrate and air interfaces. There are two basic ways in which a temperature differential can induce convective flow—through either surface tension or buoyancy forces. Both surface tension and buoyancy driven convection can create convection rolls and cells of circulating liquid.

Zhang, Yang, and Duh found the convective flow in evaporating drops to look like that pictured in the model shown in the topmost image of Fig. 3.[9-11] They cite four distinct regions of convective flow.

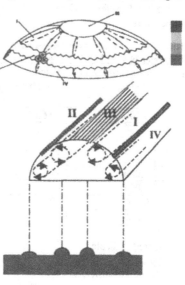

Region III is a stationary cap in which there no flow exists; Region I is a convection roll in which the liquid flows away from the stationary cap at the surface of the drop; Region IV is another convection roll in which the liquid flows away from the contact line at the surface of the drop; Region II is a ring of convection cells that move towards the stationary cap as the drop evaporates,

Figure 3: Flow model discovered by Zhang, Yang, and Duh. The top image shows their model, derived for the spherical cap, or drop. The bottom model diagrams the flow in cylindrical caps, or ribbons. Region III is a stationary cap in the drop and strip in the ribbon. Regions IV and I are convection rolls, forming rings in the drop or a pair of rolls in the ribbon. Region II is made up of convection cells, forming a ring in the drop, and a pair of lines in the ribbon. Below the ribbon is a cross-sectional cartoon of the final deposition.

and eventually merge into the stationary cap (as the cells move towards the stationary cap, Region IV grows at the expense of Region I). Since the flow model is symmetric about the cross-section, and the cross-section of a spherical cap is the same as that of a cylindrical cap, one can extrapolate what the flow would be in a ribbon. Each of the convection rolls and cells, forming a ring in the drop, form a pair in the ribbon. That model is depicted below the drop in Fig. 3.

In order to visualize the flow, we tracked either silica beads or aluminum powder in pure chloroform drops—using a microscope and video to watch and record the evaporation. We were unable to visualize the flow for the ribbons because they evaporated too fast. As discussed above, however, since the flow is symmetric, one can apply the results of the drops to the ribbons. We were also unable to visualize the flow in the solutions containing CuHex because they were too colored by the copper to see the particle flow clearly. From these videos, however, we were able to determine sites for the simultaneous solute accretion at the center and edges of ribbons and drops. Still images of the videos describe our results.

RESULTS
Ribbon Deposition

The following data can be directly read off of the surface profiles in Figure 2. The least concentrated solution, C=0.02 g/ml, formed the widest ribbon (780μm) that split into two line pairs upon drying—Fig. 2 shows clear internal and external pairs. The highest concentration, C=

0.1 g/ml, formed instead the narrowest ribbon (544 μm) that split into a single, dominant external line pair. The concentration C= 0.06 g/ml exhibited a behavior intermediate between C= 0.10 g/ml and C= 0.02 g/ml—it produced a 605 μm wide ribbon which splits into a strong external pair, but leaves a jagged middle area indicating an incipient internal line pair. For the least concentrated solution, the external pair averaged 0.8 μm in height and 30 μm in width and the internal pair averaged 1.7 μm in height and 45 μm in width. For the highest concentrated solution, the single line pair averaged 4.1 μm in height and 104 μm in width. (The line widths refer to width at half-maximum as measured above the highest neighboring baseline value.)

TableII.1: Parameter values for the writing of copper lines by the capillary method. Profiles corresponding to entries 1a-1c are shown in Figure II.1, 2a-2c in Figure II.2, and 3a-3c in Figure II.4.

Expt.	Concentration (C= g/ml)	Viscosity ($\mu= 10^{-4}$ Pa-s)	Volume/length (V/L= 10^{-5} cm^2)	Angle w/ Slide (degrees)**	Line Width (μm)
control	0.00	5.37	---	8.0	---
1a	0.100	30.65	15.5	17.8	544
1b	0.060	10.67	16.7	15.5	605
1c	0.021	6.56	22.3	12.5	780
2a	0.102	30.65	3.78	6.9	434
2b	0.063	10.67	6.27	4.0	741
2c	0.021	6.56	8.70	3.3	955
3a	0.062	10.7	3.9	7.0	436
3b	0.062	10.7	48.0	14.4	1070

Listed in Table 1 are the initial concentration, viscosity, volume/length, contact angle, and ribbon width for eight runs. We found that with decreasing concentration, the solution becomes less viscous, the volume per unit length dispensed increases, the contact angle decreases, and the ribbon width increases. The associate parameter values for the dried solute profiles depicted in Fig. 2 are listed in entries 1a-1c of Table 1. The entries 2a-2c of Table 1 refer to ribbons written in the same way as the ribbons reported in Figure 2, but using a new capillary.

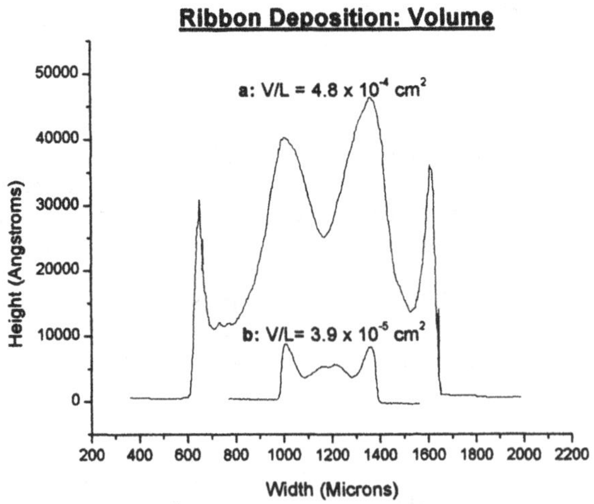

Ribbon Deposition: Volume

a: V/L = 4.8 x 10^{-4} cm^2

b: V/L= 3.9 x 10^{-5} cm^2

Figure 4: Surface profiles of dried CuHex from precursor liquid ribbons with Volume/Length: V/L = 4.8 10^{-4} and 3.9 10^{-5} cm^2. Each ribbon was written with a different sized capillary. Other relevant parameter values are listed in entries 3a-3b of TableII.1.

Their surface profiles are not incorporated since they essentially yield the same information as Figure 2—two line pairs at low concentrations, a single line pair at high concentrations.

In addition to viscosity, we measured other intrinsic parameters of the solutions for the different concentrations—density, surface tension, and evaporation rate. With increasing concentration, the density and surface tension decrease very slightly, while the evaporation rate increases, but also only slightly. The only intrinsic solution parameter changing by a significant amount is viscosity—increasing by a factor of five from the lowest to highest concentration.

The volume/length trend reported in Table 1 confirms what one would expect to find in the presence of a significant change in viscosity. The more highly concentrated, viscous solutions poured, spread, and de-wetted on the glass slide more slowly—thereby decreasing the volume per unit length dispensed. One guess as to why the contact angle increases with concentration is that solute inside the solution, by introducing surface roughness, causes the contact line to pin earlier. Since the contact angles are low overall, the ribbons are fairly flat for all concentrations. Therefore increasing volume/length dispensed creates ribbons that are wider rather than thicker. This is confirmed by the significant change in ribbon width and little change in ribbon height over the three solution concentrations. For the ribbons of Fig. 2, the ribbon heights were calculated by dividing volume/length by line width—for 1a thru 1c, the heights are 28.6, 27.6, and 28.5 μm. Clearly, the propensity to form four parallel lines can be ascribed to a viscosity-induced increase in either volume/length dispensed or ribbon width (while maintaining constant ribbon height).

In an effort to decouple the affects of viscosity from that of volume/length, studies were preformed at constant concentration but different sized capillaries. Figure 4 shows transverse scans of dried solute profiles for the parameters listed in 3a-3b of Table 1. At 48 ml/cm, there developed four distinct solid lines, while at 3.9 ml/cm, there only formed one exterior line pair and a single, unresolved central peak.

In summary, then, a relatively higher solution viscosity, lower volume per unit length dispensed, and narrower ribbon leads to the formation of one line pair at the contact line. Conversely, a relatively lower solution viscosity, higher volume per unit length dispensed, and wider ribbon leads to the formation of both an interior and exterior line pair.

Flow Visualization

The flow visualization experiments both confirmed the flow model described by Zhang, Yang, and Duh and determined sites of simultaneous solute accretion at the center and edges of drops and ribbons. The bottom image in Figure 3 depicts the sites of resulting solute accretion.

We found that as the drop evaporates, the solute (either silica beads or aluminum powder) is most dense at the boundary of the stationary cap, in the convection cells, and at the contact line. While the accretion at all three locations takes place simultaneously, the solute first deposits at the first de-pinning of the contact line. Solute also accretes at the junction between the stationary cap (region III) and the inner convection roll (region I), forming a ring around the stationary cap. Because the ring of convection cells merge into the stationary cap, the solute that accretes in them eventually adds to that already accreted at the stationary cap. The solute at the stationary cap finally deposits on the slide when the drop has de-wetted the slide almost completely and the contact line surrounds the stationary cap—almost simultaneously, the solute deposits on the slide and the rest of the drop dries.

Since multiple pinnings and de-pinnings may take place during the evaporation, multiple rings may be formed due to the contact line. However the stationary cap may only guide the accretion of the innermost ring. In summary, it is clear that the contact line guides the formation of the outer rings and the stationary cap guides the formation of the inner ring. In the ribbon, the contact line guides the external line pairs, while the stationary strip guides the inner line pair.

Figure 5a   **Figure 5b**

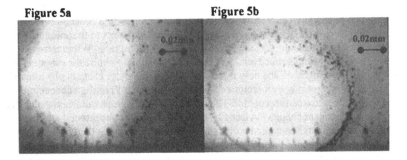

Figure 5: Still images of the flow visualization of pure chloroform drops. The dark particles are silica beads. The total lifetime of the drop is ~37 seconds. Fig. 5a is a still image of the video captured at the 30th second. Figure 5b is a still image of the video captured at the 33rd second. Taken together, the photographs depict the accumulation of solute at the boundary of the stationary cap—over time, the solute moves from being inside the inner convection roll (region I in the model) to accreting at the boundary of the stationary cap (region III in the model).

Figure 5 shows two sequential still images of the video that, together, describe solute accretion at the boundary of the stationary cap. At the earlier time (pic. 5a), solute is primarily in the convection roll, while only a thin ring exists around the stationary cap. At the later time (pic. 5b)), little solute is left in the convection roll, and a thicker ring exists around the stationary cap. The photos are both taken towards the end of the drop lifetime, after the convection cells have already merged into the stationary cap (and only one convection roll exists). The convection cells may in fact be a stabilization factor in the formation of the inner ring or line pair—the accretion at the boundary of the stationary cap is significantly clearer after the convection cells have already merged into the cap.

Thermocapillary Driven Convection

In studying thermally driven convection, one should always determine which of the two possible mechanisms—surface tension and buoyancy—dominate. Two numbers must be calculated—the Marangoni number (Ma) indicates the strength of the surface tension forces, while the Rayleigh number (Ra) indicates the strength of buoyancy forces. We found that for the ribbons, the convection is clearly Marangoni (or surface-tension) driven, the Ma/Ra ratio being on the order of 10^6. The Ma/Ra ratio increases by a factor of 6 with increasing concentration, primarily due to the increase in viscosity.

CONCLUSION

In summary, solute in drying ribbons of solution accretes into either one or two pairs of parallel lines with a definite transition phase in between. We found that a viscosity-induced change in volume/length (or equivalently ribbon width) deposited determines the mode of solute segregation. We have demonstrated that the simultaneous solute accretion correlates with the boundary of the stationary region and the contact line in the flow model; we determined that Marangoni-driven convection dominates.

ACKNOWLEDGEMENTS

The authors would like to thank Scott Miller and Anton Darbhur for their advice and expertise.

REFERENCES

1. R.D. Deegan, O. Bakajin, T.F. Dupont, G. Huber, S.R. Nagel and T.A. Witten, *Nature* **389**, 827 (1997).
2. R.D. Deegan, *Phys. Rev. E* **61**, 475 (2000).
3. J. Conway, H. Korns, and M. R. Fisch, *Langmuir* **13**, pp. 426 (1997).
4. S. Maenosono, C. D. Dushkin, S. Saita, and Y. Yamaguchi, *Langmuir* **15**, 957 (1999).
5. E. Adachi, A. S. Dimitrov, and K. Nagayama, *Langmuir* **11**, 1057 (1995).
6. P. G. deGennes, *Reviews of Modern Physics* **57**, 827 (1985).
7. J.F. Joanny and P.G. deGennes, *J. Chem. Phys.* **81**, 552 (1984).
8. C.M. Hong, H. Gleskova and S. Wagner, *Mat. Res. Soc. Symp. Proc.* **471**, 35 (1997).
9. N. Zhang and W. Yang, *Journal of Heat Transfer* **104**, 656 (1982).
10. N. Zhang and W. Yang, *Journal of Heat Transfer* **105**, 908 (1983).
11. J. C. Duh and W. Yang, *Numerical Heat Transfer, Part A* **16**, 129 (1989).

CALCULATION OF HAMAKER CONSTANTS IN NONAQUEOUS FLUID MEDIA

Nelson Bell and Duane Dimos
Ceramic Materials Department, Sandia National Laboratories, 1515 Eubank Blvd. SE, MS 1411, Albuquerque, NM 87123

ABSTRACT

Calculations of the Hamaker constants representing the van der Waals interactions between conductor, resistor and dielectric materials are performed using Lifshitz theory. The calculation of the parameters for the Ninham-Parsegian relationship for several non-aqueous liquids has been derived based on literature dielectric data. Discussion of the role of van der Waals forces in the dispersion of particles is given for understanding paste formulation. Experimental measurements of viscosity are presented to show the role of dispersant truncation of attractive van der Waals forces.

INTRODUCTION

Thick film pastes are complicated examples of colloidal processing and engineering. The desired rheological characteristics for thick film printing include a shear thinning viscosity to allow flow during printing and a yield stress to maintain printed feature definition. These properties are achieved through control of the range and magnitude of interparticle forces. All similar materials experience attractive forces due to permanent or induced dipolar interactions, and the effect of these interactions between materials is expressed macroscopically in terms of the Hamaker constant, A_{132}. This attractive force generates an agglomerated particle network that resists flow. In order to make a suspension fluid, a stabilizing mechanism must be employed that controls the magnitude of the attractive forces between the particles to give the desired rheological behavior. By keeping particles at a fixed separation distance, the strength of their attraction can be controlled. The Hamaker constant provides a baseline for understanding how much separation is required. Understanding the strength of the Hamaker constant helps determine the type and properties of the stabilizing mechanism needed to form an effective thick film paste composition.

The Hamaker constant was first calculated by summing the interactions between the dipole-pairs in a material [1]. This laborious procedure was greatly simplified when Lifshitz described how the dielectric response function of a material could be used to perform the same function more accurately as it incorporates many-body effects directly into the calculation [2]. Assuming materials to be continuous media, the Ninham-Parsegian (N-P) imaginary function can be used to represent the dielectric response as a function of frequency [3-5]. This N-P representation has no direct physical basis but can be constructed based on knowledge of the static dielectric constant, index of refraction, and infrared, ultraviolet and microwave absorption spectra. One of the properties of the N-P relationship is that it is evaluated at discrete imaginary frequencies which are distributed so that the ultraviolet and (to a lesser degree) infrared spectra dominate the determination of the Hamaker constant. Accuracy of the calculation therefore depends greatly on the accuracy of the IR and UV spectra used to construct the N-P function for the material.

Mat. Res. Soc. Symp. Proc. Vol. 624 © 2000 Materials Research Society

Most calculations of Hamaker constants have focused on aqueous media. Much of the reason for attention being paid to water is its environmentally benign nature. Yet, the majority of thick film pastes used in the electronics industry are in nonaqueous solvents, and the availability of information for these systems are not common in the literature. It is the drive of this work to fill the gap in the literature of the functions required to calculate Hamaker constants in nonaqueous media. In doing so, the role of dispersants in these media to control rheological response will be improved, and the possibility of tailoring solid and liquid to provide esired rheological properties can be explored.

EXPERIMENTAL

Rheological measurements were performed with a Bohlin CS-10 rheometer[1]. Silver slurries[2] were prepared at 40 volume % in methanol and isopropanol. Pluronic F68[3] dispersant was added at 0.4 weight % to powder to evaluate the control of interparticle force on rheological response. This dispersant is a triblock copolymer of polyethylene oxide stabilizing blocks bonded to an adsorbing block of polypropylene oxide. Experiments were performed using a 60 second preshear at 400 rpm followed by a hold time of 30 seconds. Tests were run from low shear rate to high and back to zero.

The parameters for the Ninham-Parsegian expression for liquids can be found in a variety of literature sources. Water has been most extensively characterized, and the spectral parameters for water are given in [5]. Six infra-red and five ultraviolet damped oscillators with a single Debye microwave relaxation represent the dielectric behavior very well. However, this level of detail is hard to compile for most nonaqueous solvents. The static dielectric constant and index of refraction at the sodium D line are commonly tabulated [6], and the infra-red adsorption spectra for many liquids are relatively easy to find [7]. The microwave parameters for several liquids are referenced in the compilation of Buckley and Maryott [8]. Determination of the ultraviolet characteristics for several solvents has proven to be more difficult. This is a serious complication as the ultraviolet terms are the most critical for calculating the Hamaker constant. In the absence of a full spectra, UV relaxation is commonly represented by a single oscillator of magnitude equal to $n^2 - 1$. The UV adsorption edge was chosen as the critical frequency of this oscillator and damping terms were omitted [6]. Further searches are being performed to determine the UV spectra for these nonaqueous fluids and refine the parameters for Hamaker constant calculation.

From the collected data, the oscillator strengths were determined by the changes in the value of the real dielectric constant between spectral regions. The microwave strength can be determined directly from [8]. Groups of peaks located very closely together in the infra-red spectrum are represented by a single oscillator and damping terms were omitted. The index of refraction was used to determine the magnitude of the transition between the microwave and visible region due to adsorption in the IR, and individual oscillator strengths were assigned based on the area of each IR adsorption. The ultraviolet adsorption was represented by a single oscillator at the UV adsorption edge. This collected data is presented in Table 1.

The optical parameters for solids have been tabulated in several sources that can be used to fit the oscillator models for dielectric behavior [5,9,10]. Most metal oxides can be fit in a

---

[1] Bohlin Instruments Inc., Suite 1, 11 Harts Lane, East Brunswick, NJ 08816.
[2] Superior Micropowders, 3740 Hawkins NE, Albuquerque, NM 87109.
[3] BASF Corporation, 3000 Continental Drive North, Mount Olive, NJ 07828-1234.

Table 1. Spectral Parameters for the Ninham-Parsegian expression of $\varepsilon(i\xi)$ for Nonaqueous Solvents.

Name	Type	Dielectric Constant	C_{MW}	ω_{MW} (rad/sec)	$C_{IR}(\#)$	$\omega_{IR}(\#)$ (rad/sec)	n	C_{UV}	ω_{UV} (rad/sec)
Methanol	Amphiprotic	33.64	27.64	2×10^{12}	0.254 0.123 1.042 2.801	3.0579×10^{13} 4.149×10^{13} 8.682×10^{13} 1.007×10^{14}	1.3288	0.76571	9.1885×10^{15}
1-Propanol	Amphiprotic	20.8	17.4	3×10^{9}	0.2134 0.1778 0.3808 0.70977	3.148×10^{13} 4.317×10^{13} 8.8019×10^{13} 9.977×10^{13}	1.3850	0.91823	8.9698×10^{15}
2-Propanol	Amphiprotic	20.18	17.12	1×10^{10}	0.0198 0.05927 0.1441 0.1801 0.2208 0.53462	2.4343×10^{13} 2.842×10^{13} 3.4656×10^{13} 4.125×10^{13} 8.8739×10^{13} 9.9172×10^{13}	1.3776	0.89778	9.1885×10^{15}
1-Butanol	Amphiprotic	17.84	14.84	2×10^{9}	0.13545 0.13233 0.26674 0.50743	3.1778×10^{13} 4.2871×10^{13} 8.7840×10^{13} 9.9172×10^{13}	1.3993	0.95804	8.7612×10^{15}
2-Butanol	Amphiprotic	17.26	13.32	2×10^{9}	0.2582 0.2185 0.4767 0.9931	2.9979×10^{13} 4.1132×10^{13} 8.9039×10^{13} 1.0037×10^{14}	1.3978	0.95385	7.2448×10^{15}
Benzene	Inert	2.28	N/A	N/A	0.00969 0.00561 0.0114	2.0146×10^{13} 4.437×10^{13} 9.144×10^{13}	1.5011	1.25330	6.7757×10^{15}
Toluene	Inert	2.38	N/A	N/A	0.0642 0.02295 0.0545	2.1825×10^{13} 4.4849×10^{13} 9.0778×10^{13}	1.4961	1.23832	6.6326×10^{15}

similar manner to the liquids. Values for alumina and barium titanate were taken from the compilation of Bergstrom [5]. Parameters for silver were taken from Parsegian and Weiss [11]. The parameters have been converted from eV to rad/sec. The first term has no critical frequency, and it represents the hyperbolic dependence of the conduction electrons.

Table 2. Spectral Parameters for the Ninham-Parsegian expression of $\varepsilon(i\xi)$ for Solid Materials.

Name	Dielectric Constant	Index of Refraction	C_j	ω_j (rad/sec)	g_j (rad/sec)	Source
Silver	∞	--	2.1209×10^{32}	--	--	16
			1.51997	7.8997×10^{15}	2.8864×10^{15}	
			0.54527	2.3547×10^{15}	8.20349×10^{15}	
			0.17327	3.4333×10^{16}	5.46899×10^{15}	
			2.24565	5.2563×10^{16}	1.43105×10^{17}	
Alumina	10.1	1.753	2.072	2.0×10^{16}	--	5
			7.03	1×10^{14}	--	
Barium Titanate	o 3600	2.284	4.218	0.841×10^{16}	--	5
			3595	$0.7\text{-}1.0 \times 10^{14}$	--	

RESULTS

Using the parameters from Tables 1 and 2, the Hamaker constant between identical particles in various liquid media was determined by the N-P method [3-5]. These values are given in Table 3.

Table 3. Hamaker Constants between Identical Particles in Various Solvents

Solvent	Silver (zJ)	Alumina (zJ)	Barium Titanate (zJ)
Water	149.1	36.9	106.8
Methanol	175.2	84.4	142.8
1-Propanol	168	76.0	134.4
2-Propanol	224.5	149	203.9
1-Butanol	167.3	75.0	133.2
2- Butanol	175.4	84.7	142.5
Benzene	166.4	77.0	131.5
Toluene	168	78.6	133.3

From the values in Table 3, there is a general trend regardless of solvent that ranks the Hamaker constants in the order silver > barium titanate > alumina. This trend relates to the material dielectric constant which results from Keesom and Debye electrostatic interactions. The London (dispersive) interactions that affect the IR and UV adsorption cannot compensate for the difference in static properties. The high magnitudes of the dielectric constant for barium titanate (3600) and of silver (infinity) makes it unlikely that any solvent exists which will match the dielectric properties of these materials and cause a minimum in the Hamaker constant. Alumina however has a low dielectric constant, so the possiblity to choose a solvent that minimizes van der Waals interactions is available.

In comparison of the solvents, the values for each solid material with water as the solvent can be expected to be more accurate than the nonaqueous solvents due to the higher degree of characterization of the dielectric spectra available for water. No solvent that was evaluated has a Hamaker constant lower than water, and in general they are significantly higher. Most of the nonaqueous values are of comparable magnitude with the exception of 2-propanol. The high value of A_{132} for 2-propanol predicts that with a similar dispersing mechanism, values of shear stress and viscosity will be higher in 2-propanol versus another solvent.

To test this hypothesis and the accuracy of the parameters used in calculation, silver dispersions were made in methanol and 2-propanol using a nonionic dispersant recommended for use in water and alcohols. The viscosity data was normalized for the intrinsic solvent viscosity, and the samples are compared in Figure 1. Contrary to the Hamaker constant prediction, the attractive forces in methanol seem higher than in 2-propanol. Reasons for the discrepancy include: the Hamaker values for 2-propanol are too high, there may be electrostatic forces present in the 2-propanol, or the dispersant may have different solvation characteristics in each solvent. The viscosity curves have the same qualitative behavior and exhibit little hysteresis between the rising and falling shear rate test. This suggests that the dispersant is behaving similarly between the two solvents, but it does not rule out the possibility that there may be differing adsorbed amount of polymer or a difference in the extension of the polymer from the surface. However, the estimations used in the ultraviolet spectra of 2-propanol may need to be corrected to give the most correct values. Further examination of the surface chemistry of the silver powder in each solvent and the behavior of the dispersant need to be performed before definite answers can be concluded.

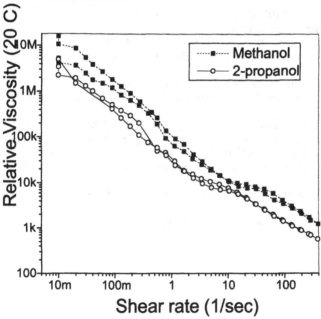

Figure 1. Relative viscosity comparison of silver powder dispersed with Pluronic F68 in methanol and isopropanol.

CONCLUSIONS

Terms needed for calculating the Hamaker constant of several nonaqueous liquids have been collected from the literature, and the Hamaker constant for three materials systems have been calculated for identical particles in these solvents and in water. Within each solid system, the Hamaker constant scales with dielectric constant. The nonaqueous solvents were calculated to have larger Hamaker constants than water, but their accuracy could be improved by using more accurate representations of the ultraviolet adsorption spectra. The examination of two nonaqueous solvents for differences in van der Waals attractive forces did not agree with values of the Hamaker constant. Reasons for the discrepancy require more investigation of each system.

ACKNOWLEDGEMENTS

Sandia is a multiprogram laboratory operated by Sandia Corporation, a Lockheed Martin Company, for the United States Department of Energy under contract DE-AC04-94AL85000. This work was performed under the DARPA MICE program under the leadership of CMS Technitronics. Special thanks go to Superior Micropowders for supplying the silver powder.

REFERENCES

1. H.C. Hamaker, Physica, **4** 1058 (1937).
2. E.M. Lifshitz, Soviet Physics, **2** 73-83 (1956).
3. D.B. Hough and L.R. White, Adv. Colloid Interface Sci., **14** 3-41 (1980).
4. J. Israelachvili, *Intermolecular and Surface Forces*, 2nd Ed., (Academic Press, London, 1995) p. 176-209.
5. L. Bergstrom, Adv. Colloid Interface Sci., **70** 125-169 (1997).
6. D.R. Lide, Ed., *CRC Handbook of Chemistry and Physics*, 78th Ed., (CRC Press, New York, 1997) p. **6**-139-172, **8**-113.
7. C.J. Pouchert, *The Aldrich Library of Infrared Spectra*, 3rd Ed., Aldrich Chemical Co., 1981.
8. F. Buckley and A.A. Maryott, *Tables of Dielectric Dispersion Data for Pure Liquids and Dilute Solutions*, NBS Circular 589, 1958.
9. E.D. Palik, *Handbook of the Optical Constants of Solids*, (Academic Press, Orlando, Fl, 1985).
10. E.D. Palik, *Handbook of the Optical Constants of Solids II*, (Academic Press, New York, 1991).
11. V.A. Parsegian and G.H. Weiss, J. Colloid Interface Sci., **81** 285-289 (1981).

AUTHOR INDEX

Abraham, M.H., 79
Alleman, J., 59
Aucoin, Richard, 9
Auyeung, R.C.Y., 29, 143

Beach, Joseph D., 87
Bell, Nelson, 275
Berry, G.J., 249
Bertagnolli, E., 163
Boyd, Ian W., 115
Bradford, William C., 87
Brinker, C. Jeffrey, 231
Bulthaup, Colin, 225
Burns, M.J., 9

Cairns, J.A., 249
Chou, S.Y., 219
Chrisey, Douglas B., 29, 135, 143
Chung, R., 143
Church, Kenneth H., 3, 135, 195
Cicoira, F., 171
Coleman, Steven M., 53
Collins, Reuben T., 87
Courter, J., 17
Cuk, T., 267
Culver, James, 195
Curtis, C.J., 59

Darhuber, Anton A., 47, 261
Davidson, M.R., 249
Derby, B., 65
Detig, Robert H., 71
Dimos, Duane, 275
Doppelt, P., 171
Duignan, M.T., 99
Dwir, B., 171

Eason, Steve, 195
Ehrlich, D.J., 9
Eichhorn, Bryan W., 35

Fan, Hongyou, 231
Fan, Y.C., 249
Feeley, Terry, 3
Fitzgerald, A.G., 249
Fitz-Gerald, J.M., 29, 143
Fore, Charlotte, 3
Fuqua, P.D., 79
Fzea, A.H., 249

Gambino, R.J., 181, 189
Gamota, Daniel R., 41
Gao, Hongjun, 205

Geiculescu, A.C., 29
Ginley, D.S., 59
Giridharan, M.G., 255
Gonen, Z. Serpil, 35
Goswami, R., 189
Greenlaw, R., 181
Gritsch, M., 163

Hansen, W.W., 79
Hebner, Thomas R., 211
Helvajian, H., 79
Herman, H., 181, 189
Herndon, Mary K., 87
Hoffmann, P., 171
Hollingsworth, Russell E., 87
Hong, C.M., 219, 267
Hubert, Brian, 225
Hull, Robert, 157
Hutter, H., 163

Jacobson, Joe, 225

Kapon, E., 171
Kar, A., 127
Knobbe, Edward T., 53
Kydd, Paul H., 135

Lakeou, S., 143
Langfischer, H., 163
Leifer, K., 171
Liu, Q., 17
Lobban, J., 249
Longo, David M., 157
López, Gabriel P., 231
Lowry, S., 243, 255
Lugstein, A., 163

Madigan, Conor F., 211
Manos, Dennis M., 205
Martelé, Y., 199
McGivern, P., 249
Megaridis, C.M., 23
Miedaner, A., 59
Miller, Scott M., 47, 261
Mohdi, R., 143

Nill, Kenneth, 9

Orme, M., 17

Parkhill, Robert L., 53, 243
Patel, A., 181
Piqué, A., 29, 143

Quick, N.R., 127

Rack, H.J., 29
Rack, P.D., 29
Rahman, K.M.A., 99
Ranade, M.B. (Arun), 35
Register, Richard A., 211
Reis, N., 65
Renn, Michael J., 107
Richard, David L., 135
Ridley, Brent, 225
Rivkin, T., 59

Sampath, S., 181, 189
Schacht, E.H., 199
Schulz, D.L., 59
Sengupta, D.K., 127
Shaikh, W., 249
Sheu, J.C., 243, 255
Shmagin, Irina, 41
Silverman, Scott, 9
Skinner, James, 41
Smith, R., 17
Stewart, Robert, 243
Sturm, J.C., 211
Sun, X., 219
Szczech, John B., 23, 41

Tan, S.Y., 189
Taylor, D.P., 79
Taylor, Robert M., 53, 195
Thomson, J., 249
Tormey, E., 181
Troian, Sandra M., 47, 211, 261, 267

Utke, I., 171

van Aert, H., 199
Van Damme, M., 199
Vansteenkiste, S.O., 199
Vermeersch, J., 199

Wagner, Sigurd, 47, 219, 261, 267
Wanzenboeck, H.D., 163
Wells, D.N., 99
Wilhelm, Eric, 225
Wu, H.D., 143
Wu, Lingling, 205

Young, H.D., 143

Zhang, Jie, 23, 41
Zhang, Jun-Ying, 115
Zhu, J., 17

Printed in the United States
By Bookmasters